Handbook of Chloroplast (Cell Biology Essentials)

Handbook of Chloroplast (Cell Biology Essentials)

Edited by Eldon Dixon

www.statesacademicpress.com

States Academic Press,
109 South 5th Street,
Brooklyn, NY 11249, USA

Visit us on the World Wide Web at:
www.statesacademicpress.com

ISBN: 978-1-63989-734-6

Cataloging-in-Publication Data

Handbook of chloroplast (cell biology essentials) / edited by Eldon Dixon.
   p. cm.
Includes bibliographical references and index.
ISBN 978-1-63989-734-6
1. Cytology. 2. Cytology--Technique. 3. Chloroplasts. I. Dixon, Eldon.
QK725 .H36 2023
581.072 4--dc23

Table of Contents

Preface

Chloroplast refers to a type of membrane-bound organelle that conducts photosynthesis in plant and algal cells. Chlorophyll is the photosynthetic pigment that absorbs the energy from sunlight. It then converts and stores this energy in the energy-storage molecules, ATP (adenosine triphosphate) and NADPH (nicotinamide adenine dinucleotide phosphate). There are various parts within the structure of chloroplast, which include membrane envelope, intermembrane space, stroma, thylakoid system, grana and chlorophyll. Grana and stroma are the two distinct regions present inside a chloroplast. The grana are the functional units of chloroplasts and consist of chlorophyll pigments. The stroma is an aqueous, alkaline and protein-rich fluid that contains various enzymes, ribosomes, DNA, and other substances present inside the chloroplast membrane surrounding the grana. This book includes some of the vital pieces of works being conducted across the world, on various topics related to the chloroplast. It will provide interesting topics for research, which interested readers can take up. The book is a resource guide for experts as well as students.

This book is a result of research of several months to collate the most relevant data in the field.

When I was approached with the idea of this book and the proposal to edit it, I was overwhelmed. It gave me an opportunity to reach out to all those who share a common interest with me in this field. I had 3 main parameters for editing this text:

1. Accuracy – The data and information provided in this book should be up-to-date and valuable to the readers.

2. Structure – The data must be presented in a structured format for easy understanding and better grasping of the readers.

3. Universal Approach – This book not only targets students but also experts and innovators in the field, thus my aim was to present topics which are of use to all.

Thus, it took me a couple of months to finish the editing of this book.

I would like to make a special mention of my publisher who considered me worthy of this opportunity and also supported me throughout the editing process. I would also like to thank the editing team at the back-end who extended their help whenever required.

Editor

Comparative Analysis of Complete Chloroplast Genome Sequences of Wild and Cultivated *Bougainvillea* (Nyctaginaceae)

Mary Ann C. Bautista [1,2,3], **Yan Zheng** [2], **Zhangli Hu** [4], **Yunfei Deng** [1,3,*] **and Tao Chen** [2,3,*]

1   South China Botanical Garden, Chinese Academy of Sciences, Guangzhou 510650, China; bautista.maryann@gmail.com
2   Fairy Lake Botanical Garden, Chinese Academy of Sciences, Shenzhen 518004, China; yanzheng@szbg.ac.cn
3   Graduate School, University of Chinese Academy of Sciences, Beijing 100049, China
4   School of Life Sciences and Oceanology, Shenzhen University, Shenzhen 518060, China; huzl@szu.edu.cn
*   Correspondence: yfdeng@scbg.ac.cn (Y.D.); taochen@szbg.ac.cn (T.C.)

Abstract: *Bougainvillea* (Nyctaginaceae) is a popular ornamental plant group primarily grown for its striking colorful bracts. However, despite its established horticultural value, limited genomic resources and molecular studies have been reported for this genus. Thus, to address this existing gap, complete chloroplast genomes of four species (*Bougainvillea glabra*, *Bougainvillea peruviana*, *Bougainvillea pachyphylla*, *Bougainvillea praecox*) and one *Bougainvillea* cultivar were sequenced and characterized. The *Bougainvillea* cp genomes range from 153,966 bp to 154,541 bp in length, comprising a large single-copy region (85,159 bp–85,708 bp) and a small single-copy region (18,014 bp–18,078 bp) separated by a pair of inverted repeats (25,377–25,427 bp). All sequenced plastomes have 131 annotated genes, including 86 protein-coding, eight rRNA, and 37 tRNA genes. These five newly sequenced *Bougainvillea* cp genomes were compared to the *Bougainvillea spectabilis* cp genome deposited in GeBank. The results showed that all cp genomes have highly similar structures, contents, and organization. They all exhibit quadripartite structures and all have the same numbers of genes and introns. Codon usage, RNA editing sites, and repeat analyses also revealed highly similar results for the six cp genomes. The amino acid leucine has the highest proportion and almost all favored synonymous codons have either an A or U ending. Likewise, out of the 42 predicted RNA sites, most conversions were from serine (S) to leucine (L). The majority of the simple sequence repeats detected were A/T mononucleotides, making the cp genomes A/T-rich. The contractions and expansions of the IR boundaries were very minimal as well, hence contributing very little to the differences in genome size. In addition, sequence variation analyses showed that *Bougainvillea* cp genomes share nearly identical genomic profiles though several potential barcodes, such as *ycf*1, *ndh*F, and *rpo*A were identified. Higher variation was observed in both *B. peruviana* and *B. pachyphylla* cp sequences based on SNPs and indels analysis. Phylogenetic reconstructions further showed that these two species appear to be the basal taxa of *Bougainvillea*. The rarely cultivated and wild species of *Bougainvillea* (*B. pachyphylla*, *B. peruviana*, *B. praecox*) diverged earlier than the commonly cultivated species and cultivar (*B. spectabilis*, *B. glabra*, *B.* cv.). Overall, the results of this study provide additional genetic resources that can aid in further phylogenetic and evolutionary studies in *Bougainvillea*. Moreover, genetic information from this study is potentially useful in identifying *Bougainvillea* species and cultivars, which is essential for both taxonomic and plant breeding studies.

Keywords: *Bougainvillea*; Nyctaginaceae; chloroplast genome; phylogeny

1. Introduction

The family Nyctaginaceae, distributed primarily in the tropics and subtropics, contains around 400 species of trees, shrubs, and herbs classified in ca. 31 genera [1,2]. Nyctaginaceae has been well-recognized as one of the core groups of Caryophyllales (Centrospermae) based on the presence of betalain pigments, free-central placentation, p-type sieve tube elements, perisperm, and molecular evidence [1,3]. One of the most popular genera in Nyctaginaceae is *Bougainvillea*, a tropical and subtropical shrubby vine cultivated primarily for its colorful showy bracts. Their vibrant structures often mistaken as "flowers" are actually bracts or specialized leaves (ca. 0.5–2-inch long), in which the true flowers are attached at the mid-rib [4]. The true perfect flowers are normally small, tubular, white or yellowish in color, and surrounded by colorful petaloid bracts [4]. Due to *Bougainvillea*'s growth habit and attractive bracts, it became a widely known plant for landscaping [4]. It is commonly used in gardens as hedges or barriers, topiaries, and as ground cover on banks.

The horticulturally important species, such as *Bougainvillea glabra* and *Bougainvillea spectabilis*, are native to South America, but were brought and introduced to various countries all over the world. There are approximately 18 species of *Bougainvillea*, but only the above two species are well-known for cultivation [1,5]. The purpose of planting *Bougainvillea* is mostly ornamental, but recently a number of studies have explored the potential use of *Bougainvillea* as a medicinal plant and as a pollution-mitigating plant in industrial areas [6–11]. Recent research studies have tapped into the potential of *Bougainvillea* as an anti-inflammatory, anticancer, antioxidant, antimicrobial, and antihyperglycemic plant [7–11]. Specifically, *Bougainvillea spectabilis* has been well-known for its ability to lower blood sugar and improve liver function, while *Bougainvillea* cv. *buttiana* and *Bougainvillea glabra* exhibit significant anti-inflammatory activities [7,9,10]. The plant group has garnered research attention from the horticultural and pharmaceutical industries, and even in environmental studies. However, no recent studies have focused on the taxonomy of *Bougainvillea*, particularly regarding the wild species not utilized for cultivation.

Most of the publications on *Bougainvillea* focused on taxonomic descriptions were published decades ago [12–14], and no actual revision of the topic has been attempted since then. Molecular studies based on short-fragment sequences centered on Nyctaginaceae include only a few sequences of *Bougainvillea*. Phylogenetic studies on family Nyctaginaceae based on *ndh*F, *rps*16, *rpl*16, and nrITS revealed that *Bougainvillea glabra* and *Bougainvillea infesta* are actually closely related to *Belemia* and *Phaeoptilum* [15]. Similarly, the phylogenomic study about the order Caryophyllales includes only *Bougainvillea spectabilis* as representative of Bougainvilleeae [16]. To date, no phylogenetic study has been published focusing specifically on *Bougainvillea* species.

Even though next generation sequencing (NGS) has instigated a rapid increase in the available complete organelle genome sequences of plants that can be used for phylogenetic studies, sequences of *Bougainvillea* are rarely deposited in the database. The commercial value of *Bougainvillea* as an ornamental plant has overshadowed the need for genetic studies. Hence, the majority of the available sequences in the GenBank are from the commercially cultivated *Bougainvillea*. Short-genome announcement papers have presented the gene content and structure of *B. glabra* and *B. spectabilis* [17,18]. It was also shown that *Bougainvillea* is clustered together with *Acleisanthes*, *Mirabilis*, and *Nyctaginia*; however, specific relationships cannot be inferred due to the lack of available sequences [17]. Currently, there is very little information available on the genetic structures of *Bougainvillea*, particularly on their plastome features and specific phylogenetic placements.

It has been established that plastomes, particularly chloroplast genomes, are useful in phylogenetic studies due to their ability to self-replicate, their conservative structure, and their slow evolutionary rate [19]. Investigating the genome organization and genetic information for plastid genomes provide scientists relevant data that can be utilized for species conservation, phylogenetic reconstruction, molecular marker development, genomic evolution studies, and for solving taxonomic complexities in different taxa [19–21]. Therefore, sequencing additional cp genomes of *Bougainvillea* species is an initial step that is needed to fill the gap in genomic resources before conducting further studies. Thus, this study aimed to characterize the cp genome sequences of several cultivated and wild *Bougainvillea* species (*Bougainvillea glabra*, *Bougainvillea peruviana*, *Bougainvillea pachyphylla*, *Bougainvillea praecox*, and *Bougainvillea* cv.). Specifically, the study aimed to discuss the structures and features of the five newly sequenced cp genomes, including the codon usage, RNA editing, simple sequence repeats, tandem repeats, IR contractions and expansion, divergent regions, SNPs, and indels. Phylogenetic analyses were also conducted to determine the relationship among cultivated and wild *Bougainvillea* in the family Nyctaginaceae.

2. Results and Discussion

2.1. General Features of Bougainvillea Chloroplast Genomes

The assembled chloroplast genome sequences of *Bougainvillea glabra*, *Bougainvillea peruviana*, *Bougainvillea pachyphylla*, *Bougainvillea praecox*, and *Bougainvillea* cv., together with the available sequence of *Bougainvillea spectabilis*, range from 153,966 bp to 154,541 bp. Similar to most sequenced Nyctaginaceae cp genomes, they all exhibit the typical quadripartite structure, consisting of a large single-copy (LSC) region (85,159–85,708 bp), a small single-copy (SSC) region (18,014–18,078 bp), and a pair of inverted repeats (25,377–25,427 bp) (Figure 1, Table 1). The commonly cultivated *Bougainvillea spectabilis* and *Bougainvillea glabra* have the largest cp genomes, but despite the differences in size, all cp genomes have a total of 131 genes, including 86 protein-coding genes, eight rRNA, and 37 tRNA. Of the identified genes, there are seven protein-coding (*ndh*B, *rpl*2, *rpl*23, *rps*7, *rps*12, *ycf*1, *ycf*2), four rRNA (*rrn*4.5, *rrn*5, *rrn*16, *rrn*23), and seven tRNA (*trn*I-CAU, *trn*L-CAA, *trn*V-GAC, *trn*I-GAU, *trn*A-UGC, *trn*R-ACG, *trn*N-GUU) genes duplicated in the IR regions. Thus, there are 79 unique protein-coding genes that function primarily in photosynthesis and transcription–translation processes, while the remaining are transfer RNA genes (30), and ribosomal RNA genes (four) (Table 2). The numbers of rRNA and tRNA genes are highly conserved among Nyctaginaceae, but the number of protein-coding genes differs due to acetyl-CoA carboxylase subunit D gene (*accD*) loss in some genera. *Bougainvillea* cp genomes have an intact *accD* gene, whereas *Nycataginia capitata* (MH286318.1) and *Pisonia aculeata* (MK397866.1) lack the *accD* gene. In general, the genome structure and features of the *Bougainvillea* cp genomes are highly similar to most Nyctaginaceae chloroplast genomes—no significant changes in the gene order or gene content have been observed.

Table 1. Complete chloroplast genome features of five *Bougainvillea* species and one cultivar.

Features	B. glabra	B. peruviana	B. pachyphylla	B. praecox	B. cultivar	B. spectabilis
Genome size (bp)	154,536	153,966	154,062	154,306	154,520	154,541
LSC length (bp)	85,708	85,159	85,181	85,474	85,688	85,694
SSC length (bp)	18,038	18,025	18,027	18,014	18,078	18,077
IR length (bp)	25,395	25,391	25,427	25,409	25,377	25,385
Total no. of genes	131	131	131	131	131	131
Protein-coding genes	86	86	86	86	86	86
rRNA	8	8	8	8	8	8
tRNA	37	37	37	37	37	37
Overall GC content (%)	36.5%	36.6%	36.5%	36.5%	36.5%	36.4%
GC content in LSC (%)	34.2%	34.3%	34.3%	34.3%	34.2%	34.2%
GC content in SSC (%)	29.5%	29.6%	29.6%	29.5%	29.5%	29.5%
GC content in IR (%)	42.8%	42.8%	42.8%	42.8%	42.8%	42.7%

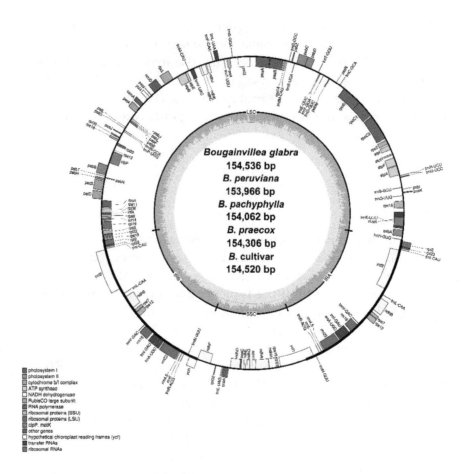

Figure 1. Circular gene map of five newly sequenced *Bougainvillea* chloroplast genomes. The genes drawn outside the circle are transcribed clockwise, while the genes on the inside are transcribed counterclockwise. Genes belonging to different functional groups are color-coded. The dark gray plot in the inner circle represents the GC (Guanine-Cytosine) content, whereas the light-grey corresponds to the AT (Adenine-Thymine) content.

In addition, there are also 17 intron-containing genes identified in all six cp genomes, of which 15 genes (*rps*16, *atp*F, *rpo*C1, *pet*B, *pet*D, *rpl*16, *rpl*2, *ndh*B, *ndh*A, *trn*I-GAU, *trn*A-UGC, *trn*V-UAC, *trn*L-UAA, *trn*G-UCC, *trn*K-UUU) have one intron, while *clp*P and *ycf*3 have two introns each (Table 2). The *trn*K-UUU gene has the longest intron (2508–2524 bp), which encodes the *mat*K ORF. This is normally perceived in published cp genomes, since plastid *trn* introns are relatively longer in comparison to nuclear tRNA introns [22]. A trans-splicing event also occurs in the *rps*12 gene, with the 5' exon positioned in the LSC region, while the 3' exons are duplicated in the IR regions. The aforementioned genes with introns can be categorized into three types—genes for electron transfer, protein synthesis, and ATP synthesis. Most introns in general do not code for proteins, but recent studies have shown that they can enhance gene expression and regulation in specific locations [23,24]; hence, they are potential sites for efficient processing of native or foreign transcripts that can improve particular horticultural traits in plants such as *Bougainvillea*.

Overall, the GC contents of the six *Bougainvillea* cp genomes are almost identical, ranging from 36.4% to 36.6%. Interestingly, the GC contents of the IR regions (42.7–42.8%) are higher compared to the LSC (34.2–34.3%) and SSC regions (29.5–29.6%). Higher GC contents in IR regions are typically linked to the presence of rRNA genes in the IRs or to the GC-biased conversion (gBGC) [20]. The gBGC is a preferential fixation of AT to GC mutations over GC to AT mutations, thus increasing the GC contents in recombination hotspots such as IR regions [25–27].

Table 2. List of genes encoded by six *Bougainvillea* cp genomes.

Gene Category	Gene Names
ATP Synthase	*atp*A, *atp*B, *atp*E, *atp*F *, *atp*H, *atp*I
NADH dehydrogenase	*ndh*A *, *ndh*B$^{(X2)}$ *, *ndh*C, *ndh*D, *ndh*E, *ndh*F, *ndh*G, *ndh*H, *ndh*I, *ndh*J, *ndh*K
Cytochrome b/f complex	*pet*A, *pet*B *, *pet*D *, *pet*G, *pet*L, *pet*N
Photosystem I	*psa*A, *psa*B, *psa*C, *psa*I, *psa*J
Photosystem II	*psb*A, *psb*B, *psb*C, *psb*D, *psb*E, *psb*F, *psb*H, *psb*I, *psb*J, *psb*K, *psb*L, *psb*M, *psb*N, *psb*T, *psb*Z
RubisCO large subunit	*rbc*L
Ribosomal protein genes (large subunits)	*rpl*2$^{(X2)}$ *, *rpl*14, *rpl*16 *, *rpl*20, *rpl*22, *rpl*23$^{(X2)}$, *rpl*32, *rpl*33, *rpl*36
Ribosomal protein genes (small subunits)	*rps*2, *rps*3, *rps*4, *rps*7$^{(X2)}$, *rps*8, *rps*11, *rps*12$^{(X2)}$ #, *rps*14, *rps*15, *rps*16 *, *rps*18, *rps*19
RNA Polymerase	*rpo*A, *rpo*B, *rpo*C1 *, *rpo*C2
Ribosomal RNA genes	*rrn*4.5$^{(X2)}$, *rrn*5$^{(X2)}$, *rrn*16$^{(X2)}$, *rrn*23$^{(X2)}$
Transfer RNA genes	*trn*I-CAU$^{(X2)}$, *trn*L-CAA$^{(X2)}$, *trn*V-GAC$^{(X2)}$, *trn*I-GAU$^{(X2)}$ *, *trn*A-UGC$^{(X2)}$ *, *trn*R-ACG$^{(X2)}$, *trn*N-GUU$^{(X2)}$, *trn*L-UAG, *trn*P-UGG, *trn*W-CCA, *trn*M-CAU, *trn*V-UAC *, *trn*F-GAA, *trn*L-UAA *, *trn*T-UGU, *trn*S-GGA, *trn*fM-CAU, *trn*G-GCC, *trn*S-UGA, *trn*T-GGU, *trn*E-UUC, *trn*Y-GUA, *trn*D-GUC, *trn*C-GCA, *trn*R-UCU, *trn*G-UCC *, *trn*S-GCU, *trn*Q-UUG, *trn*K-UUU *, *trn*H-GUG
ATP-dependent protease	*clp*P **
Maturase	*mat*K
Hypothetical chloroplast reading frames	*ycf*1$^{(X2)}$, *ycf*2$^{(X2)}$, *ycf*3 **, *ycf*4
Acetyl-CoA carboxylase	*acc*D
C-type cytochrome synthesis gene	*ccs*A
Envelope membrane protein	*cem*A
Translational initiation factor	*inf*A

Note: ^{X2} duplicated genes; * genes with one intron; ** genes with two introns; # trans-spliced gene.

2.2. Codon Usage Analysis

Approximately 48–52% of the six *Bougainvillea* cp genomes are comprised of protein-coding genes with 24,557–26,717 codons. Of these codons, leucine (10.50–10.64%) and isoleucine (8.67–8.76%) are the most abundant amino acids, whereas cysteine (1.10–1.15%) has the lowest frequency (Figure 2). The high leucine frequency can be attributed to the fact that leucine biosynthesis is greatly needed in chloroplasts, due to its important function in photosynthesis-related metabolism [28]. On the other hand, cysteine is quite reactive and considered toxic if it is allowed to accumulate above a certain level [29]. It is also highly susceptible to changes in biological conditions [30]. This pattern is highly uniform in most angiosperm cp genomes [21,31].

In addition, codon usage analysis also revealed that there are particular amino acid codons that are more frequently used or preferred [32]. This codon usage bias has been commonly observed in plant genomes, and it is assumed that preferred codons are normally utilized in highly expressed genes [33]. Based on the relative synonymous codon usage values (RSCU), all amino acid codons found in the six *Bougainvillea* cp genomes exhibit codon preferences, except for tryptophan (UGG) and methionine (AUG) (RSCU = 1). There are 30 codons that are highly favored (RSCU > 1) and 32 codons that are less preferred (RSCU < 1). Moreover, it can be observed that out of the 30 preferred codons, 29 are A/U-ending codons, meaning that C/G-ending codons are less common in the chloroplast genomes. Several reported cp genomes from other families such as Zingiberaceae, Euphorbiaceae, and Asparagaceae have constantly reported the same occurrence [31,34,35]. The bias towards high A/U occurrences in the third nucleotide position of codons appears to be conserved among higher plants.

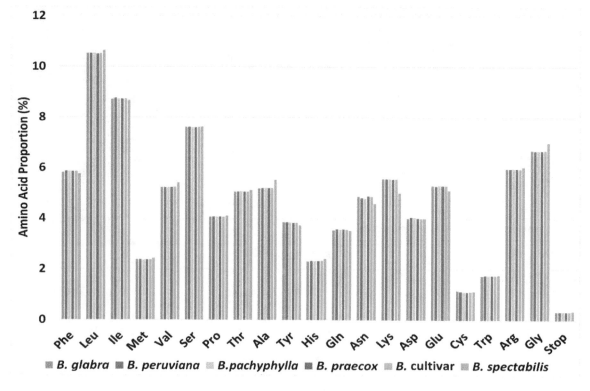

Figure 2. Percentages of amino acid in the protein-coding regions of six *Bougainvillea* chloroplast genomes.

2.3. RNA Editing Sites

Nucleotide sequences in cpDNAs are commonly altered at the transcript level through RNA editing and RNA splicing. Thus, determining the RNA editing sites is necessary in order to understand coding information in the cp genomes. Putative RNA editing sites in the six *Bougainvillea* cp genomes were predicted using the Predictive RNA Editor for Plants (PREP) software, which identified a total of 42 RNA editing sites. Parallel to other seed plants, the annotated RNA editing sites in the *Bougainvillea* cp genomes were C to U conversions located in the first or second position of the codons (Table S2). C to U editing mainly occurred in chloroplast and mitochondrial genomes of angiosperms and gymnosperms, although U to C conversions were also observed in ferns and bryophytes [36]. Consistent with prior studies, most of the editing sites were observed to be distributed mainly in the *ndh* genes (*ndh*A, *ndh*B, *ndh*D, *ndh*F, *ndh*G) of the cp genomes, particularly in *ndh*B (12 sites) [36,37]. In addition, most of the RNA editing amino acids have a tendency to be converted from serine (S) to Leucine (L), and the majority of the changes were from hydrophilic to hydrophobic. These results indicate that RNA editing increases hydrophobicity, which might influence the proteins' secondary or tertiary structures [36,38]. RNA editing can actually lead to re-establishment of conserved amino acid residues, increases in hydrophobicity, and regulation of protein expression [36,38].

2.4. Simple Sequence Repeats and Tandem Repeat Analyses

The analysis of repetitive sequences identified a total of 80, 83, 84, 80, 89, and 89 chloroplast simple sequence repeats (cpSSRs) in *Bougainvillea glabra*, *Bougainvillea peruviana*, *Bougainvillea pachyphylla*, *Bougainvillea praecox*, *Bougainvillea* cv., and *Bougainvillea spectabilis*, respectively (Figure 3A–C). *Bougainvillea spectabilis* and *Bougainvillea* cv. have the highest numbers of cpSSRs, while *Bougainvillea glabra* and *Bougainvillea praecox* have the lowest numbers of cpSSRs. In agreement with other studies, about 71.25–78.3% of the identified cpSSRs are A/T mononucleotides, and most of the dinucleotides, trinucleotides, tetranucleotides, and pentanucleotides contain A/T, hence contributing to the AT richness of *Bougainvillea* cp genomes (Figure 3D). Trinucleotides are rare in *Bougainvillea* cp genomes—only

B. glabra has trinucleotides detected in its cp genome. Chloroplast simple sequence repeats are also primarily found in the intergenic spacers of the LSC region (Figure 3A,B). The frequency of cpSSRs in coding regions is relatively lower, since the rate of mutation in cpSSRs is higher and might affect gene expression [39]. As mentioned in earlier studies, cpSSRs in the non-coding regions typically exhibit intraspecies differences in repeat number [40]. SSRs are also considered as a popular chloroplast marker due to their high level of polymorphism, co-dominance mode of inheritance, and multi-allelic nature [41]. Therefore, identified cpSSRs can be potentially useful for population studies or phylogeographic studies of *Bougainvillea* species and their cultivars. The initial cpSSR study conducted in fifty cultivars of *Bougainvillea* verified that SSRs can be used for molecular characterization and identification of cultivars [41].

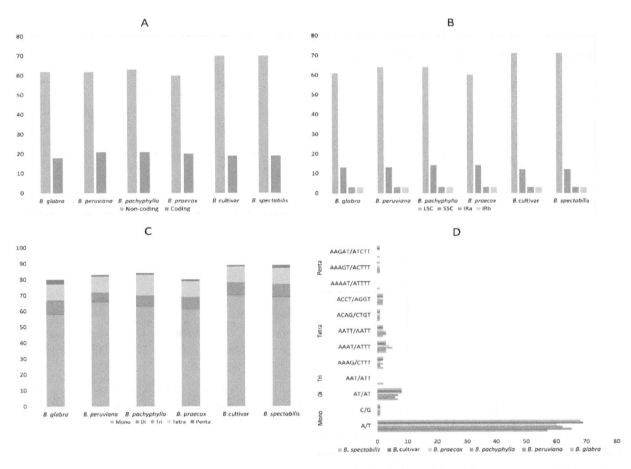

Figure 3. Analysis of simple sequence repeats (SSRs) in six *Bougainvillea* chloroplast genomes. (**A**) Simple sequence repeats detected in coding and non-coding regions of six *Bougainvillea* cp genomes. (**B**) Simple sequence repeats distributions in the LSC, SSC, and IR regions of *Bougainvillea* cp genomes. (**C**) Numbers of different types of SSRs identified in the *Bougainvillea* cp genomes. (**D**) Frequency of various SSR types identified in six *Bougainvillea* cp genomes.

In addition to SSRs, tandem repeats (TRs) were also determined in the six cp genomes using Tandem Repeats Finder v4.04. *Bougainvillea glabra* (26) has the highest number of identified tandem repeats, whereas *Bougainvillea peruviana* (17) and *Bougainvillea pachyphylla* (19) have the lowest. The remaining three (*B. praecox, B. spectabilis, B. cultivar*) have the same number of tandem repeats. Similar to SSR, TRs in the six cp genomes are mostly distributed in the non-coding areas of the LSC region (Figure 4A,B). The lengths of the identified TRs range from 11 to 32 bp, but they are mainly around 15–18 bp (Figure 4C). All TRs found in the coding regions of six genomes are located in the hypothetical chloroplast reading frame *ycf*2. A higher density of TRs in non-coding regions is quite common in angiosperms, since mutations in TRs situated in the known protein-coding regions can result in protein function changes [42]. TR mutations occur due to modifications in the number of repeating units, which can possibly cause unfavorable phenotypes [43].

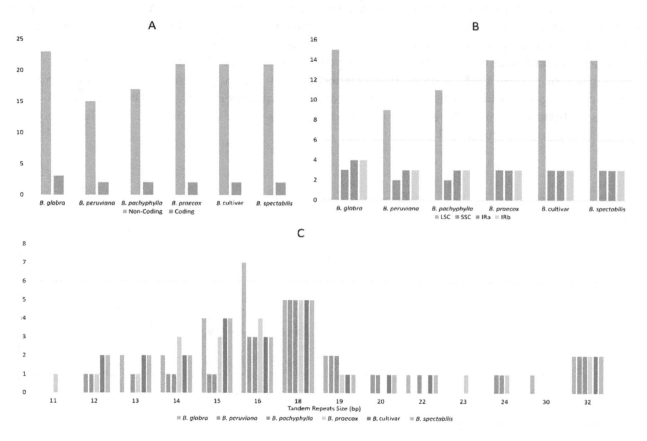

Figure 4. Tandem repeat analysis in six *Bougainvillea* cp genomes. (**A**) Frequency of tandem repeats in the non-coding and coding regions of six cp genomes. (**B**) Distributions of the detected tandem repeats in LSC, SSC, and IR regions. (**C**) Lengths of the identified tandem repeats in all six *Bougainvillea* cp genomes.

2.5. IR Contraction and Expansion

Inverted repeat regions are generally conserved among land plants. They all contain four rRNA, five tRNA genes, and a few protein-coding genes as a result of some expansion and contraction in the IR junctions [44]. Sequences flanking the IR junctions may vary among different species, which might result in genome size variation. In *Bougainvillea* cp genomes, the gene contents and arrangements are highly similar, however there are few contractions and expansions in the IR boundaries (Figure 5). For instance, the IRb–LSC junction (JLB) is situated within the *rps*19 gene in all *Bougainvillea* cp genomes, thus *rps*19 has a 114 bp extension in the IRb region. In addition, all taxa have the *ycf*1 gene in the IRa–SSC junction (JSA), hence producing long fragments of *ycf*1 in the IRb–SSC junction (JSB). It is also evident that the partial copies of *ycf*1 in the IRb regions of all cp genomes overlap with the *ndh*F gene. However, in five *Bougainvillea* cp genomes (*B. glabra*, *B. peruviana*, *B. pachyphylla*, *B. praecox*, *B. cv.*), the *ycf*1 fragments have two bp extension in the IRb–SSC junction (JSB), suggesting an infinitesimal expansion of the IR. This IR contraction and expansion pattern is usually observed in most angiosperms [31,44,45]. All of the sequences used in this study belong to the same genus, so the IR boundary shifts are relatively minor, hence contributing very little to the observed differences in genome size.

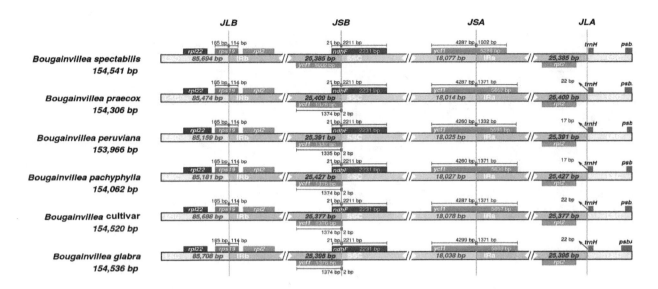

Figure 5. Comparisons of LSC, SSC, and IRs junctions among the six chloroplast genomes.

2.6. Sequence Variation Analyses among Bougainvillea cp Genomes

Sequence divergence among *Bougainvillea* cp genomes was compared through multiple sequence alignment carried out in mVISTA. Generally, no significant rearrangements were observed among *Bougainvillea* cp genomes, but several regions displayed higher variation than others. Resulting alignment analysis using *Bougainvillea glabra* as a reference showed that the coding regions of the *Bougainvillea* cp genomes are less divergent compared to the non-coding regions, whereas the non-coding regions are more variable than the coding regions (Figure 6). Likewise, the IR regions have lower divergence and more conserved compared to LSC and SSC regions. Specifically, protein-coding genes such as *ycf*1 and *ndh*F and non-coding regions such as start-*psb*A, *rps*16-*psb*K, *psb*I-*atp*A, *psa*A-*ycf*3, *pet*D-*rpo*A, and *ndh*F-*rpl*32 are considered to be highly divergent regions.

To further examine the sequence divergence in the *Bougainvillea* cp genomes, nucleotide diversity (Pi) values were calculated using DnaSP v5.10 (Figure 7). Similar standards to those employed in the Zingiberaceae family were used to determine the divergence hotspots in coding and non-coding regions [31]. For the 79 unique protein-coding genes, the nucleotide diversity values range from 0 to 0.2282, with an average of 0.00307. Eleven protein-coding genes (*pet*N, *psa*I, *psb*J, *pet*G, *rpo*A, *rps*8, *rps*11, *rpl*22, *rps*19, *ycf*1, *ndh*F) positioned at the single copy regions exhibit higher Pi values (>0.005) (Figure 7A). On the other hand, the Pi values of the non-coding regions range from 0 to 0.03629, with an average of 0.00798. Among these regions, 10 regions (start-*psb*A, *rps*16 CDS1-*psb*K, *psb*K-*psb*I, *psb*I-*atp*A, *cem*A-*pet*A, *pet*D CDS2-*rpo*A, *ndh*F-*rpl*32, *rpl*32-*ccs*A, *ndh*E-*ndh*G, *ndh*I-*ndh*A CDS2) have high diversity values (>0.012) (Figure 7B). Most of these divergence hotspots are located in the LSC and SSC regions, signifying that IR regions are less variable. These results also exemplify that most hypervariable regions shown in the mVISTA alignment have higher nucleotide diversity (Pi) values as well.

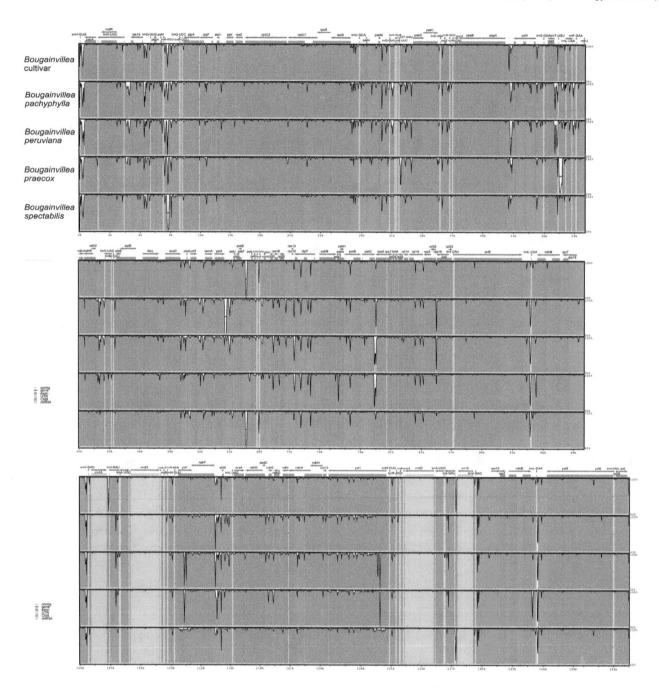

Figure 6. Sequence identity plots (mVISTA) among *Bougainvillea* species. Alignments of the five *Bougainvillea* plastomes, with *Bougainvillea glabra* as the reference genome. Genes are color-coded, whereby pink regions represent conserved non-coding sequences (CNS) and purple regions indicate protein-coding sequences. Grey arrows above the alignments indicate gene directions. The y-axis denotes the percentages of identity, ranging between 50% and 100%.

The overall variation among the *Bougainvillea* sequences was analyzed through mVISTA and DnaSP v5.10, however to further elucidate the differences among taxa, SNP/indel analysis was also conducted using MUMmer 4 and Geneious 2020.2. Using *Bougainvillea glabra* as a reference, the single-nucleotide polymorphisms (SNPs) and indels (insertions deletions) were identified in the other five *Bougainvillea* cp genomes. General results revealed that more SNPs and indels were detected in *B. pachyphylla* and *B. peruviana* in contrast to the other three plastomes. When aligned to *B. glabra*, *B. pachyphylla* has 571 SNPs in the non-coding regions and 317 SNPs in the coding regions, whilst *B. peruviana* has 545 SNPs in the non-coding regions and 320 SNPs coding regions (Figure 8A).

Lesser SNPs were found in *B. praecox* (309, 200), *B. spectabilis* (283, 195), and *B. cultivar* (282, 163), indicating that these three have higher sequence similarities to *B. glabra* (Figure 8A). There are also around 45–58 protein-coding regions with SNPs; specifically, *ycf*1, *ndh*F, *rpo*C2, and *rpo*C1 have higher numbers of SNPs (Figure 8B, Table S3). In all *Bougainvillea*, the *ycf*1 gene has the highest numbers of both synonymous and non-synonymous SNPs. The RNA polymerase genes (*rpo*A, *rpo*B, *rpo*C2, *rpo*C1), particularly *rpo*C1 and *rpo*C2, predominantly contain synonymous and non-synonymous SNPs (Table S3). Similar to the mVISTA alignment and nuclear diversity analyses, *ycf*1 and *ndh*F are also considered highly variable due to their high numbers of SNPs.

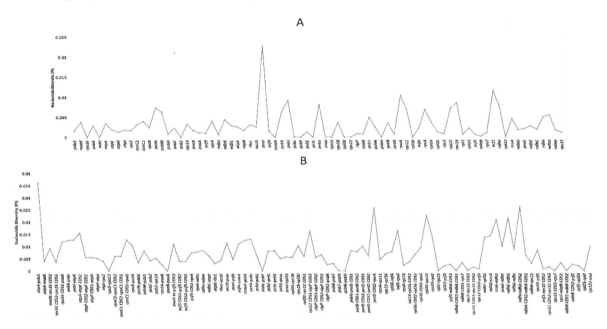

Figure 7. Nucleotide diversity (Pi) of various regions in *Bougainvillea* chloroplast genomes. **(A)** Nucleotide diversity values in the protein-coding regions. **(B)** Nucleotide diversity values in the non-coding regions.

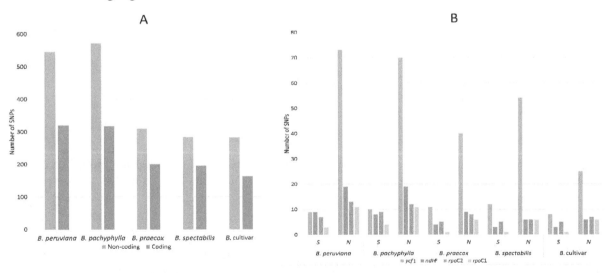

Figure 8. Summary of SNPs detected in the five *Bougainvillea* chloroplast genomes. **(A)** Frequency of SNPs in the coding and non-coding regions. **(B)** Protein-coding genes with highest numbers of synonymous and non-synonymous SNPs.

A similar trend can be observed in *Bougainvillea* indels, *B. peruviana* (366), and *B. pachyphylla* (357), which have more indels in comparison to *B. praecox* (218), *B. spectabilis* (171), and *B. cultivar* (171) (Figure 9A). Both *B. spectabilis* and *B. cultivar* have lesser indels, indicating less differences from

B. glabra. The presence of the large deletion in the *clp*P intron of *B. spectabilis* (55 bp), *B. cultivar* (55 bp), *B. pachyphylla* (47 bp), and *B. peruviana* (29 bp) mainly differentiates these four from *B. glabra* (Figure S1). On the other hand, both *B. peruviana* and *B. pachyphylla* differ from *B. glabra* by having a 43 bp deletion in the spacer in between *rpl*22 and *rps*19 (Figure S2). Although small indels are quite common in all *Bougainvillea* cp genomes, more one-bp indels (116) are discovered in *B. peruviana* and *B. pachyphylla* (Figure 9C). The presence of copious amounts of small indels and several large indels results in high sequence variation in *B. peruviana* and *B. pachyphylla* (Figure 9A,C). In addition, the majority of the indels detected in the five cp genomes are in the non-coding regions—only *ycf*1, *ycf*2, *cem*A, *ndh*K, *ndh*D, *rpo*A, *psb*N, and *rps*15 have indels (Figure 9B). Again, *ycf*1 and one of the RNA polymerase genes (*rpo*A) exhibit a high degree of variation due to the presence of indels.

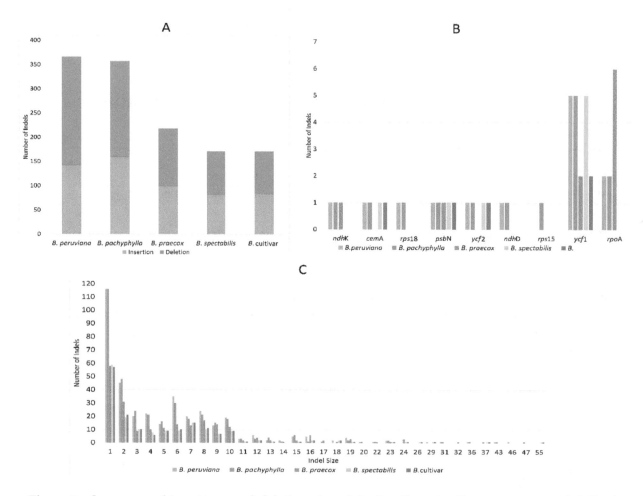

Figure 9. Summary of insertions and deletions found in five *Bougainvillea* cp genomes. (**A**) Total number of indels in five *Bougainvillea* species. (**B**) Numbers of indels located in the protein-coding genes. (**C**) Lengths of indels identified in five cp genomes.

Consolidating the results from various analyses, regions such as *ycf*1, *ndh*F, and *rpo*A are potential molecular markers. Consistent with all analyses, *ycf*1 has high sequence divergence, high diversity value, and has the most SNPs and indels (Figure 6, Figure 7A, Figure 8B, and Figure 9B). Another good barcode candidate is the *ndh*F gene, which displays high sequence divergence, high Pi value, and high SNP density (Figure 6, Figure 7A, and Figure 8B). Likewise, the nucleotide diversity analysis also showed that *rpo*A is considered a variable region (Figure 7A). Additionally, it also has good amounts of detected SNPs and indels (Table S2, Figure 9B). Therefore, to check whether these regions can actually differentiate *Bougainvillea* species, phylogenetic analyses was also conducted and compared to the phylogenetic tree inferred from all the protein-coding regions in *Bougainvillea* cp genomes.

2.7. Phylogenetic Analysis

Several of the phylogenetic reconstructions focusing on the Nyctaginaceae family were based on a single gene or a few gene regions from plastid or nuclear DNA [15,46], but none of them explicitly discussed the relationships among species of *Bougainvillea*. With the advent of next-generation sequencing, several cultivated *Bougainvillea spectabilis* and *Bougainvillea glabra* genomes have been published, but due to the lack of other available sequences, phylogenetic analyses have focused only on *Bougainvillea's* placement within the Nyctaginaceae family [17,18]. Thus, in this study, the phylogeny of *Bougainvillea* was reconstructed using eight complete chloroplast genomes of *Bougainvillea* (including the sequences from GenBank) and six other species from the Nyctaginaceae family. Species from Petiveriaceae were used as outgroups. In addition, the highly variable regions from the same dataset were extracted and used to construct a phylogenetic tree. As a whole, the resulting ML and BI trees based on complete cp genomes have a consistent and well-supported topology (Figure 10). Likewise, a similar topology was observed for the ML tree generated from the potential markers obtained from the sequence variation analyses (Figure S3).

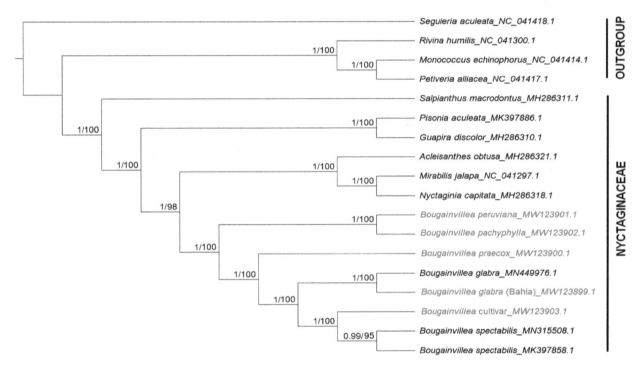

Figure 10. Maximum Likelihood (ML) and Bayesian Inference (BI) consensus tree based on the 79 concatenated protein-coding regions of 14 Nyctaginaceae cp genomes. Species from Petiveriaceae were used as outgroups. Numbers on each node represent bootstrap support and Bayesian posterior probability (BPP) values. Branches with bootstrap values > 75 and BPP values > 95 are considered as highly supported.

Bougainvilleeae, as represented by the *Bougainvillea* species, has higher affinity to the Nyctagineae tribe (*Acleisanthes, Mirabilis, Nyctaginia*) compared to other tribes within Nyctaginaceae. Within the *Bougainvillea* genus, *Bougainvillea peruviana* and *Bougainvillea pachyphylla* appear to be the basal taxa, while *Bougainvillea praecox* is a sister to two distinct subclades: the "glabra" and "spectabilis" subclades.

Based on its morphology, *B. peruviana* was assumed to be closely related to either *B. glabra* [12,47] or *B. pachyphylla* [48], but molecular data from cp genomes revealed that it has a closer relationship to *B. pachyphylla*. The sequence variation analysis discussed earlier also showed that *B. peruviana* and *B. pachyphylla* are the two genomes that differ most from *B. glabra*. In terms of morphological structure, *B. pachyphylla* is not too distinct from *B. peruviana*—it differs only by having thick and leathery leaves and a densely puberulent perianth in comparison to *B. peruviana* [48,49]. *B. peruviana* is considered one of the most sTable Species, as there is less variation in the shapes of the bracts and leaves [4]. This species is also not as vigorous as the cultivars and hybrids [4]. Therefore, the striking morphological and molecular similarities of *B. pachyphylla* to *B. peruviana* must be taken into consideration in future taxonomic revisions for the genus *Bougainvillea*.

On the other hand, *Bougainvillea praecox* is actually distinct from other *Bougainvillea* by being sparsely spiny or unarmed [12]. It is not usually used for cultivation, but it has showy white bracts that become greenish when dried. Based on the tree, *B. praecox* is the sister to two distinct groups containing *Bougainvillea glabra* and *Bougainvillea spectabilis*.

The close relationship between *Bougainvillea glabra* and *Bougainvillea* spectabilis is not surprising, as morphologically *B. glabra* is highly similar to *B. spectabilis* [10]. *B. glabra* differs only by having puberulent to glabrate branches and leaves, while *B. spectabilis* has fulvous–villous branches and densely villous abaxial leaf surfaces [12,49]. In addition, horticulturists have observed that *B. spectabilis* has stouter spines and wavy bracts. The sequences from GenBank are mostly cultivated and are either classified as *B. glabra* or *B. spectabilis*. Similarly, the *Bougainvillea* cultivar clusters together with *B. spectabilis* and *B. glabra*, since most of the cultivars are actually crosses between *B. spectabilis* and *B. glabra*. Many crosses between the two species have produced new hybrids and horticultural cultivars. As presented earlier, less sequence variation was observed among these three species, so closer relationships are also evident in the resulting trees.

In general, it can be observed that rarely cultivated and wild species of *Bougainvillea* (*B. pachyphylla*, *B. peruviana*, *B. praecox*) diverged earlier than the commonly cultivated species of *Bougainvillea*. These results showed that information inferred from highly variable regions and complete cp genome sequences resulted in consistent phylogeny. The potential barcodes can also successfully differentiate *Bougainvillea* species and cultivars, meaning more samples can be sequenced to give a broader view of the evolutionary relationships within this genus.

3. Materials and Methods

3.1. Plant Samples and DNA Extraction

The leaf samples of the four *Bougainvillea* species and one cultivar used in this study were collected from Brazil, Peru, Ecuador, and China. Two samples were obtained from Brazil, namely *Bougainvillea glabra* (M.B.M. da Cruz 0001 NY) from Ilhéus, Bahia; and *Bougainvillea praecox* (Chen T. 2012063001 SZG) from Jardim Botanico Plantarum, Nova Odes. *Bougainvillea pachyphylla* (Sagastegui A. et al. 15924 MO) was from Chota, Cajamarca, Peru, while *Bougainvillea peruviana* (Chen T. et al. 2014052606 SZG) was collected on the way from Loja to Macara, Ecuador. Lastly, the *Bougainvillea* cultivar (Chen T. 2020031204 SZG), which is originally from India, was acquired from Fairy Lake Botanical Garden, Shenzhen, China. Fresh leaves of *B. glabra*, *B. praecox*, *B. peruviana*, and *B. cultivar* were dried in silica gel, while the *B. pachyphylla* leaf sample was obtained from herbarium material. Total genomic DNA was extracted from each sample through the modified CTAB (cetyl trimethylammonium bromide) method [50], then the DNA quality was checked through agarose gel electrophoresis, nanodrop method, and Qubit 2.0.

3.2. Chloroplast Genome Sequencing, Assembly, and Annotation

After the DNA quality assessment, DNA was sheared to fragments using a Covaris ultrasonic disruptor. Short-insert (350–400 bp) libraries were constructed using the Nextera XT DNA Library Preparation Kit. Sequencing was performed in Illumina Novseq 600 platform, and for each sample around 10.0 Gb of raw data were generated, with an average read length of 150 bp and sequencing depths of 263.8X–1627.3X. The Illumina raw sequence reads were filtered for adaptor sequences, undersized inserts, duplicated reads, and low-quality reads using the NGS-QC (Next Generation Sequencing Quality Control) toolkit [51]. High-quality reads were assembled into contigs via the de novo assembler SPAdes 3.11.0, using a k-mer set of 93, 105, 117, 121 [52].

The assembled *Bougainvillea* cp genomes were annotated using Plann software [53] and the online annotation tools cpGAVAS [54] and DOGMA [55]. Protein-coding gene annotation was also verified by BlastN searches of the non-redundant database at the National Center for Biotechnology Information (NCBI). RNAmmer 1.2 [56] and tRNAscan-SE v2.0 [57] were used to annotate rRNA and tRNA genes, respectively. The genome maps of the *Bougainvillea* cp genomes were generated using the online program OGDRAW v1.3.1 [58], then the cp genome sequences were deposited in National Center for Biotechnology Information (NCBI) GenBank with the accession numbers MW123899–MW123903.

3.3. Codon Usage and RNA Editing Sites Prediction

Relative synonymous codon usage (RSCU) was determined for all protein-coding genes using MEGAX software [59]. Amino acid frequency values were also obtained from MEGA X [59] and manually verified. In addition, probable RNA editing sites in six *Bougainvillea* cp genomes were identified using Predictive RNA Editor for Plants (PREP) suite (http://prep.unl.edu/), with a cutoff value of 0.8 [60]. The default settings (35 coding sequences) were used to predict putative RNA editing sites.

3.4. Repeat Analysis

Simple sequence repeats (SSRs) in the *Bougainvillea* cp genomes were identified using MISA (http://pgrc.ipk-gatersleben.de/misa/) [61], with motif sizes of one to six nucleotides and thresholds of 10, 5, 5, 3, 3, and 3. Similar parameters had been used in other angiosperm cp genomes. Aside from SSRs, tandem repeats were also analyzed with the aid of Tandem Repeats Finder Program v4.04 [62] using default parameters. All identified repeats were manually filtered and redundant results were excluded. For tandem repeats, repeats with more than 90% sequence identity were included.

3.5. Genome Comparison and Divergence Analyses

The five newly sequenced *Bougainvillea* cp genomes were compared to the available cpDNA sequence of *Bougainvillea spectabilis*. For comparison, sequence alignment was carried out in mVISTA (http://genome.lbl.gov/vista/mvista/about.shtml) using Shuffle-LAGAN mode [63]. *Bougainvillea glabra* was used as the reference. In addition, the nucleotide diversity (Pi) values of the cp genomes were determined to detect various divergence hotspots. Using MAFFT v7.388 [64], *Bougainvillea* cp sequences were aligned, then DnaSP v5.10 [65] was utilized to compute the nucleotide diversity (Pi) values. Sliding window analysis was used, with a window length of 600 bp and step size of 200 bp. The expansions and contractions in the IR junctions were also depicted using the online tool IRscope (https://irscope.shinyapps.io/irapp/) [66]. To further examine variations among the *Bougainvillea* cp genomes, SNPs and indels were also identified and located using MUMmer 4 [67] and Geneious Prime 2020.2 [68]. The *Bougainvillea glabra* cp genome was also used as the reference.

3.6. Phylogenetic Analyses

Together with the five newly sequenced *Bougainvillea* cp genomes, three available *Bougainvillea* sequences and six Nyctaginaceae sequences deposited in NCBI GenBank were included in the dataset. Four complete cp genomes from the closely related family Petiveriaceae were used as outgroups. From these sequences, 79 protein-coding regions were extracted using Geneious Prime 2020.2. Multiple alignments of these extracted regions were conducted in MAFFT v7.388 [64]. Ambiguous regions in the alignments were manually removed, then the filtered alignments were concatenated using Geneious Prime 2020.2 [68], which generated a final alignment of 66,426 bp. Aside from the 79 concatenated protein-coding regions, sequenced alignment was also conducted in the selected highly divergent regions to test their barcode effectivity. The genes *ndh*F, *rpo*A, and *ycf*1 were extracted and concatenated using Geneious Prime 2020.2 [68]

Maximum likelihood (ML) analyses were performed using RAxML 8.2.11 [69] and IQ-TREE v6.10 [70] with the GTR+I+G (General Time Reversible + Invariable Sites + Gamma Distribution) nucleotide substitution model. The best fit model was determined through jModelTest2, executed in

CIPRES Gateway [71]. The bootstrap consensus tree was inferred from 1000 replicates. Furthermore, Bayesian inference tests were also conducted in MrBayes 3.2.6 [72] with the general time reversible (GTR) DNA substitution model and the gamma distribution rate variation across sites. Bayesian Inference analyses were carried out in CIPRES Gateway [71], with four Markov chain Monte Carlo (MCMC) running for one million generations, with sampling every 1000 generations and the first 25% being discarded as burn-in. The resulting branches with bootstrap support > 75% for maximum likelihood and Bayesian posterior probabilities (BPP) > 0.95 for BI were considered as significantly supported.

4. Conclusions

Bougainvillea is deemed to be one of the most popular genera in the Nyctaginaceae family, as it is used mainly as an ornamental plant. However, despite its horticultural value, its molecular and phylogenetic aspects are not well-researched. Hence, in this paper, the complete chloroplast genomes of four species (B. glabra, B. peruviana, B. pachyphylla, B. praecox) and one cultivar were sequenced and analyzed. In general, all Bougainvillea cp genomes newly sequenced from GenBank, including Bougainvillea spectabilis, have similar genome structures and features. They all display the typical quadripartite structure and have the same numbers of genes (131) and introns (17). Highly similar patterns were also obtained from codon usage, RNA editing, and repeat analyses. Although the Bougainvillea cp genomes are highly conserved and no rearrangement was observed, several highly divergent regions were identified. Moreover, phylogenetic analyses revealed the early divergence of the rarely cultivated or wild Bougainvillea (B. peruviana, B. pachyphylla, B. praecox). B. pachyphylla and B. peruviana were shown to be the basal taxa of Bougainvillea. Close relationships among Bougainvillea glabra, Bougainvillea spectabilis, and Bougainvillea cultivar were also confirmed. These results show that chloroplast genomes can provide sufficient information that can be used in phylogenetic studies. However, more cp genome sequences are needed to further elucidate relationships among Bougainvillea species and cultivars.

Supplementary Materials:
Table S1: Codon usage. Table S2: RNA editing sites. Table S3: List of synonymous and non-synonymous SNPs. Figure S1: Large deletions in clpP introns of Bougainvillea cp genomes. Figure S2: Large deletions in Bougainvillea peruviana and Bougainvillea pachyphylla. Figure S3: Maximum likelihood tree based on potential Bougainvillea barcodes.

Author Contributions: All analyses, M.A.C.B.; writing of the manuscript and preparation of figures, tables, and graphs, M.A.C.B.; manuscript revision, M.A.C.B. and T.C.; conceptualization, planning, and supervision, T.C. and Y.D.; sample collection, T.C. and Y.Z.; resources and funding acquisition, T.C. and Z.H. All authors have read and agreed to the published version of the manuscript.

Acknowledgments: The authors would like to show sincere gratitude to Aimee Caye G. Chang and Sadaf Adnan for providing valuable suggestions related to the analyses; and to Nora Oleas of Technological University Indoamerica and Harri Lorenzi of the Jardim Botânico Plantarum for their kind assistance in the sample collection.

References

1.	Mabberley, D.J. The Plant Book; Cambridge Univercity Press: Cambridge, UK, 1987; pp. 1–706.
2.	Bittrich, V.; Kühn, U. Nyctaginaceae. In The Families and Genera of Flowering Plants; Kubitzki, K., Rohwer, J.G., Bittrich, V., Eds.; Springer: Berlin, Germany, 1993; Volume 2, pp. 473–486.
3.	Bremer, B.; Bremer, K.; Chase, M.W.; Fay, M.F.; Reveal, J.L.; Soltis, D.E.; Soltis, P.S.; Stevens, P.F.; Anderberg, A.A.; Moore, M.J.; et al. An update of the Angiosperm Phylogeny Group classification for the orders and families of flowering plants: APG II. Bot. J. Linn. Soc. **2003**, 141, 399–436.
4.	Kobayashi, K.D.; McConnell, J.; Griffis, J. Bougainvillea. In Ornamentals and Flowers; College of Tropical Agriculture and Human Resources, University of Hawaii: Honolulu, HI, USA, 2007; Volume OF-38, pp. 1–12.

5. Plants of the World Online. Facilitated by the Royal Botanic Gardens, Kew. Available online: http://www.plantsoftheworldonline.org/ (accessed on 10 October 2020).

6. Kulshreshtha, K.; Rai, A.; Mohanty, C.S.; Roy, R.K.; Sharma, S.C. Particulate pollution mitigating ability of some plant species. *Int. J. Environ. Res.* **2009**, *3*, 137–142.

7. Chauhan, P.; Mahajan, S.; Kulshrestha, A.; Shrivastava, S.; Sharma, B.; Goswamy, H.M.; Prasad, G.B.K.S. *Bougainvillea spectabilis* Exhibits Antihyperglycemic and Antioxidant Activities in Experimental Diabetes. *J. Evid. Based Complementary Altern. Med.* **2016**, *21*, 177–185. [CrossRef] [PubMed]

8. Abarca-Vargas, R.; Petricevich, V.L. *Bougainvillea* Genus: A Review on Phytochemistry, Pharmacology, and Toxicology. *J. Evid. Based Complementary Altern. Med.* **2018**, *2018*, 9070927. [CrossRef]

9. Ogunwande, I.A.; Avoseh, O.N.; Olasunkanmi, K.N.; Lawal, O.A.; Ascrizzi, R.; Guido, F. Chemical composition, anti-nociceptive and anti-inflammatory activities of essential oil of *Bougainvillea glabra*. *J. Ethnopharmacol.* **2019**, *232*, 188–192. [CrossRef]

10. Abarca-Vargas, R.; Petricevich, V.L. Extract from *Bougainvillea × buttiana* (Variety Orange) Inhibits Production of LPS-Induced Inflammatory Mediators in Macrophages and Exerts a Protective Effect In Vivo. *Biomed. Res. Int.* **2019**, *2019*, 2034247. [CrossRef]

11. Rauf, M.A.; Oves, M.; Rehman, F.U.; Khan, A.R.; Husain, N. *Bougainvillea* flower extract me.diated zinc oxide's nanomaterials for antimicrobial and anticancer activity. *Biomed. Pharmacother.* **2019**, *116*, 108983. [CrossRef]

12. Standley, P.C. *The Nyctaginaceae and Chenopodiaceae of Northwestern South America*; Field Museum of Natural History-Botanical Series: Chicago, IL, USA, 1931; Volume 11, pp. 73–114.

13. Heimerl, A. Nyctaginaceae. In *Engler & Prantl, Naturl. Pflanzenfam*, 2nd ed.; W. Engelmann: Leipzig, Germany, 1934; Volume 4, pp. 86–134.

14. Toursarkissian, M. Las Nictaginaceas argentinas. Revista Museo Argentino de Ciencias Naturales Bernardino Rivadavia. *Botanica* **1975**, *5*, 1–83.

15. Douglas, N.A.; Manos, P.S. Molecular phylogeny of Nyctaginaceae: Taxonomy, biogeography and characters associated with a radiation of xerophytic genera in North America. *Am. J. Bot.* **2007**, *94*, 856–872. [CrossRef]

16. Yao, G.; Jin, J.J.; Lia, H.T.; Yanga, J.B.; Mandalad, V.S.; Croleyd, M.; Mostowd, R.; Douglas, N.A.; Chase, M.W.; Christenhuszg, M.J.M.; et al. Plastid phylogenomic insights into the evolution of Caryophyllales. *Mol. Phylogenet. Evol.* **2019**, *134*, 74–86. [CrossRef]

17. Ni, J.; Lee, S.Y.; Hu, X.; Wang, W.; Zhang, J.; Ruan, L.; Dai, S.; Liu, G. The complete chloroplast genome of a commercially exploited ornamental plant, *Bougainvillea glabra* (Caryophyllales: Nyctaginaceae). *Mitochondrial DNA Part B* **2019**, *4*, 3390–3391. [CrossRef]

18. Wang, N.; Qiu, M.Y.; Yang, Y.; Li, J.W.; Xou, X.X. Complete chloroplast genome sequence of *Bougainvillea spectabilis* (Nyctaginaceae). *Mitochondrial DNA Part B* **2019**, *4*, 4010–4011. [CrossRef]

19. Moore, M.J.; Soltis, P.S.; Bell, C.D.; Burleigh, J.G.; Soltis, D.E. Phylogenetic analysis of 83 plastid genes further resolves the early diversification of eudicots. *Proc. Natl. Acad. Sci. USA* **2010**, *107*, 4623–4628. [CrossRef]

20. Li, B.; Li, Y.; Cai, Q.; Lin, F.; Huang, P.; Zheng, Y. Development of chloroplast genomic resources for Akebia quinata (Lardizabalaceae). *Conserv. Genet. Resour.* **2016**, *8*, 447–449. [CrossRef]

21. Munyao, J.N.; Dong, X.; Yang, J.X.; Mbandi, E.M.; Wanga, V.O.; Oulo, M.A.; Saina, J.K.; Musili, P.M.; Hu, G.W. Complete Chloroplast Genomes of *Chlorophytum comosum* and *Chlorophytum gallabatense*: Genome Structures, Comparative and Phylogenetic Analysis. *Plants* **2020**, *9*, 296. [CrossRef] [PubMed]

22. Kwon, E.C.; Kim, J.H.; Kim, N.S. Comprehensive genomic analyses with 115 plastomes from algae to seed plants: Structure, gene contents, GC contents, and introns. *Genes Genome* **2020**, *42*, 553–570. [CrossRef] [PubMed]

23. Xu, J.; Feng, D.; Song, G.; Wei, X.; Chen, L.; Wu, X.; Li, X.; Zhu, Z. The first intron of rice EPSP synthase enhances expression of foreign gene. *Sci. China Life Sci.* **2003**, *46*, 561–569. [CrossRef]

24. Daniell, H.; Lin, C.; Yu, M.; Chang, W. Chloroplast genomes: Diversity, evolution, and applications in genetic engineering. *Genome Biol.* **2016**, *17*, 134. [CrossRef]

25. Vu, H.T.; Tran, N.; Nguyen, T.H.; Vu, Q.L.; Bui, M.H.; Le, M.T.; Le, L. Complete chloroplast genome of *Pahiopedilum delenatii* and phylogenetic relationships among Orchidaceae. *Plants* **2020**, *9*, 61. [CrossRef]

26. Wu, C.S.; Chaw, S.M. Evolutionary stasis in Cycad plastomes and the first case of plastome GC-biased gene conversion. *Genome Biol. Evol.* **2015**, *7*, 2000–2009. [CrossRef]

27. Niu, Z.; Xue, Q.; Wang, H.; Xie, X.; Zhu, S.; Liu, W.; Ding, X. Mutational Biases and GC-Biased Gene Conversion AT to GC Content in the Plastomes of *Dendrobium* Genus. *Int. J. Mol. Sci.* **2017**, *18*, 2307.

28. Knill, T.; Reichelt, M.; Paetz, C.; Gershenzon, J.; Binder, S. *Arabidopsis thaliana* encodes a bacterial-type heterodimeric isopropylmalate isomerase involved in both Leu biosynthesis and the Met chain elongation pathway of glucosinolate formation. *Plant Mol. Biol.* **2009**, *71*, 227–239. [CrossRef]

29. Hildebrandt, T.M.; Nunes Nesi, A.; Araujo, W.L.; Braun, H.P. Amino Acid Catabolism in Plants. *Mol. Plant.* **2015**, *8*, 1563–1579. [CrossRef] [PubMed]

30. Marino, S.M.; Gladyshev, V.N. Analysis and functional prediction of reactive cysteine residues. *J. Biol. Chem.* **2012**, *287*, 4419–4425. [CrossRef] [PubMed]

31. Li, D.M.; Zhu, G.F.; Xu, Y.C.; Ye, Y.J.; Liu, J.M. Complete Chloroplast Genomes of Three Medicinal *Alpinia* Species: Genome Organization, Comparative Analyses and Phylogenetic Relationships in Family Zingiberaceae. *Plants* **2020**, *9*, 286. [CrossRef] [PubMed]

32. Wu, X.M.; Wu, S.F.; Ren, D.M.; Zhu, Y.P.; He, F.C. The analysis method and progress in the study of codon bias. *Yi Chuan = Hered.* **2007**, *29*, 420–426. [CrossRef]

33. Zhou, Z.; Dang, Y.; Zhou, M.; Li, L.; Yu, C.H.; Fu, J.; Chen, S.; Liu, Y. Codon usage is an important determinant of gene expression levels largely through its effects on transcription. *Proc. Natl. Acad. Sci. USA* **2016**, *113*, E6117–E6125. [CrossRef]

34. Wang, Z.; Xu, B.; Li, B.; Zhou, Q.; Wang, G.; Jiang, X.; Wang, C.; Xu, Z. Comparative analysis of codon usage patterns in chloroplast genomes of six *Euphorbiaceae* species. *PeerJ* **2020**, *8*, e8251. [CrossRef]

35. Lee, S.R.; Kim, K.; Lee, B.Y.; Lim, C.E. Complete chloroplast genomes of all six *Hosta* species occurring in Korea: Molecular structures, comparative, and phylogenetic analyses. *BMC Genom.* **2019**, *20*, 833. [CrossRef]

36. He, P.; Huang, S.; Xiao, G.; Zhang, Y.; Yu, J. Abundant RNA editing sites of chloroplast protein-coding genes in *Ginkgo biloba* and an evolutionary pattern analysis. *BMC Plant Biol.* **2016**, *16*, 257. [CrossRef]

37. Sasaki, T.; Yukawa, Y.; Miyamoto, T.; Obokata, J.; Sugiura, M. Identification of RNA Editing Sites in Chloroplast Transcripts from the Maternal and Paternal Progenitors of Tobacco (*Nicotiana tabacum*): Comparative Analysis Shows the Involvement of Distinct Trans-Factors for *ndh*B Editing. *Mol. Biol. Evol.* **2003**, *20*, 1028–1035. [CrossRef] [PubMed]

38. Yura, K.; Go, M. Correlation between amino acid residues converted by RNA editing and functional residues in protein three-dimensional structures in plant organelles. *BMC Plant Biol.* **2008**, *8*, 79. [CrossRef] [PubMed]

39. Vieira, M.L.; Santini, L.; Diniz, A.L.; de Freitas Munhoz, C. Microsatellite markers: What they mean and why they are so useful. *Genet. Mol. Biol.* **2016**, *39*, 312–328. [CrossRef] [PubMed]

40. Ebert, D.; Peakall, R. Chloroplast simple sequence repeats (cpSSRs): Technical resources and recommendations for expanding cpSSR discovery and applications to a wide array of plant species. *Mol. Ecol. Resour.* **2009**, *9*, 673–690. [CrossRef] [PubMed]

41. Kumar, P.; Janakiram, T.; Bhat, K.V.; Jain, R.; Prasad, K.V.; Prabhu, K.V. Molecular characterization and cultivar identification in *Bougainvillea* spp. using SSR markers. *Indian J. Agric. Sci.* **2014**, *84*, 1024–1030.

42. Zhao, Z.; Guo, C.; Sutharzan, S.; Li, P.; Echt, C.S.; Zhang, J.; Liang, C. Genome-wide analysis of tandem repeats in plants and green algae. *G3 (Bethesda)* **2014**, *4*, 67–78. [CrossRef]

43. Verstrepen, K.J.; Jansen, A.; Lewitter, F.; Fink, G.R. Intragenic tandem repeats generate functional variability. *Nat. Genet.* **2005**, *37*, 986–990. [CrossRef]

44. Li, J.; Zhang, D.; Ouyang, K.; Chen, X. The complete chloroplast genome of the miracle tree *Neolamarckia cadamba* and its comparison in Rubiaceae family. *Biotechnol. Biotechnol. Equip.* **2018**, *32*, 1314–3530. [CrossRef]

45. Feng, S.; Zheng, K.; Jiao, K.; Cai, Y.; Chen, C.; Mao, Y.; Wang, L.; Zhan, X.; Ying, Q.; Wang, H. Complete chloroplast genomes of four *Physalis* species (Solanaceae): Lights into genome structure, comparative analysis, and phylogenetic relationships. *BMC Genom.* **2020**, *20*, 242. [CrossRef]

46. Douglas, N.; Spellenberg, R. A new tribal classification of Nyctaginaceae. *Taxon* **2010**, *59*, 905–910. [CrossRef]

47. Tripathi, S.; Singh, S.; Roy, R.K. Pollen morphology of *Bougainvillea* (Nyctaginaceae): A popular ornamental plant of tropical and sub-tropical gardens of the world. *Rev. Palaeobot. Palynol.* **2017**, *239*, 31–46. [CrossRef]

48. Standley, P.C. *Studies of American Plants*; Field Museum of Natural History-Botanical Series: Chicago, IL, USA, 1931; Volume 8, pp. 44–48.

49. Standley, P.C. Nyctaginaceae. In *Flora of Peru*; Macbride, J.F., Ed.; Field Museum of Natural History-Botanical Series: Chicago, IL, USA, 1937; Volume 13, pp. 518–546.

50. Murray, M.G.; Thompson, W.F. Rapid isolation of high molecular weight plant DNA. *Nucleic Acids Res.* **1980**, *8*, 4321–4326. [CrossRef]

51. Patel, R.K.; Jain, M. NGS QC Toolkit: A toolkit for quality control of next generation sequencing data. *PLoS ONE* **2012**, *7*, e30619. [CrossRef]

52. Bankevich, A.; Nurk, S.; Antipov, D.; Gurevich, A.A.; Dvorkin, M.; Kulikov, A.S.; Lesin, V.M.; Nikolenko, S.I.; Pham, S.; Prjibelski, A.D.; et al. SPAdes: A new genome assembly algorithm and its applications to single-cell sequencing. *J. Comput. Biol.* **2012**, *19*, 455–477. [CrossRef]

53. Huang, D.I.; Cronk, Q. Plann: A command-line application for annotating plastome sequences. *Appl. Plant Sci.* **2015**, *3*, 1500026. [CrossRef]

54. Liu, C.; Shi, L.; Zhu, Y.; Chen, H.; Zhang, J.; Lin, X.; Guan, X. CpGAVAS, an integrated web server for the annotation, visualization, analysis, and GenBank submission of completely sequenced chloroplast genome sequences. *BMC Genom.* **2012**, *13*, 715. [CrossRef]

55. Wyman, S.K.; Jansen, R.K.; Boore, J.L. Automatic annotation of organellar genomes with DOGMA. *Bioinformatics* **2004**, *20*, 3252–3255. [CrossRef]

56. Lagesen, K.; Hallin, P.F.; Rodland, E.; Staerfeldt, H.H.; Rognes, T.; Ussery, D.W. RNAmmer: Consistent annotation of rRNA genes in genomic sequences. *Nucleic Acids Res.* **2007**, *35*, 3100–3108. [CrossRef]

57. Chan, P.P.; Lowe, T.M. tRNAscan-SE: Searching for tRNA genes in genomic sequences. *Methods Mol. Biol.* **2019**, *1962*, 1–14.

58. Greiner, S.; Lehwark, P.; Bock, R. OrganellarGenomeDRAW (OGDRAW) version 1.3.1: Expanded toolkit for the graphical visualization of organellar genomes. *Nucleic Acids Res.* **2019**, *47*, W59–W64. [CrossRef]

59. Kumar, S.; Stecher, G.; Li, M.; Knyaz, C.; Tamura, K. MEGA X: Molecular Evolutionary Genetics Analysis across computing platforms. *Mol. Biol. Evol.* **2018**, *35*, 1547–1549. [CrossRef] [PubMed]

60. Mower, J.P. The PREP suite: Predictive RNA editors for plant mitochondrial genes, chloroplast genes and user-defined alignments. *Nucleic Acids Res.* **2009**, *37*, W253–W259. [CrossRef] [PubMed]

61. Beier, S.; Thiel, T.; Münch, T.; Scholz, U.; Mascher, M. MISA-web: A web server for microsatellite prediction. *Bioinformatics* **2017**, *33*, 2583–2585. [CrossRef] [PubMed]

62. Benson, G. Tandem repeats finder: A program to analyze DNA sequences. *Nucleic Acids Res.* **1999**, *27*, 573–580. [CrossRef] [PubMed]

63. Frazer, K.A.; Pachter, L.; Poliakov, A.; Rubin, E.M.; Dubchak, I. VISTA: Computational tools for comparative genomics. *Nucleic Acids Res* **2004**, *32*, W273–W279. [CrossRef]

64. Katoh, K.; Standley, D.M. MAFFT multiple sequence alignment software version 7: Improvements in performance and usability. *Mol. Biol. Evol.* **2013**, *30*, 772–780. [CrossRef]

65. Librado, P.; Rozas, J. DnaSP v5: A software for comprehensive analysis of DNA polymorphism data. *Bioinformatics* **2009**, *25*, 1451–1452. [CrossRef]

66. Amiryousefi, A.; Hyvönen, J.; Poczai, P. IRscope: An online program to visualize the junction sites of chloroplast genomes. *Bioinformatics* **2018**, *34*, 3030–3031. [CrossRef]

67. Marcais, G.; Delcher, A.L.; Phillippy, A.M.; Coston, R.; Salzberg, S.L.; Zimin, A. MUMmer4: A fast and versatile genome alignment system. *PLoS Comput. Biol.* **2018**, *14*, e1005944. [CrossRef]

68. Kearse, M.; Moir, R.; Wilson, A.; Stones-Havas, S.; Cheung, M.; Sturrock, S.; Buxton, S.; Cooper, A.; Markowitz, S.; Duran, C.; et al. Geneious Basic: An integrated and extendable desktop software platform for the organization and analysis of sequence data. *Bioinformatics* **2012**, *28*, 1647–1649. [CrossRef]

69. Stamatakis, A. RAxML Version 8: A tool for phylogenetic analysis and post-analysis of large phylogenies. *Bioinformatics* **2014**, *30*, 1312–1313. [CrossRef] [PubMed]

70. Trifinopoulos, J.; Nguyen, L.T.; von Haeseler, A.; Minh, B.Q. W-IQ-TREE: A fast online phylogenetic tool for maximum likelihood analysis. *Nucleic Acids Res.* **2016**, *44*, W232–W235. [CrossRef] [PubMed]

71. Miller, M.A.; Pfeiffer, W.; Schwartz, T. Creating the CIPRES science gateway for inference of large phylogenetic trees. In Proceedings of the Gateway Computing Environments Workshop (GCE), New Orleans, LA, USA, 14 November 2010; pp. 1–8.

72. Ronquist, F.; Huelsenbeck, J.P. MrBayes 3: Bayesian phylogenetic inference under mixed models. *Bioinformatics* **2003**, *19*, 1572–1574. [CrossRef] [PubMed]

The Complete Chloroplast Genome of the Vulnerable *Oreocharis esquirolii* (Gesneriaceae): Structural Features, Comparative and Phylogenetic Analysis

Li Gu [1,2,3], **Ting Su** [1,2,3], **Ming-Tai An** [4] **and Guo-Xiong Hu** [1,2,*]

[1]   College of Life Sciences, Guizhou University, Guiyang 550025, China; guligz@163.com (L.G.); sutingyo@163.com (T.S.)
[2]   The Key Laboratory of Plant Resources Conservation and Germplasm Innovation in Mountainous Region Ministry of Education, Guizhou University, Guiyang 550025, China
[3]   Institute of Agro-Bioengineering, Guizhou University, Guiyang 550025, China
[4]   College of Forestry, Guizhou University, Guiyang 550025, China; mtan@gzu.edu.cn
*   Correspondence: gxhu@gzu.edu.cn

Abstract: *Oreocharis esquirolii*, a member of Gesneriaceae, is known as *Thamnocharis esquirolii*, which has been regarded a synonym of the former. The species is endemic to Guizhou, southwestern China, and is evaluated as vulnerable (VU) under the International Union for Conservation of Nature (IUCN) criteria. Until now, the sequence and genome information of *O. esquirolii* remains unknown. In this study, we assembled and characterized the complete chloroplast (cp) genome of *O. esquirolii* using Illumina sequencing data for the first time. The total length of the cp genome was 154,069 bp with a typical quadripartite structure consisting of a pair of inverted repeats (IRs) of 25,392 bp separated by a large single copy region (LSC) of 85,156 bp and a small single copy region (SSC) of 18,129 bp. The genome comprised 114 unique genes with 80 protein-coding genes, 30 tRNA genes, and four rRNA genes. Thirty-one repeat sequences and 74 simple sequence repeats (SSRs) were identified. Genome alignment across five plastid genomes of Gesneriaceae indicated a high sequence similarity. Four highly variable sites (*rps16-trnQ*, *trnS-trnG*, *ndhF-rpl32*, and *ycf 1*) were identified. Phylogenetic analysis indicated that *O. esquirolii* grouped together with *O. mileensis*, supporting resurrection of the name *Oreocharis esquirolii* from *Thamnocharis esquirolii*. The complete cp genome sequence will contribute to further studies in molecular identification, genetic diversity, and phylogeny.

Keywords: Gesneriaceae; next-generation sequencing; complete chloroplast genome; *Oreocharis*; *Thamnocharis*

1. Introduction

Traditionally, *Oreocharis* Benth. was a genus of the Gesneriaceae including 27 species [1,2]. Phylogenetic researches showed that *Oreocharis* was not monophyletic and up to 10 other genera were transferred to the genus [3–5]. Recently, an increasing number of new species of *Oreocharis* have been discovered and now approximately 135 species are recorded within this genus [6,7]. *Oreocharis* is mainly distributed in the tropical and subtropical areas in the south and southwest of China with a few extending to neighboring countries, such as Vietnam, Thailand, and Japan [3,8–11].

Oreocharis esquirolii H. Lév. was first established by Augustin Abel Hector Léveillé in 1911 based on a collection (*Esquirol 628*) from Guizhou, southwestern China [12]. Based on the character of actinomorphic corolla, Wang [13] transferred this species to his newly established genus, namely *Thamnocharis esquirolii* (H. Lév.) W. T. Wang. However, molecular phylogenetic results showed that

Thamnocharis was embedded into *Oreocharis* [3,4], and Möller et al. [4] resurrected *Oreocharis esquirolii* with regarding *Thamnocharis* as a synonym of *Oreocharis*.

Oreocharis esquirolii is endemic to Zhenfeng and Xingren County, Guizhou, southwestern China with a narrow distribution [2]. The species grows in thicket or hilly forest at an altitude of about 1500–1600 m. Due to excessive deforestation, serious vegetation damage and habitat degradation or loss, the population of the species decreased significantly with a risk of extinction. Based on restriction in habitat coupled with other threats, *O. esquirolii* was evaluated as vulnerable (VU) in IUCN's Red List of Threatened Species (http://www.iucnredlist.org/) and was listed as a national grade-I protected plant by China's government in 1999.

In plants, chloroplast (cp) genome is highly conserved in gene order, gene content, and genome organization [14,15] with a typical quadripartite structure consisting of a large single copy region (LSC), a small single-copy region (SSC), and a pair of inverted repeats (IRs) [16,17]. In most angiosperm chloroplasts, the cp genome ranges from 72 kb to 217 kb [18]. Chloroplast genome usually codes for 110–130 genes, including about 80 protein-coding genes, four rRNA genes, and about 30 tRNA genes [19]. The highly conserved structure of cp genome makes it often used to infer evolutionary relationships of higher taxa [20]. Currently, cp genome has also been demonstrated to be effective to solve species-level phylogenetic relationships in some taxa [21–23]. Comparative analysis of cp genome can provide valuable information for understanding structural and organizational changes of plant cp genome, and effectively help to reveal processes of plant molecular evolution and diversification [16,24,25]. Although cp genomes of some taxa of Gesneriaceae have been reported [26–29], the cp genome of *O. esquirolii* is not included. In this study, we sequenced the cp genome of *O. esquirolii* for the first time, aiming to present the genomic features of *O. esquirolii* and compare its structure and gene organization within Gesneriaceae. In addition, based on available cp genomes in GenBank, we inferred its phylogenetic position in Gesneriaceae.

2. Results and Discussion

2.1. Gene Content and Structure of Chloroplast Genome of Oreocharis esquirolii

Generally, the angiosperm cp genome is considered to be conserved [30]. In this study, we sequenced the cp genome of *Oreocharis esquirolii* and compared its features with other species from Gesneriaceae. The cp genome features of *O. esquirolii* were similar to other reported species in the Gesneriaceae concerning gene content, order, and orientation [28,31]. The whole cp genome of *O. esquirolii* was found to be 154,069 bp in length with a typical quadripartite structure, comprising a pair of inverted repeats (IRa and IRb) of 25,392 bp separated by a LSC region of 85,156 bp and a SSC region of 18,129 bp (Figure 1). Additionally, comparisons of length and GC content with the other 16 species from Gesneriaceae showed that their lengths ranged from 152,373 bp (*Primulina eburnea*) to 154,069 bp (*O. esquirolii*) and the GC content from 37.40% (*O. mileensis*) to 37.59% (*Primulina huaijiensis*) (Table S1). Notably, *O. esquirolii*, has the longest overall length (154,069 bp) but the shortest IR regions (25,392 bp), which may be related to the contraction of the IR regions.

Gene annotation revealed that the cp genome of *O. esquirolii* contained 114 unique genes, including a set of 80 protein-coding genes, 30 tRNA genes, and four rRNA genes. Amongst them, 19 genes were duplicated in the IR regions, comprising eight protein-coding genes (*ndhB*, *ycf1*, *ycf2*, *ycf15*, *rpl2*, *rps7*, *rpl23*, and *rps12*), four rRNA genes (*rrn4.5*, *rrn23*, *rrn5*, and *rrn16*), and seven tRNA genes (*trnA*$^{-UGC}$, *trnI*$^{-CAU}$, *trnI*$^{-GAU}$, *trnL*$^{-CAA}$, *trnN*$^{-GUU}$, *trnV*$^{-GAC}$, and *trnR*$^{-ACG}$) (Table 1). Fourteen intron-containing genes were detected, including nine protein-coding genes (*atpF*, *ndhA*, *ndhB*, *rpl2*, *rpl16*, *rps16*, *clpP*, *rpoC1*, and *ycf3*) and five tRNA genes (*trnA*$^{-UGC}$, *trnI*$^{-GAU}$, *trnK*$^{-UUU}$, *trnL*$^{-UAA}$, and *trnV*$^{-UAC}$). Of the 14 genes, two (*clpP* and *ycf3*) harbored two introns and the other 12 contained only one intron with the *trnK*$^{-UUU}$ including the largest intron (2,497 bp) and the *trnL*$^{-UAA}$ having the smallest intron (476 bp) (Table 2). Content (%) of the four bases was T (31.67%) > A (30.83%) > C (19.04%) > G (18.45%). Similarly to previous reports [26,30], the GC content in the IR regions of *O. esquirolii* (43.21%) was higher than that in the LSC (35.43%) and SSC (31.16%) (Table 3), which could be attributed to the presence of the eight rRNA sequences in IR regions [32].

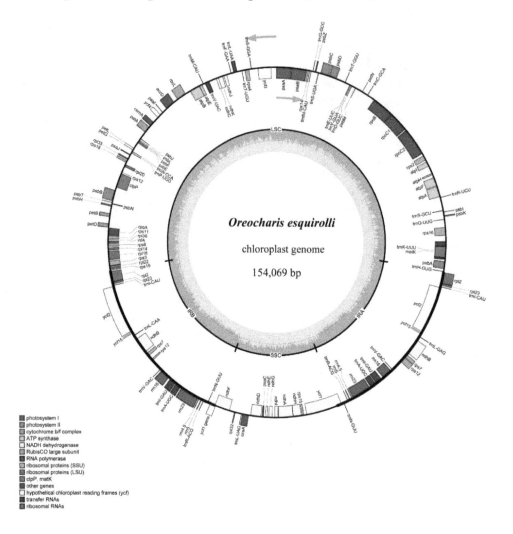

Figure 1. Gene map of chloroplast genome of *Oreocharis esquirolii*. Genes outside the circle are transcribed in counterclockwise direction and those inside in clockwise direction. LSC indicates large single copy; SSC small single copy, and IR inverted repeat.

Table 1. Genes present in chloroplast genome of *Oreocharis esquirolii*.

Category	Gene Group	Gene Names
Photosynthesis	Subunits of ATP synthase	*atpA, atpB, atpE, atpF *, atpI, atpH*
	Subunits of NADH dehydrogenase	*ndhA *, ndhB * (×2), ndhC, ndhD, ndhE, ndhF, ndhG, ndhH, ndhI, ndhJ, ndhK,*
	Subunits of cytochrome	*petA, petB, petD, petG, petL, petN*
	Subunits of photosystem I	*psaA, psaB, psaC, psaJ, psaI*
	Subunits of photosystem II	*psbA, psbB, psbC, psbD, psbE, psbH, psbK, psbN, psbJ, psbF, psbL, psbI, psbM, psbT, psbZ*
	Subunit of rubisco	*rbcL*
Other genes	Subunit of Acetyl-CoA-carboxylase	*accD*

Table 1. *Cont.*

Category	Gene Group	Gene Names
	c-type cytochrome synthesis gene	*ccsA*
	Envelop membrane protein	*cemA*
	Protease	*clpP* **
	Translational initiation	*infA*
	Maturase	*matK*
Self-replication	Large subunit of ribosome	*rpl2* * (×2), *rpl14*, *rpl16* *, *rpl20*, *rpl22*, *rpl23* (×2), *rpl32*, *rpl33*, *rpl36*
	DNA dependent RNA polymerase	*rpoA*, *rpoC2*, *rpoB*, *rpoC1*
	Small subunit of ribosome	*rps12* ** (×2), *rps2*, *rps3*, *rps4*, *rps7* (×2), *rps8*, *rps11*, *rps14*, *rps15*, *rps16* *, *rps18*, *rps19*
	rRNA Genes	*rrn4.5* (×2), *rrn5* (×2), *rrn16* (×2), *rrn23* (×2)
	tRNA Genes	*trnK-UUU* *, *trnI-GAU* * (×2), *trnA-UGC* * (×2), *trnV-UAC* *, *trnL-UAA* *, *trnS-UGA*, *trnS-GCU*, *trnS-GGA*, *trnY-GUA*, *trnL-CAA* (×2), *trnL-UAG*, *trnL-GAG*, *trnM-CAU*, *trnR-ACG* (×2), *trnP-UGG*, *trnW-CCA*, *trnD-GUC*, *trnH-GUG*, *trnF-GAA*, *trnT-UGU*, *trnE-UUC*, *trnN-GUU* (×2), *trnV-GAC* (×2), *trnT-GGU*, *trnQ-UUG*, *trnR-UCU*, *trnG-GCC*, *trnC-GCA*, *trnI-CAU* (×2), *trnfM-CAU*
Unknown function	Conserved open reading frames	*ycf1* (×2, ψ), *ycf2* (×2), *ycf3* **, *ycf4*, *ycf15* (×2)

(×2) gene in two copies, * gene which contains one intron, ** gene which contains two introns, ψ one of two duplicated genes is a pseudogene.

Table 2. Length of exons and introns within intron-containing genes in the chloroplast genome of *Oreocharis esquirolii*.

Gene	Region	Exon1 (bp)	Intron1 (bp)	Exon2 (bp)	Intron2 (bp)	Exon3 (bp)
atpF	LSC	144	707	411		
ndhA	SSC	552	1062	540		
ndhB	IR	777	679	756		
rpl2	IR	390	673	435		
rpl16	LSC	9	824	399		
rps16	LSC	42	921	210		
clpP	LSC	69	814	291	644	228
rpoC1	LSC	453	812	1611		
trnA-UGC	IR	38	807	35		
trnI-GAU	IR	37	941	35		
trnK-UUU	LSC	37	2497	36		
trnL-UAA	LSC	37	476	48		
trnV-UAC	LSC	38	586	35		
ycf3	LSC	126	692	228	714	153

Table 3. AT and GC content in different regions in the chloroplast genome of *Oreocharis esquirolii*.

Region	Length (bp)	A (%)	T (%)	G (%)	C (%)	GC (%)
LSC	85,156	31.54	33.03	17.30	18.13	35.43
SSC	18,129	34.34	34.50	15.00	16.16	31.16
IRA	25,392	28.38	28.41	22.42	20.79	43.21
IRB	25,392	28.38	28.41	22.42	20.79	43.21
CDS	79,650	30.70	31.59	20.10	17.61	37.71
Total genome	154,069	30.83	31.67	18.45	19.04	37.49

2.2. *Codon Usage Bias Analysis*

Codon usage refers to an organism's use of similar codons when encoding amino acids. Non-random use of synonymous codons is widespread both within and between organisms [33]. Many studies have shown that there are species-specific patterns of codon usage due to various factors such as codon hydrophilicity, gene length, expression levels, and protein secondary structure base composition [34,35]. The frequency of codons in the cp genome of *Oreocharis esquirolii* was calculated based on protein-coding genes. In total, all genes were encoded by 26,550 codons, of which AUU (Ile) was the most frequent (1111 codons) and UGC (Cys) was the least frequent (90 codons). Among the amino acids encoded by these codons, Leucine (2,784 codons, 10.49%), with the highest coding rate, was the most frequent. However, Cysteine (309 codons, 1.16%) was found less due to their high sensitivity to changes in physiological and environmental conditions [36] (Table S2). If the relative value of synonymous codon usage (RSCU) is greater than one, the codon usage is highly preferred, indicating that the codon is used more often than expected but not preferred if the value is equal to one and less preferred with values of less than one [36,37]. Codon usage analysis showed that codon usage was biased towards T and A at the third codon position in the cp genome of *O. esquirolii*. Furthermore, 30 highly preferred codons were detected in the *O. esquirolii* with an RSCU value greater than 1.0. Of the 30 codons, except for UUG ending with G, all codons terminated with A or T, and no C was found in the third position (Figure 2, Table S2).

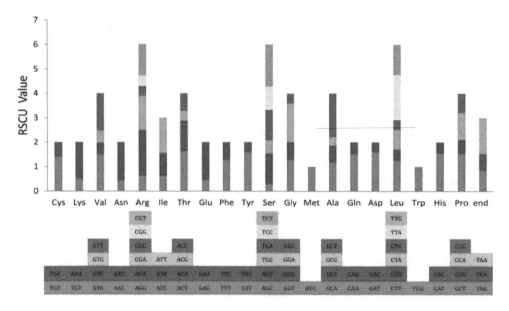

Figure 2. Amino acid frequencies and RSCU value of the protein-coding sequences of *Oreocharis esquirolii*.

2.3. *SSRs Analysis*

Simple sequence repeats (SSRs) are tandemly repeats of DNA sequences, comprising one to six (mono-, di-, tri-, tetra-, penta-, and hexa-) repeat nucleotide units. Being highly reliable, reproducible, and highly polymorphic, SSRs have been widely applied in molecular identification, genetic diversity, and population genetic studies [22,38–40]. In this study, SSRs of both *Oreocharis esquirolii* and *O. mileensis* were analyzed. A total of 74 SSRs were found in *O. esquirolii*, of which 54 were in the LSC regions, 12 in the IR and eight in the SSC regions. Comparatively, in the congeneric *O. mileensis*, 76 SSRs were detected with 55, 12, and nine SSRs distributed in the LSC, IR, and SSC regions, respectively (Figure 3). Besides, 27 SSRs were discovered in the coding sequences (CDS), 38 in the intergenic spacers (IGS), and nine in the intron regions of the *O. esquirolii* cp genome, whereas the values in the *O. mileensis* were 29 in CDS, 38 in IGS and nine in intron regions (Table S3). In terms of repeat unit, total five types of repeats (mono-, di-, tri-, tetra-, and penta-) were detected in *O. esquirolii* and *O. mileensis* cp genomes. Dinucleotide repeats were the most frequent, accounting for 55.41% (41) and 53.95% (41), respectively, followed by mononucleotide with 32.43% (24) and 31.98% (24), tetranucleotide with 10.81% (8) and 10.53% (8), and the least frequent trinucleotide with 1.35% (1) and 1.32% (1). It is worth noting that the pentanucleotide repeats (2, 2.63%) were only detected in *O. mileensis*, (Figure 3A,B, Table S3). Among the identified repeat units, dinucleotide repeat unit (AG/CT and AT/TA) was the most abundant. This finding supports the view that cp SSRs are generally composed of short polyadenine (polyA) or polythymine (polyT) repeats and rarely contain tandem guanine (G) or cytosine (C) repeats [40,41]. In addition, rarity or absence of pentanucleotide and hexanucleotide repeats in these two species demonstrated again that the two types of repeat unit are rather rare among cp SSRs [26,40].

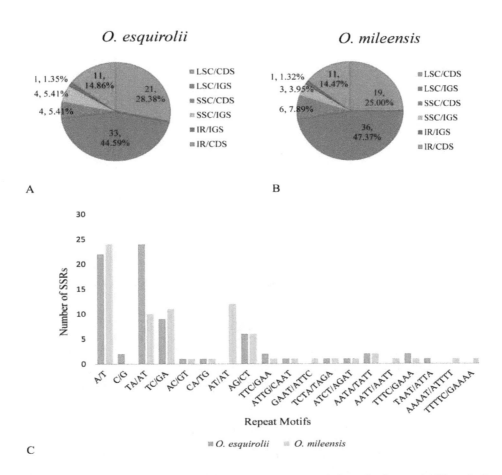

Figure 3. Simple Sequence Repeats (SSRs) in cp genomes of *Oreocharis esquirolii* and *O. mileensis*. (**A,B**) Frequencies of identified SSRs in LSC, IR, and SSC regions; (**C**) Numbers of SSRs.

2.4. Analysis of Repeat Sequences

Thirty-one repeat sequences were identified in both cp genomes of *Oreocharis mileensis* and *O. esquirolii*. In *O. esquirolii*, 13 (41.94%) forward repeats, 17 (54.84%) palindromic repeats, and one (3.23%) reverse repeats were identified. Similarly, in *O. mileensis*, palindromic repeats (19, 61.29%) are the most frequent, followed by forward repeats (12, 38.71%). However, none reverse repeats were identified in *O. mileensis* (Figure 4C, Table S4). Additionally, in the cp genome of *O. esquirolii*, the repeat sequence length ranged from 30 bp to 56 bp, while in *O. mileensis*, the length varied from 30 bp to 137 bp. Further analysis of the percentage of repeats in LSC, SSC and IR regions of *O. esquirolii*, and *O. mileensis* revealed that the LSC contained the largest number of repeats, accounting for 58.06%, and 61.29%, respectively, followed by the IR region with 35.48% and 35.48%, and the SSC region with 6.46% and 3.23% (Figure 4A,B).

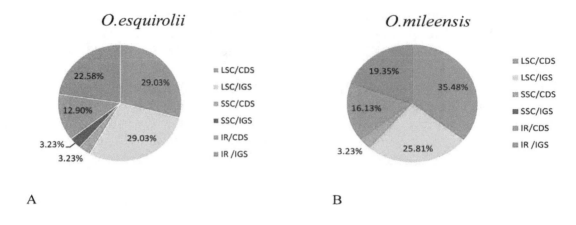

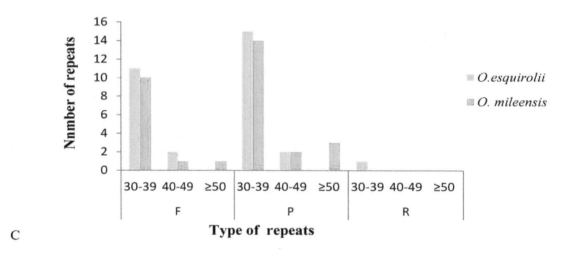

Figure 4. Repeat sequences in the chloroplast genome of *Oreocharis esquirolii* and *O. mileensis*. (**A,B**) Percentages of repeats in LSC, IR, and SSC regions; (**C**) Numbers of repeat types detected (F: forward, P: palindrome, R: reverse).

2.5. Comparisons of Chloroplast Genome among Oreocharis esquirolii and Closely Related Species

Expansion and contraction of the IR region, contributing to variation of cp genome size, plays a crucial role in the evolution of plants [42,43]. Junctions between single copy regions and IR regions among closely related species of *Lysionotus pauciflorus, Petrocodon jingxiensis, Primulina huaijiensis, Oreocharis esquirolii*, and *O. mileensis* were compared in this study. These genomes showed a bit variances at the junctions, but the general gene structures, contents, and orientations were the same. The LSC/IRb junction had expanded to *rps19* gene in four species (*Lysionotus pauciflorus*, 35 bp, *Oreocharis mileensis*, 31 bp, *Petrocodon jingxiensis*, 32 bp, and *Primulina huaijiensis*, 25 bp). However, in *O. esquirolii*, the *rps19* gene did not span the LSC/IRb junction (44 bp away from the junction), suggesting that the IR regions of *O. esquirolii* underwent significant contraction compared with the other four species. This phenomenon was also observed in *Streptocarpus* [31]. A pseudogenized *ycf1* occurred at the IRb/SSC junctions in all species as a result of the extension of SSC/IRa junction into the *ycf1* gene, with variable extensions of the gene into the SSC region observed in the five species. In contrast, *ycf1* was mainly located in the SSC region ranging from 4752 bp to 4266 bp. An overlap of Ψ*ycf1* and *ndhF* genes was observed in all five species: *Lysionotus pauciflorus* (137 bp), *O. mileensis* (42 bp), *Primulina huaijiensis* (88 bp), *O. esquirolii* (109 bp), and *Petrocodon jingxiensis* (109 bp) (Figure 5).

Mauve was used to check for possible rearrangements within the cp genomes of five species (*Lysionotus pauciflorus, Orecharis esquirolii, O. mileensis, Petrocodon jingxiensis*, and *Primulina huaijiensis*).

The results indicated that the organization of the five Gesneriaceae cp genome was highly conserved, without translocations or inversions detected (Figure 6).

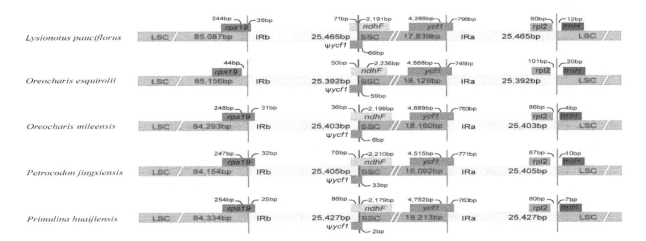

Figure 5. Comparisons of LSC, SSC, and IR border regions among five chloroplast genomes of Gesneriaceae.

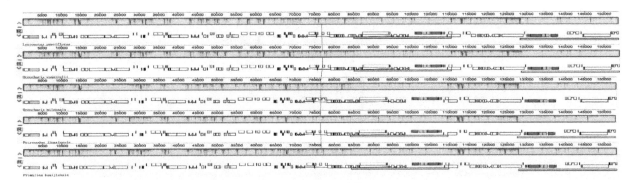

Figure 6. Mauve multiple alignment of five chloroplast genomes of Gesneriaceae, with *Oreocharis esquirolii* as the reference.

A sliding window analysis was used to estimate the level of variation across regions in the five Gesneriaceae cp genomes. The nucleotide diversity (Pi) values ranged from 0.00000 to 0.09606, with a mean of 0.01381. All highly divergent sequences were restricted to the single copy (SC) regions, with the highest peak occurring in the SSC region. Four hyper-variable regions were identified with nucleotide diversity values higher than 0.05, of which three were intergenic spacers (*rps16-trnQ*, *trnS-trnG*, and *ndhF-rpl32*), and the remaining one was *ycf1* gene (Figure 7). Generally, the intergenic regions exhibit higher nucleotide diversity than the coding regions. As expected, of the four hypervariable regions detected in five Gesneriaceae cp genomes, three were in intergenic regions, while only one in genic region. Similar result was also found in recent cp genome analysis [31,43]. Although not commonly used because of large number of primer pairs needed to sequence the entire region, as a hypervariable gene detected here, *ycf1* could be regarded as a potential marker in phylogenetic analysis of Gesneriaceae, and it have been demonstrated to be effective in Orchidaceae and Lamiaceae [44,45].

The pairwise cp genomic alignment between *O. esquirolii* and its closely related species was analyzed using mVISTA with the annotation of *O. mileensis* as a reference. Results showed that IR regions were found to be more conserved than the single copy regions, so were genic regions, coding regions, and exons compared with intergenic regions, non-coding and introns. Highly divergent regions among the five species of cp genomes were mainly located in the intergenic spacers, such as *trnH*[-GUG]*-psbA*, *rps16-trnQ*[-UUG], *atpH-atpI*, *trnL*[-UAG]*-ccsA*, and *ycf4-cemA*, and few (*rpl16* and *ycf1*) were distributed in protein-coding regions (Figure 8). These regions can provide phylogenetic information as well as serve as unique barcodes for DNA.

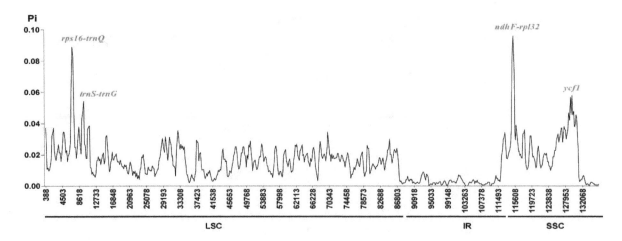

Figure 7. Nucleotide diversity (Pi) in the complete chloroplast genomes of five species of Gesneriaceae. Sliding window analysis with a window length of 600 bp and a step size of 200 bp.

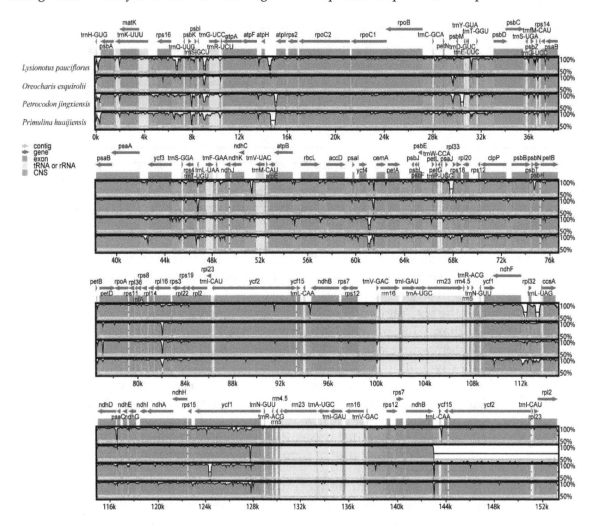

Figure 8. Visualization of genome alignment of five chloroplast genomes of Gesneriaceae using *Oreocharis mileensis* as reference.

2.6. Phylogenetic Position of Oreocharis esquirolii

Based on whole cp genome sequences of 26 taxa within Lamiales, the phylogenetic relationship of Gesneriaceae was inferred using Bayesian inference (BI) and maximum likelihood (ML) analyses. As topology of BI and ML trees were identical, the two trees were combined with addition of bootstrap values of ML and posterior probabilities values of BI. Phylogenetic results showed Gesneriaceae was monophyletic, and *O. esquirolii* grouped with *O. mileensis* (Figure 9). As bearing actinomorphic corolla, Wang [13] transferred *O. esquirolii* to *Thamnocharis esquirolii*. Together with other genera such as *Bournea*, *Tengia*, and *Conandron*, *Thamnocharis* was classed into tribe Ramondieae that is sometimes considered to be primitive in Gesneriaceae [2]. However, phylogenetic analysis showed that actinomorphic genera are scattered over clades with zygomorphic corolla, and hypothesized that flora actinomorphy has evolved in a convergent manner [13]. In addition, phylogenetic studies also indicated that *Oreocharis* is non-monophyletic with several genera including *Thamnocharis* embedded [3,4,46], and finally, Möller et al. [4] regarded *Thamnocharis esquirolii* as a synonym of *Oreocharis esquirolii*. Although the sampling is very limited in our analysis, the sister relationship between *Oreocharis esquirolii* and *O. mileensis* support resurrection of the name *Oreocharis esquirolii* from *Thamnocharis esquirolii*.

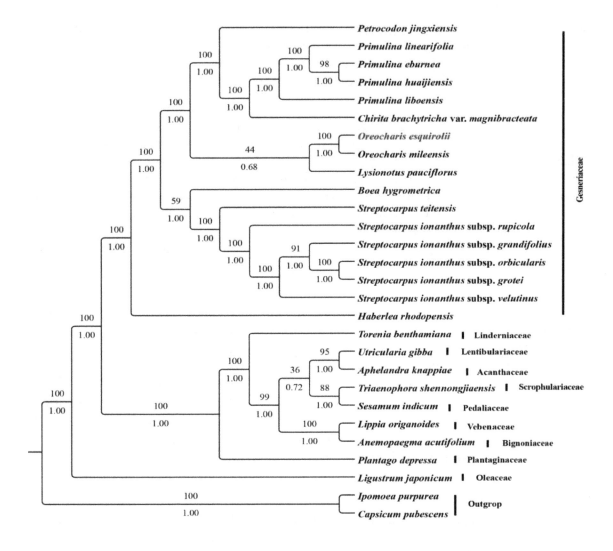

Figure 9. Phylogenetic relationships of 28 species based on complete chloroplast genome sequence. ML bootstrap values are given above branches and posterior probabilities are indicated below.

3. Materials and Methods

3.1. Plant Material, DNA Extraction, Sequencing, and Assembly

Young leaves of *Oreocharis esquirolii* were collected from Longtoudashan Natural Reverse, Zhenfeng, Guizhou, Southwestern China, and were put into silica gel to preserve. Total genomic DNA was extracted from about 100 mg of dried leaf material according to a modified CTAB method [47]. DNA integrity was assessed by electrophoresis on a 1% agarose gel and its concentration and yield was determined and calculated with Qubit. The DNA sample meeting the requirements of sequencing was sent to the BGI-Wuhan and Illumina HiSeq 2500 platform was used for sequencing. After filtering the low-quality data and adaptors, clean data were obtained. Then, GetOrganelle [48], a fast toolkit for accurate *de novo* assembly of organelle genomes which was jointly completed by SPAdes [49], Bowtie2 [50], and BLAST+ [51], was used to assemble the cp genome of *O. esquirolii* with *O. mileenis* (MK342624) [28] as a reference. Assembly graph was visualized using Bandage v.8.0 [52] and then a whole circular cp genome was generated.

3.2. Genome Annotation and Sequence Submission

The cp genome was annotated using program PGA [53] with *Oreocharis mileensis* [28] as a reference, then coupled with manual adjustment using Geneious v.10.1.3 [54]. MEGA 6.06 [55] was used to analyze AT and GC contents. Finally, the circular genome map was generated with OGDRAW v.1.2 [56] and submitted to NCBI GenBank under Accession Number MT612436.

3.3. Codon Usage, Repetitive Sequence, and SSR Analysis

The codon usage frequency was calculated based on protein-coding genes using CodonW 1.4.2. [57]. REPuter [58] was used to identify repeat sequences, including direct (forward), inverted (palindromic), complement, and reverse repeats. The repeat sizes were limited to a minimum of 30 bp and a maximum of 300, with sequence identities greater than 90% (Hamming distance of 3). MISA [59], an SSR motif scanning tool written in Perl, was adopted to detect SSRs. The minimum thresholds were set to 10 repeat units for mononucleotide SSRs, four repeat units for dinucleotide and trinucleotide SSRs, and three repeat units for tetranucleotide, pentanucleotide, and hexanucleotide SSRs [60].

3.4. Genome Comparison

Based on previous phylogenetic results together with the current reported data, five Gesneriaceae cp genomes (*Lysionotus pauciflorus*, *Petrocodon jingxiensis*, *Primulina huaijiensis*, *Oreocharis mileensis*, and *O. esquirolii*) were selected for comparative analysis. To explore the expansion and contraction of IR regions of *Oreocharis esquirolii*, comparison of boundaries between IRs and single copy regions was performed in Geneious v.10.1.3 [54]. The mVISTA [61] was used to assess the similarity among the five cp genomes, and the default parameters were utilized to align the cp genomes in Shuffle-LAGAN mode. Chloroplast genome sequence alignment was carried out with the Mauve program [62] to check the gene order and sequence variations. Sliding window analysis of nucleotide variability in the cp genome was conducted using DnaSP [63]. The step size was set to 200 bp, with a 600 bp window length.

3.5. Phylogenetic Analyses

To explore the phylogenetic position of *Oreocharis esquirolii* among the limited number of species available across Gesneriaceae, complete cp genomes of 26 species within Lamiales were selected to conduct analyses, using *Ipomoea purpurea* and *Capsicum pubescens* from Solanales as outgroups [64,65] (Table S5). Multiple sequence alignment of cp genome sequences were performed using MAFFT [66], and poorly aligned positions and regions with a too-high divergence were excluded from the alignment using Gblocks v0.91 [67]. Bayesian inference (BI) and Maximum likelihood (ML) methods were adopted for phylogenetic analyses. ML analysis was performed using RAxML–HPC2 on XSEDE v.8.2.12 as implemented on the CIPRES Science Gateway (http://www.phylo.org/) [68] under the GTRGAMMA model. Bootstrap iteration (–#|–N) was set to 1000, and other parameters followed default settings. BI analysis was performed in MrBayes v3.2.6 [69] as implemented in PhyloSuite [70] with the ModelFinder [71] used to select the best model. Under the Akaike information criterion (AIC), the GTR+F+I+G4 model was selected for the data matrix. The Markov Chain Monte Carlo (MCMC) algorithm was calculated for 2,000,000 generations with two parallel searches using four chains, each starting with a random tree. The convergence was reached with the average standard deviation of split frequencies (ASDFs) following 0.01. Trees were sampled at every 1000 generations with the first 25% discarded as burn-in, and the remaining trees were used to construct majority-rule consensus trees.

4. Conclusions

Oreocharis esquirolii, also known as *Thamnocharis esquirolii*, is categorized under IUCN criteria as vulnerable. We assembled and characterized the complete cp genome of *O. esquirolii* for the first time. The cp genome features of *O. esquirolii* were similar to other reported species of Gesneriaceae concerning gene content, order, and orientation. SSRs analysis supports the view that cp SSRs are generally composed of short polyA or polyT, and pentanucleotide and hexanucleotide repeats are rather rare. Comparative analyses revealed that no arrangements occurred in Gesneriaceae, intergenic regions were more variable than coding regions, and some hypervariable regions such as *rps16-trnQ*, *trnS-trnG*, *ndhF-rpl32* and *ycf1* may be applied to address phylogenetic issues of Gesneriaceae. Phylogenetic analysis supported synonymizing *Thamnocharis esquirolii* as *Oreocharis esquirolii*. The complete cp genome sequence will contribute to further studies in molecular identification, genetic diversity, and phylogeny.

Supplementary Materials:
Table S1: Comparison of the features of *Oreocharis esquirolii* with other Gesneriaceae chloroplast genomes. Table S2: Comparative analysis of chloroplast codon usage bias of *Oreocharis esquirolii*. Table S3: Distribution of simple sequence repeats (SSRs) loci in the chloroplast genome *Oreocharis esquirolii* and *O. mileensis*. Table S4: List of repeated sequences and their locations in chloroplast genome of *Oreocharis esquirolii* and *O. mileensis*. Table S5: Taxa used in phylogenetic analysis in this study.

Author Contributions: Conceptualization and supervision were done by G.-X.H.; Investigation, G.-X.H., M.-T.A., L.G., and T.S.; Data curation was done by L.G. and T.S.; formal analysis was performed by L.G. and T.S.; Writing—original draft, L.G.; revision and manuscript editing was done by G.-X.H.; resources and funding acquisition was provided by G.-X.H. and M.-T.A. All authors have read and agreed to the published version of the manuscript.

Acknowledgments: We thank Jia-Xin Yang for the help with data analysis. We also thank anonymous reviewers for helpful comments and precious remarks to improve the manuscript.

References

1. Bentham, G. Gesneriaceae. In *Genera Plantarum*; Bentham, G., Hooker, J.D., Eds.; Lovell Reeve & Co.: London, UK, 1876; pp. 990–1025.
2. Wang, W.T.; Pan, K.Y.; Li, Z.Y.; Weitzman, A.L.; Skog, L.E. Gesneriaceae. In *Flora of China*; Wu, C.Y., Raven, P.H., Hong, D.Y., Eds.; Science Press: Beijing, China; Missouri Botanical Garden Press: St. Louis, MO, USA, 1998; Volume 18, pp. 244–401.
3. Wang, Y.Z.; Liang, R.H.; Wang, B.H.; Li, J.M.; Qiu, Z.J.; Li, Z.Y.; Weber, A. Origin and phylogenetic relationships of the Old World Gesneriaceae with actinomorphic flowers inferred from *ITS* and *trnL-trnF* sequences. *Taxon* **2010**, *59*, 1044–1052. [CrossRef]
4. Möller, M.; Middleton, D.; Nishii, K.; Wei, Y.G.; Sontag, S.; Weber, A. A new delineation for *Oreocharis* incorporating an additional ten genera of Chinese Gesneriaceae. *Phytotaxa* **2011**, *23*, 1–36. [CrossRef]

5.   Möller, M.; Forrest, A.; Wei, Y.G.; Weber, A. A molecular phylogenetic assessment of the advanced Asiatic and Malesian didymocarpoid Gesneriaceae with focus on non-monophyletic and monotypic genera. *Plant Syst. Evol.* **2011**, *292*, 223–248. [CrossRef]

6.   Möller, M. Species discovery in time: An example from Gesneriaceae in China. *Guihaia* **2019**, *26*, 1–16. [CrossRef]

7.   Wen, F.; Li, S.; Xin, Z.B.; Fu, L.F.; Hong, X.; Cai, L.; Qin, J.Q.; Pan, B.; Pan, F.G.; Wei, Y.G. The updated plant list of Gesneriaceae in China under the new Chinese naming rules. *Guihaia* **2019**, *26*, 37–63. [CrossRef]

8.   Middleton, D.J.; Möller, M. *Tribounia*, a new genus of Gesneriaceae from Thailand. *Taxon* **2012**, *61*, 1286–1295. [CrossRef]

9.   Do, T.V.; Wei, Y.G.; Wen, F. *Oreocharis caobangensis* (Gesneriaceae), a new species from Cao Bang Province, northern Vietnam. *Phytotaxa* **2017**, *302*, 65–70. [CrossRef]

10. Chen, W.H.; Middleton, D.J.; Nguyen, H.Q.; Nguyen, H.T.; Averyanov, L.V.; Chen, R.Z.; Nguyen, K.S.; Möller, M.; Shui, Y.M. Two new species of *Oreocharis* (Gesneriaceae) from Northwest Vietnam. *Gard. Bull.* **2017**, *69*, 295–305. [CrossRef]

11. Chen, W.H.; Nguyen, Q.H.; Chen, R.Z.; Nguyen, T.H.; Nguyen, S.K.; Nguyen, V.T.; Möller, M.; Middleton, D.J.; Shui, Y.M. Two new species of *Oreocharis* (Gesneriaceae) from Fan Si Pan, the highest mountain in Vietnam. *Phytokeys* **2018**, *94*, 95–106. [CrossRef]

12. Léveillé, H. Decades plantarum novarum. LIV–LVII. *Repert. Spec. Nov. Regni Veg.* **1911**, *9*, 321–330.

13. Wang, W.T. Genus novum primitivum Gesneriacearum e Sina. *Acta Phytotax. Sin.* **1981**, *19*, 485–489.

14. Chumley, T.W.; Palmer, J.D.; Mower, J.P.; Fourcade, H.M.; Calie, P.J.; Boore, J.L.; Jansen, R.K. The complete chloroplast genome sequence of *Pelargonium × hortorum*: Organization and evolution of the largest and most highly rearranged chloroplast genome of land plants. *Mol. Biol. Evol.* **2006**, *23*, 2175–2190. [CrossRef] [PubMed]

15. Wicke, S.; Schneeweiss, G.M.; dePamphilis, C.W.; Mueller, K.F.; Quandt, D. The evolution of the plastid chromosome in land plants: Gene content, gene order, gene function. *Plant Mol. Biol.* **2011**, *76*, 273–297. [CrossRef] [PubMed]

16. Wang, M.; Cui, L.; Feng, K.; Deng, P.; Du, X.; Wan, F.; Song, W.; Nie, X. Comparative analysis of Asteraceae chloroplast genomes: Structural organization, RNA editing and evolution. *Plant Mol. Biol. Rep.* **2015**, *33*, 1526–1538. [CrossRef]

17. Palmer, J.D. Plastid chromosomes: Structure and evolution. In *The Molecular Biology of Plastids*; Bogorad, L., Vasil, I.K., Eds.; Academic Press: Cambridge, MA, USA, 1991; pp. 5–53. [CrossRef]

18. Chen, Q.; Wu, X.; Zhang, D. Phylogenetic analysis of *Fritillaria cirrhosa* D. Don and its closely related species based on complete chloroplast genomes. *PeerJ* **2019**, *7*, e7480. [CrossRef]

19. Wyman, S.K.; Jansen, R.K.; Boore, J.L. Automatic annotation of organellar genomes with DOGMA. *Bioinformatics* **2004**, *20*, 3252–3255. [CrossRef]

20. Givnish, T.J.; Zuluaga, A.; Spalink, D.; Soto Gomez, M.; Lam, V.K.Y.; Saarela, J.M.; Sass, C.; Iles, W.J.D.; de Sousa, D.J.L.; Leebens-Mack, J.; et al. Monocot plastid phylogenomics, timeline, net rates of species diversification, the power of multi-gene analyses, and a functional model for the origin of monocots. *Am. J. Bot.* **2018**, *11*, 1888–1910. [CrossRef]

21. He, Y.; Xiao, H.; Deng, C.; Xiong, L.; Yang, J.; Peng, C. The complete chloroplast genome sequences of the medicinal plant *Pogostemon cablin*. *Int. J. Mol. Sci.* **2016**, *17*, 820. [CrossRef]

22. Li, Y.; Zhang, Z.; Yang, J.; Lv, G. Complete chloroplast genome of seven *Fritillaria species*, variable DNA markers identification and phylogenetic relationships within the genus. *PLoS ONE* **2018**, *13*, e0194613. [CrossRef]

23. Zong, D.; Zhou, A.; Zhang, Y.; Zou, X.; Li, D.; Duan, A.; He, C. Characterization of the complete chloroplast genomes of five *Populus* species from the western Sichuan plateau, southwest China: Comparative and phylogenetic analyses. *PeerJ* **2019**, *7*, e6386. [CrossRef]

24. Rivas, D.L.; Lozano, J.; Ortiz, J.J.; Angel, R. Comparative analysis of chloroplast genomes: Functional annotation, genome-based phylogeny, and deduced evolutionary patterns. *Genome Res.* **2002**, *12*, 567–583. [CrossRef] [PubMed]

25. Liu, Y.; Huo, N.X.; Dong, L.L.; Wang, Y.; Zhang, S.X.; Young, H.A.; Feng, X.X.; Gu, Y.Q. Complete chloroplast genome sequences of Mongolia medicine *Artemisia frigida* and phylogenetic relationships with other plants. *PLoS ONE* **2013**, *8*, e57533. [CrossRef] [PubMed]

26. Ivanova, Z.; Sablok, G.; Daskalova, E.; Zahmanova, G.; Apostolova, E.; Yahubyan, G.; Baev, V. Chloroplast genome analysis of resurrection tertiary relict *Haberlea rhodopensis* highlights genes important for desiccation stress response. *Front. Plant Sci.* **2017**, *8*, 204. [CrossRef] [PubMed]

27. Hou, N.; Wang, G.; Li, C.R.; Luo, Y. Characterization of the complete chloroplast genomes of three *Chirita* species (*C. brachytricha, C. eburnea* & *C. liboensis*) endemic to China. *Conserv. Genet. Resour.* **2018**, *10*, 597–600. [CrossRef]

28. Meng, J.; Zhang, L.; He, J. Complete plastid genome of the endangered species *Paraisometrum mileense* (Gesneriaceae) endemic to China. *Mitochondrial DNA B Resour.* **2019**, *4*, 3585–3586. [CrossRef]

29. Xin, Z.B.; Fu, L.F.; Fu, Z.X.; Li, S.; Wei, Y.G.; Wen, F. Complete chloroplast genome sequence of *Petrocodon jingxiensis* (Gesneriaceae). *Mitochondrial DNA B Resour.* **2019**, *4*, 2771–2772. [CrossRef]

30. Kyalo, C.M.; Gichira, A.W.; Li, Z.Z.; Saina, J.K.; Malombe, I.; Hu, G.W.; Wang, Q.F. Characterization and comparative analysis of the complete chloroplast genome of the critically endangered species *Streptocarpus teitensis* (Gesneriaceae). *Biomed. Res. Int.* **2018**, *2018*, 1507847. [CrossRef]

31. Kyalo, C.M.; Li, Z.Z.; Mkala, E.M.; Malombe, I.; Hu, G.W.; Wang, Q.F. The first glimpse of *Streptocarpus ionanthus* (Gesneriaceae) phylogenomics: Analysis of five subspecies' chloroplast genomes. *Plants* **2020**, *9*, 456. [CrossRef]

32. Qian, J.; Song, J.; Gao, H.; Zhu, Y.; Xu, J.; Pang, X.; Yao, H.; Sun, C.; Li, X.E.; Li, C.; et al. The complete chloroplast genome sequence of the medicinal plant *Salvia miltiorrhiza*. *PLoS ONE* **2013**, *8*, e57607. [CrossRef]

33. Liu, Q.; Dou, S.; Ji, Z.; Xue, Q. Synonymous codon usage and gene function are strongly related in *Oryza sativa*. *Biosystems* **2005**, *80*, 123–131. [CrossRef]

34. Srivastava, D.; Shanker, A. Identification of Simple Sequence Repeats in chloroplast genomes of Magnoliids through bioinformatics approach. *Interdiscip. Sci.* **2015**, *8*, 327–336. [CrossRef] [PubMed]

35. Li, Y.; Kuang, X.J.; Zhu, X.X.; Zhu, Y.J.; Sun, C. Codon usage bias of *Catharanthus roseus*. *Zhongguo Zhong Yao Za Zhi China J. Chin. Mater. Med.* **2016**, *41*, 4165–4168. [CrossRef]

36. Marino, S.M.; Gladyshev, V.N. Analysis and functional prediction of reactive Cysteine residues. *J. Biol. Chem.* **2012**, *287*, 4419–4425. [CrossRef] [PubMed]

37. Sharp, P.M.; Li, W.H. The codon Adaptation Index-a measure of directional synonymous codon usage bias, and its potential applications. *Nucleic Acids Res. Suppl.* **1987**, *15*, 1281–1295. [CrossRef]

38. Raza, A.; Mehmood, S.S.; Ashraf, F.; Khan, R.S.A. Genetic diversity analysis of *Brassica* species using PCR-based SSR markers. *Gesunde Pflanz.* **2018**, *71*, 1–7. [CrossRef]

39. Torokeldiev, N.; Ziehe, M.; Gailing, O.; Finkeldey, R. Genetic diversity and structure of natural *Juglans regia* L. populations in the southern Kyrgyz Republic revealed by nuclear SSR and EST–SSR markers. *Tree Genet. Genomes* **2019**, *15*, 5. [CrossRef]

40. Kuang, D.Y.; Wu, H.; Wang, Y.L.; Gao, L.M.; Zhang, S.Z.; Lu, L. Complete chloroplast genome sequence of *Magnolia kwangsiensis* (Magnoliaceae): Implication for DNA barcoding and population genetics. *Genome* **2011**, *54*, 663–673. [CrossRef]

41. Du, Y.P.; Bi, Y.; Yang, F.P.; Zhang, M.F.; Chen, X.Q.; Xue, J.; Zhang, X.H. Complete chloroplast genome sequences of *Lilium*: Insights into evolutionary dynamics and phylogenetic analyses. *Sci. Rep.* **2017**, *7*, 5751. [CrossRef]

42. He, L.; Qian, J.; Li, X.; Sun, Z.; Xu, X.; Chen, S. Complete chloroplast genome of medicinal plant *Lonicera japonica*: Genome rearrangement, intron gain and loss, and implications for phylogenetic studies. *Molecules* **2017**, *22*, 249. [CrossRef]

43. Zhao, F.; Drew, B.T.; Chen, Y.P.; Hu, G.X.; Li, B.; Xiang, C.L. The chloroplast genome of *Salvia*: Genomic characterization and phylogenetic analysis. *Int. J. Plant Sci.* **2020**, *181*, 812–830. [CrossRef]

44. Neubig, K.M.; Whitten, W.M.; Carlsward, B.S.; Blanco, M.A.; Endara, L.; Williams, N.H.; Moore, M. Phylogenetic utility of *ycf1* in orchids: Aplastid gene more variable than *matK. Plant Syst. Evol.* **2009**, *277*, 75–84. [CrossRef]

45. Drew, B.T.; Sytsma, K.J. Testing the monophyly and placement of *Lepechinia* in the tribe Mentheae (Lamiaceae). *Syst. Bot.* **2011**, *36*, 1038–1049. [CrossRef]

46. Möller, M.; Pfosser, M.; Jang, C.G.; Mayer, V.; Clark, A.; Hollingsworth, M.L.; Barfuss, M.H.J.; Wang, Y.Z.; Kiehn, M.; Weber, A. A preliminary phylogeny of the 'didymocarpoid Gesneriaceae' based on three molecular data sets: Incongruence with available tribal classifications. *Am. J. Bot.* **2009**, *96*, 989–1010. [CrossRef] [PubMed]

47. Doyle, J.; Doyle, J. A rapid DNA isolation procedure from small quantities of fresh leaf tissues. *Phytochemistry* **1987**, *19*, 11–15.

48. Jin, J.J.; Yu, W.B.; Yang, J.B.; Song, Y.; dePamphilis, C.W.; Yi, T.S.; Li, D.Z. GetOrganelle: A fast and versatile toolkit for accurate de novo assembly of organelle genomes. *Genome Biol.* **2020**, *21*, 241. [CrossRef] [PubMed]

49. Bankevich, A.; Nurk, S.; Antipov, D.; Gurevich, A.A.; Dvorkin, M.; Kulikov, A.S.; Lesin, V.M.; Nikolenko, S.I.; Son, P.; Prjibelski, A.D.; et al. SPAdes: A new genome assembly algorithm and its applications to single-cell sequencing. *J. Comput. Biol.* **2012**, *19*, 455–477. [CrossRef]

50. Langmead, B.; Salzberg, S.L. Fast gapped-read alignment with Bowtie 2. *Nat. Methods* **2012**, *9*, 357–359. [CrossRef]

51. Camacho, C.; Coulouris, G.; Avagyan, V.; Ma, N.; Papadopoulos, J.; Bealer, K.; Madden, T.L. BLAST+: Architecture and applications. *BMC Bioinform.* **2009**, *10*, 421. [CrossRef]

52. Wick, R.R.; Schultz, M.B.; Zobel, J.; Holt, K.E. Bandage: Interactive visualization of *de novo* genome assemblies. *Bioinformatics* **2015**, *31*, 3350–3352. [CrossRef]

53. Qu, X.J.; Moore, M.J.; Li, D.Z.; Yi, T.S. PGA: A software package for rapid, accurate, and flexible batch annotation of plastomes. *Plant Methods* **2019**, *15*, 50. [CrossRef]

54. Kearse, M.; Moir, R.; Wilson, A.; Stones-Havas, S.; Cheung, M.; Sturrock, S.; Buxton, S.; Cooper, A.; Markowitz, S.; Duran, C.; et al. Geneious Basic: An integrated and extendable desktop software platform for the organization and analysis of sequence data. *Bioinformatics* **2012**, *28*, 1647–1649. [CrossRef] [PubMed]

55. Tamura, K.; Stecher, G.; Peterson, D.; Filipski, A.; Kumar, S. MEGA6: Molecular evolutionary genetics analysis version 6.0. *Mol. Biol. Evol.* **2013**, *30*, 2725–2729. [CrossRef] [PubMed]

56. Lohse, M.; Drechsel, O.; Kahlau, S.; Bock, R. OrganellarGenomeDRAW-a suite of tools for generating physical maps of plastid and mitochondrial genomes and visualizing expression data sets. *Nucleic Acids Res. Suppl.* **2013**, *41*, W575–W581. [CrossRef] [PubMed]

57. Hélène, C.; Frédérique, L.; Michel, C.; Alain, H. Codon usage and gene function are related in sequences of *Arabidopsis thaliana*. *Gene* **1998**, *209*, GC1–GC38. [CrossRef]

58. Kurtz, S.; Choudhuri, J.V.; Ohlebusch, E.; Schleiermacher, C.; Stoye, J.; Giegerich, R. REPuter: The manifold applications of repeat analysis on a genomic scale. *Nucleic Acids Res. Suppl.* **2001**, *29*, 4633–4642. [CrossRef]

59. Thiel, T.; Michalek, W.; Varshney, R.K.; Graner, A. Exploiting EST databases for the development and characterization of gene-derived SSR-markers in barley (*Hordeum vulgare* L.). *Theor. Appl. Genet.* **2003**, *106*, 411–422. [CrossRef] [PubMed]

60. Munyao, J.N.; Dong, X.; Yang, J.X.; Mbandi, E.M.; Wanga, V.O.; Oulo, M.A.; Saina, J.K.; Musili, R.M.; Hu, G.W. Complete chloroplast genomes of *Chlorophytum comosum* and *Chlorophytum gallabatense*: Genome structures, comparative and phylogenetic analysis. *Plants* **2020**, *9*, 296. [CrossRef]

61. Mayor, C.; Brudno, M.; Schwartz, J.R.; Poliakov, A.; Rubin, E.M.; Frazer, K.A.; Pachter, L.S.; Dubchak, I. VISTA: Visualizing global DNA sequence alignments of arbitrary length. *Bioinformatics* **2000**, *16*, 1046–1047. [CrossRef]

62. Kurtz, S.; Phillippy, A.; Delcher, A.L.; Smoot, M.; Shumway, M.; Antonescu, C.; Salzberg, S.L. Versatile and open software for comparing large genomes. *Genome Biol.* **2004**, *5*, R12. [CrossRef]

63. Rozas, J.; Albert, F.M.; Juan, C.S.; Sara, G.R.; Pablo, L.; Sebastian, E.R.O.; Alejandro, S.G. DnaSP 6: DNA sequence polymorphism analysis of large data sets. *Mol. Biol. Evol.* **2017**, *34*, 3299–3302. [CrossRef]

64. Liu, B.; Tan, Y.H.; Liu, S.; Olmstead, R.G.; Min, D.Z.; Chen, Z.D.; Joshee, N.; Vaidya, B.N.; Chung, R.C.K.; Li, B. Phylogenetic relationships of *Cyrtandromoea* and *Wightia* revisited: A new tribe in Phrymaceae and a new family in Lamiales. *J. Syst. Evol.* **2020**, *1*, 1–17. [CrossRef]

65. Li, H.T.; Yi, T.S.; Gao, L.M.; Ma, P.F.; Zhang, T.; Yang, J.B.; Gitzendanner, M.A.; Fritsch, P.W.; Cai, J.; Luo, Y.; et al. Origin of angiosperms and the puzzle of the Jurassic gap. *Nat. Plants* **2019**, *5*, 461–470. [CrossRef] [PubMed]

66. Katoh, K.; Standley, D.M. MAFFT multiple sequence alignment software version 7: Improvements in performance and usability. *Mol. Biol. Evol.* **2013**, *30*, 772–780. [CrossRef] [PubMed]

67. Talavera, G.; Castresana, J. Improvement of phylogenies after removing divergent and ambiguously aligned blocks from protein sequence alignments. *Syst. Biol.* **2007**, *56*, 564–577. [CrossRef]

68. Miller, M.A.; Pfeiffer, W.T.; Schwartz, T. Creating the CIPRES Science Gateway for inference of large phylogenetic trees. In Proceedings of the SC10 Workshop on Gateway Computing Environments (GCE10), New Orleans, LA, USA, 14 November 2010. [CrossRef]

69. Ronquist, F.; Teslenko, M.; van der Mark, P.; Ayres, D.L.; Darling, A.; Höhna, S.; Laget, B.; Liu, L.; Suchard, M.A.; Huelsenbeck, J.P. MrBayes 3.2: Efficient Bayesian phylogenetic inference and model choice across a large model space. *Syst. Biol.* **2012**, *61*, 539–542. [CrossRef]

70. Zhang, D.; Gao, F.L.; Jakovlić, I.; Zou, H.; Zhang, J.; Li, W.X.; Wang, G.T. PhyloSuite: An integrated and scalable desktop platform for streamlined molecular sequence data management and evolutionary phylogenetics studies. *Mol. Ecol. Resour.* **2020**, *20*, 348–355. [CrossRef]

71. Kalyaanamoorthy, S.; Minh, B.Q.; Wong, T.; Haeseler, A.; Jermiin, L.S. ModelFinder: Fast model selection for accurate phylogenetic estimates. *Nat. Methods* **2017**, *14*, 587–589. [CrossRef]

3

Comparative Genomics of the Balsaminaceae Sister Genera *Hydrocera triflora* and *Impatiens pinfanensis*

Zhi-Zhong Li [1,2,†], Josphat K. Saina [1,2,3,†], Andrew W. Gichira [1,2,3], Cornelius M. Kyalo [1,2,3], Qing-Feng Wang [1,3,*] and Jin-Ming Chen [1,3,*]

[1] Key Laboratory of Aquatic Botany and Watershed Ecology, Wuhan Botanical Garden, Chinese Academy of Sciences, Wuhan 430074, China; wbg_georgelee@163.com (Z.-Z.L.); jksaina@wbgcas.cn (J.K.S.); gichira@wbgcas.cn (A.W.G.); cmulili90@gmail.com (C.M.K.)

[2] University of Chinese Academy of Sciences, Beijing 100049, China

[3] Sino-African Joint Research Center, Chinese Academy of Sciences, Wuhan 430074, China

[*] Correspondence: qfwang@wbgcas.cn (Q.-F.W.); jmchen@wbgcas.cn (J.-M.C.)

[†] These authors contributed equally to this work.

Abstract: The family Balsaminaceae, which consists of the economically important genus *Impatiens* and the monotypic genus *Hydrocera*, lacks a reported or published complete chloroplast genome sequence. Therefore, chloroplast genome sequences of the two sister genera are significant to give insight into the phylogenetic position and understanding the evolution of the Balsaminaceae family among the Ericales. In this study, complete chloroplast (cp) genomes of *Impatiens pinfanensis* and *Hydrocera triflora* were characterized and assembled using a high-throughput sequencing method. The complete cp genomes were found to possess the typical quadripartite structure of land plants chloroplast genomes with double-stranded molecules of 154,189 bp (*Impatiens pinfanensis*) and 152,238 bp (*Hydrocera triflora*) in length. A total of 115 unique genes were identified in both genomes, of which 80 are protein-coding genes, 31 are distinct transfer RNA (tRNA) and four distinct ribosomal RNA (rRNA). Thirty codons, of which 29 had A/T ending codons, revealed relative synonymous codon usage values of >1, whereas those with G/C ending codons displayed values of <1. The simple sequence repeats comprise mostly the mononucleotide repeats A/T in all examined cp genomes. Phylogenetic analysis based on 51 common protein-coding genes indicated that the Balsaminaceae family formed a lineage with Ebenaceae together with all the other Ericales.

Keywords: Balsaminaceae; chloroplast genome; *Hydrocera triflora*; *Impatiens pinfanensis*; phylogenetic analyses

1. Introduction

The family Balsaminaceae of the order Ericales contains only two genera, *Impatiens* Linnaeus (1753:937) and *Hydrocera* Wight and Arnott (1834:140) and are predominantly perennial and annual herbs [1]. The monotypic genus *Hydrocera*, with a single species *Hydrocera triflora*, is characterized by actinomorphic flowers, a pentamerous calyx and corolla without any fusion between perianth parts, contrary to highly similar sister genus *Impatiens* whose flowers are highly zygomorphic [2]. *Impatiens*, one of the largest genera in angiosperms, consists of over 1000 species [3–6] primarily distributed in the Old World tropics, subtropics and temperate regions, but also in Europe, and central and North America [5,7]. In contrast, the sister *Hydrocera*, which is a semi-aquatic plant, is restricted to the lowlands of Indo-Malaysia [1]. Besides, the geographical regions, including south-east Asia, the eastern Himalayas, tropical Africa, Madagascar, southern India and Sri Lanka occupied by *Impatiens*, have been identified as diversity hotspots [7,8]. Recently, numerous new species have been recorded within these regions each year [9–14].

The controversial nature of classification of the genus *Impatiens* [1,15], for example different floral characters, its hybridization nature and species radiation, has made it under-studied. The species in prolific genus *Impatiens* are economically used as ornamentals, medicinal, as well as experimental research plant materials [16]. Additionally, previous studies have shown the genus *Impatiens* to possess potential anticancer compounds by decreasing patients' cancer cell count and increasing their life span and body weight [17]. The glanduliferins A and B isolated from the stem act to inhibit the growth of human cancer cells for growth inhibitory activity of human cancer cells [18]. As well, some polyphenols from *Impatiens* stems have showed antioxidant and antimicrobial activities [19].

In angiosperms, the chloroplast genome (cp) typically has a quadripartite organization consisting of a small single copy (SSC, 16–27 kb) and one large single copy (LSC) of about 80–90 kb long separated by two identical copies of inverted repeats (IRs) of about 20–88 kb with the total complete chloroplast genome size ranging from 72 to 217 kb [20–22]. Most of the complete cp genomes contains 110–130 distinct genes, with approximately 80 genes coding for proteins, 30 tRNA and 4 rRNA genes [21]. In addition, due to the highly conserved gene order and gene content, they have been used in plant evolution and systematic studies [23], determining evolutionary patterns of the cp genomes [24], phylogenetic analysis [25,26], and comparisons of angiosperm, gymnosperm, and fern families [27]. Moreover, the cp genomes are useful in genetic engineering [28], phylogenetics and phylogeography of angiosperms [29], and estimation of the diversification pattern and ancestral state of the vegetation within the family [30].

The Ericales (Bercht and Presl) form a well-supported clade (Asterid) containing more than 20 families [31]. Up to now, complete cp genomes representing approximately half of the families in the order Ericales have been sequenced including: Actinidiaceae [32,33], Ericaceae [34,35], Ebenaceae [36], Sapotaceae [37], Primulaceae [38,39] Styracaceae [40], and Theaceae, Pentaphylacaceae, Sladeniaceae, Symplocaceae, Lecythidaceae [30]. In addition the *Impatiens* and *Hydrocera* intergeneric phylogenetic relationship has been done using chloroplast *atpB-rbcL* spacer sequences [4]. However, there are no reports of complete chloroplast genomes in the family Balsaminaceae to date. This limitation of genetic information has hindered the progress and understanding in taxonomy, phylogeny, evolution and genetic diversity of Balsaminaceae. Analyses of more cp genomes are needed to provide a robust picture of generic and familial relationships of families in order Ericales.

This study aims to determine the complete sequences of the chloroplast genomes of *I. pinfanensis* (Hook. f.) and *H. triflora* using a high-throughput sequencing method. Additionally, comparisons with other published cp genomes in the order Ericales will be made in order to determine phylogenetic relationships among the representatives of Ericales.

2. Results and Discussion

2.1. The I. pinfanensis and H. triflora Chloroplast Genome Structure and Gene Content

The complete chloroplast genomes of *I. pinfanensis* and *H. triflora* share the common feature of possessing a typical quadripartite structure composed of a pair of inverted repeats (IRs) separating a large single copy (LSC) and a small single copy (SSC), similar to other angiosperm cp genomes [23]. The cp genome size of *I. pinfanensis* is 154,189 bp, with a pair of inverted repeats (IRs) of 17,611 bp long that divide LSC of 83,117 bp long and SSC of 25,755 bp long (Table 1). On the other hand, the *H. triflora* complete cp genome is 152,238 bp in length comprising a LSC region of 84,865 bp in size, a SSC of 25,622 bp size, and a pair of IR region 18,082 bp each in size. The overall guanine-cytosine (GC) contents of *I. pinfanensis* and *H. triflora* genomes are 36.8% and 36.9% respectively. Meanwhile, the GC contents in the LSC, SSC, and IR regions are 34.5%/34.7%, 29.3%/29.9%, and 43.1%/43.1% respectively.

Table 1. Comparison of the chloroplast genomes of *Impatiens pinfanensis* and *Hydrocera triflora*.

Species	Impatiens pinfanensis	Hydrocera triflora
Total Genome length (bp)	154,189	152,238
Overall G/C content (%)	36.8	36.9
Large single copy region	83,117	84,865
GC content (%)	34.5	34.7
Short single copy region	25,755	25,622
GC content (%)	29.3	29.9
Inverted repeat region	17,611	18,082
GC content (%)	43.1	43.1
Protein-Coding Genes	80	80
tRNAs	31	31
rRNAs	4	4
Genes with introns	17	17
Genes duplicated by IR	18	18

Like in typical angiosperms, both *I. pinfanensis* and *H. triflora* cp genomes encode 115 total distinct genes of which 80 are protein coding, 31 distinct tRNA and four distinct rRNA genes. Of these 62 genes coding for proteins and 23 tRNA genes were located in the LSC region, seven protein-coding genes, all the four rRNA genes and seven tRNA genes were replicated in the IR regions, while the SSC region was occupied by 11 protein-coding genes and one tRNA gene. The *ycf1* gene was located at the IR and SSC boundary region (Figures 1 and 2).

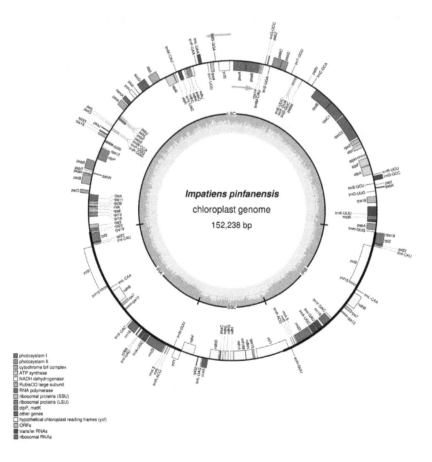

Figure 1. Gene map of the *Impatiens pinfanensis* chloroplast genome. Genes lying outside of the circle are transcribed clockwise, while genes inside the circle are transcribed counterclockwise. The colored bars indicate different functional groups. The dark gray area in the inner circle corresponds to GC content while the light gray corresponds to the adenine-thymine (AT) content of the genome.

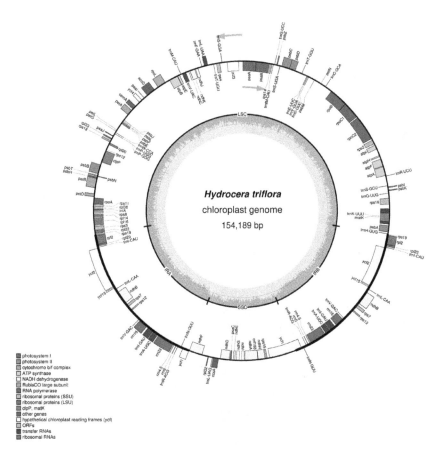

Figure 2. Gene map of the *Hydrocera triflora* chloroplast genome. Genes lying outside of the circle are transcribed clockwise, while genes inside the circle are transcribed counterclockwise. The colored bars indicate different functional groups. The dark gray area in the inner circle corresponds to (guanine cytosine) GC content while the light gray corresponds to the AT content of the genome.

Among the 115 unique genes in *I. pinfanensis* and *H. triflora* cp genomes, 14 genes contain one intron, comprised of eight genes coding for proteins (*atpF, rpoC1, rpl2, petB, rps16, ndhA, ndhB, ndhK*) and six tRNAs (*trnL-UAA, trnV-UAC, trnK-UUU, trnI-GAU, trnG-GCC* and *trnA-UGC*) (Table 2), while *ycf3, clpP* and *rps12* genes each contain two introns. These genes have maintained intron content in other angiosperms. The trans-splicing gene *rps12* has its 5′exon located in LSC, whereas the 3′exon is located in the IRs, which is similar to that in *Diospyros* species (Ebenaceae) [36,41] and *Actinidia chinensis* (Actinidiaceae) [41]. Oddly, *rps19* and *ndhD* genes in both species begin with uncommon start codons GTG and ACG respectively, which is consistent with previous reports in other plants [36]. However, the standard start codon can be restored through RNA editing process [42,43].

The complete cp genome of *I. pinfanensis* and *H. triflora* were found to be similar, although some slight variations such as genome size, gene loss and IR expansion and contraction factors were detected, despite the two species being from the same family Balsaminaceae. For instance, *H. triflora* cp genome is 1951 bp smaller than that of sister species *I. pinfanensis*. The SSC region of *I. pinfanensis* is shorter (17,611 bp) compared to that of *H. triflora*, which is 18,082 bp long. The GC content of *H. triflora* is slightly higher (36.9%) than that of *I. pinfanensis* (36.8%). Both species possess highest GC values in the IR regions (43.1%) compared to LSC and SSC region showing the lowest values (34.5%/34.7% and 29.3%/29.9%) respectively. The IR region is more conserved than the single copy region (SSC) in both species, due to presence of conserved rRNA genes in the IR region, which is also the reason for its high GC content. Both cp genomes are AT-rich with the genome organization and content of the two species almost the same and highly conserved, these results are similar to those of other recently published Ericales chloroplast genomes [34,36].

Table 2. Genes encoded in the *Impatiens pinfanensis* and *Hydrocera triflora* Chloroplast genomes.

Group of Genes	Gene Name
rRNA genes	rrn16(×2), rrn23(×2), rrn4.5(×2), rrn5(×2),
tRNA genes	trnA-UGC * (×2), trnC-GCA, trnD-GUC, trnE-UUC, trnF-GAA, trnG-UCC, trnG-GCC *, trnH-GUG, trnI-CAU(×2), trnI-GAU * (×2), trnK-UUU *, trnL-CAA(×2), trnL-UAA *, trnL-UAG, trnfM-CAU, trnM-CAU, trnN-GUU(×2), trnP-GGG, trnP-UGG, trnQ-UUG, trnR-ACG(×2), trnR-UCU, trnS-GCU, trnS-GGA, trnS-UGA, trnT-GGU, trnT-UGU, trnV-GAC(×2), trnV-UAC *, trnW-CCA, trnY-GUA
Ribosomal small subunit	rps2, rps3, rps4, rps7(×2), rps8, rps11, rps12_5'end, rps12_3'end * (×2), rps14, rps15, rps16 *, rps18, rps19
Ribosomal large subunit	rpl2 * (×2), rpl14, rpl16, rpl20, rpl22, rpl23(×2), rpl32, rpl33, rpl36
DNA-dependent RNA polymerase	rpoA, rpoB, rpoC1 *, rpoC2
Large subunit of rubisco	rbcL
Photosystem I	psaA, psaB, psaC, psaI, psaJ, ycf3 **
Photosystem II	psbA, psbB, psbC, psbD, psbE, psbF, psbH, psbI, psbJ, psbK, psbL, psbM, psbN, psbT, psbZ
NADH dehydrogenase	ndhA *, ndhB * (×2), ndhC, ndhD, ndhE, ndhF, ndhG, ndhH, ndhI, ndhJ, ndhK
Cytochrome b/f complex	petA, petB *, petD, petG, petL, petN
ATP synthase	atpA, atpB, atpE, atpF *, atpH, atpI
Maturase	matK
Subunit of acetyl-CoA carboxylase	accD
Envelope membrane protein	cemA
Protease	clpP **
Translational initiation factor	infA
c-type cytochrome synthesis	ccsA
Conserved open reading frames (ycf)	ycf1, ycf2(×2), ycf4, ycf15(×2)

Genes with one or two introns are indicated by one (*) or two asterisks (**), respectively. Genes in the IR regions are followed by the (×2) symbol.

2.2. Codon Usage

The relative synonymous codon usage (RSCU) has been divided into four models, i.e., RSCU value of less than 1.0 (lack of bias), RSCU value between 1.0 and 1.2 (low bias), RSCU value between 1.2 and 1.3 (moderately bias) and RSCU value greater than 1.3 (highly bias) [44,45]. To determine codon usage, we selected 52 shared protein-coding genes between *I. pinfanensis* and *H. triflora* with length of >300 bp for calculating the effective number of codons. As shown in (Table 3), the relative synonymous codon usage (RSCU) and codon usage revealed biased codon usage in both species with values of 30 codons showing preferences (<1) except tryptophan and methionine, with 29 having A/T ending codons. The TAA stop codon was found to be preferred.

All the protein-coding genes contained 22,900 and 22,995 codons in *I. pinfanensis* and *H. triflora* cp genomes respectively. In addition, our results indicated that 2408 and 2439 codons encode leucine while 253 and 259 encode cysteine in *I. pinfanensis* and *H. triflora* cp genomes as the most and least frequently universal amino acids respectively. The Number of codons (Nc) of the individual PCGs varied from *petD* (37.10) to *ycf3* (54.84) and *rps18* (32.11) to *rpl2* (54.24) in *I. pinfanensis* and *H. triflora* respectively (Table S1). Like recently reported in cp genomes of higher plants, our study showed that there was bias in the usage of synonymous codons except tryptophan and methionine. Our result is in line with previous findings of codon usage preference for A/T ending in other land plants [46,47].

2.3. SSR Analysis Results

Analysis of SSR occurrence using the microsatellite identification tool (MISA) detected Mono-, di-, tri-, tetra-, penta- and hexa-nucleotides categories of SSRs in the cp genomes of eight Ericales. A total of 197 and 159 SSRs were found in the *I. pinfanensis* and *H. triflora* cp genomes respectively. Not all the SSR types were identified in all the species, Penta and hexanucleotide repeats were not found in *I. pinfanensis*, *Diospyros lotus*, and *Pouteria campechiana*, while only hexanucleotides were not identified in *Ardisia polysticta* and *Barringtonia fusicarpa* (Table 4).

Among the SSR types discovered mononucleotide repeat units were highly represented, which were found 180 and 141 times in *I. pinfanensis* and *H. triflora* respectively. Most of the mononucleotide repeats consisting of A or T were most common (117–176 times), whereas C/G were less in number (1–8 times), and all the dinucleotide repeat sequences in all the species were AT repeats. This result is consistent with previous reports, which showed most angiosperm cp genome to be AT-rich [36,38,48].

Table 3. Codon usage in *Impatiens pinfanensis* and *Hydrocera triflora* chloroplast genomes.

Amino Acid	Codon	Number I. pinfanensis	Number H. triflora	RSCU I. pinfanensis	RSCU H. triflora
Phe	UUU	913	908	**1.40**	**1.38**
	UUC	387	406	0.60	0.62
Leu	UUA	854	842	**2.11**	**2.07**
	UUG	468	486	**1.16**	**1.20**
	CUU	517	503	**1.28**	**1.24**
	CUC	160	162	0.40	0.40
	CUA	310	315	0.77	0.78
	CUG	121	128	0.30	0.32
Ile	AUU	1035	1020	**1.54**	**1.52**
	AUC	359	376	0.53	0.56
	AUA	624	611	0.93	0.91
Met	AUG	547	548	**1.00**	**1.00**
Val	GUU	482	469	**1.55**	**1.52**
	GUC	134	135	0.43	0.44
	GUA	457	457	**1.47**	**1.48**
	GUG	167	174	0.54	0.56
Tyr	UAU	704	697	**1.64**	**1.65**
	UAC	155	146	0.36	0.35
TER	UAA	41	44	**1.50**	**1.63**
	UAG	23	19	0.84	0.70
His	CAU	405	421	**1.54**	**1.57**
	CAC	121	114	0.46	0.43
Gln	CAA	627	626	**1.54**	**1.53**
	CAG	186	192	0.46	0.47
Asn	AAU	885	868	**1.59**	**1.57**
	AAC	231	238	0.41	0.43
Lys	AAA	976	978	**1.55**	**1.54**
	AAG	284	289	0.45	0.46
Asp	GAU	720	737	**1.64**	**1.64**
	GAC	159	160	0.36	0.36
Glu	GAA	914	929	**1.55**	**1.55**
	GAG	264	272	0.45	0.45

Amino Acid	Codon	Number I. pinfanensis	Number H. triflora	RSCU I. pinfanensis	RSCU H. triflora
Ser	UCU	482	482	**1.69**	**1.67**
	UCC	252	264	0.88	0.92
	UCA	360	324	**1.26**	**1.12**
	UCG	142	181	0.50	0.63
Pro	CCU	376	371	**1.59**	**1.58**
	CCC	175	167	0.74	0.71
	CCA	294	290	**1.24**	**1.23**
	CCG	103	112	0.43	0.48
Thr	ACU	493	500	**1.70**	**1.74**
	ACC	198	180	0.68	0.63
	ACA	358	368	**1.24**	**1.28**
	ACG	108	104	0.37	0.36
Ala	GCU	580	593	**1.86**	**1.85**
	GCC	183	191	0.59	0.60
	GCA	346	353	**1.11**	**1.10**
	GCG	141	143	0.45	0.45
Cys	UGU	191	196	**1.53**	**1.51**
	UGC	58	63	0.47	0.49
TER	UGA	18	18	0.66	0.67
Trp	UGG	412	412	**1.00**	**1.00**
Arg	AGA	406	407	**1.81**	**1.77**
	AGG	134	143	0.60	0.62
Arg	CGU	302	299	**1.35**	**1.30**
	CGC	88	95	0.39	0.41
	CGA	317	333	**1.41**	**1.45**
	CGG	98	103	0.44	0.45
Ser	AGU	363	72	**1.27**	**1.29**
	AGC	110	108	0.39	0.37
Gly	GGU	525	525	**1.33**	**1.35**
	GGC	160	165	0.40	0.42
	GGA	639	625	**1.62**	**1.61**
	GGG	258	238	0.65	0.61

RSCU: Relative synonymous Codon Usage. RSCU > 1 are highlighted in bold.

Table 4. SSR types and amount in the *Impatiens pinfanensis* and *Hydrocera triflora* Chloroplast genomes.

SSR Type	Repeat Unit	Amount							
		Impatiens pinfanensis	*Hydrocera triflora*	*Actinidia kolomikta*	*Ardisia polysticta*	*Diospyros lotus*	*Barringtonia fusicarpa*	*Pouteria campechiana*	*Primula persimilis*
Mono	A/T	176	139	117	153	146	154	161	134
	C/G	4	2	4	4	4	8	1	4
Di	AT/AT	8	9	8	5	3	13	11	6
Tri	AAG/CTT	1	0	0	0	0	0	1	1
	AAT/ATT	3	3	2	1	1	2	4	0
	AGC/CTG	0	0	0	0	1	0	0	0
Tetra	AAAG/CTTT	1	0	3	2	1	3	1	1
	AAAT/ATTT	2	3	3	3	4	3	6	2
	AATG/ATTC	1	0	0	1	0	0	0	0
	AATT/AATT	1	0	1	0	0	0	1	0
	AGAT/ATCT	1	0	0	0	0	0	0	0
	AAGT/ACTT	0	1	0	0	0	1	0	0
	AACT/AGTT	0	0	0	1	0	0	0	0
	AATC/ATTG	0	0	2	0	1	1	0	0
	AAAC/GTTT	0	0	0	0	0	0	1	0
	AAGG/CCTT	0	0	0	0	0	0	1	0
Penta	AATAC/ATTGT	0	1	0	0	0	0	0	0
	AAAAT/ATTTT	0	0	1	0	0	0	0	0
	AAATT/AATTT	0	0	0	1	0	0	0	0
	AATGT/ACATT	0	0	0	0	0	1	0	0
	AATAT/ATATT	0	0	0	0	0	0	0	1
Hexa	AATCCC/ATTGGG	0	1	0	0	0	0	0	0
	AGATAT/ATATCT	0	0	0	0	0	0	0	1
	AAGATG/ATCTTC	0	0	1	0	0	0	0	0
Total		197	159	143	171	161	187	188	150

2.4. Selection Pressure Analysis of Evolution

The ratio of Synonymous (Ks) and non-synonymous (Ka) Substitution can determine whether the selection pressure has acted on a particular protein-coding sequence. Eighty common protein-coding genes shared by *I. pinfanensis* and *H. triflora* genomes were used. As suggested by Makałowski and Boguski [49] the Ka/Ks values are less than one in protein-coding genes as a result of less frequent non-synonymous (Ka) nucleotide Substitutions than the Synonymous (Ks) substitutions (Table S2). We found that the Ka/Ks values of the two species were low (<1) approaching zero, except for one gene *psbK* found in the LSC region, which has a ratio of 1.0259 (Figure 3). This indicates a negative selection all genes except *psbK* gene and shows that the protein-coding genes in both species are quite highly conserved (Table S2). The LSC, SSC, and IR regions average Ks values between the two species were 0.0995, 0.0314, and 0.1334 respectively. Based on Ka/Ks comparison among the regions, only *ycf1* gene in IR region and most of the genes in the LSC and SSC regions revealed higher Ks values. The higher Ks values signaled that on average more genes found in the SSC region have experienced higher selection pressures in contrast to other cp genome regions (LSC and IR). The non-synonymous (Ka) value varied from 0.005 (*psbE*) to 0.0927 (*ycf1*) while Ks ranged from 0.058 (*psbN*) to 0.2944 (*ndhE*). Based on sequence similarity among the IR, SSC and LSC regions, the IR region was more conserved. This is in agreement with previous reports that found out that IR region diverged at a slower rate than the LSC and SSC regions as a result of frequent recombinant events taking place in IR region leading to selective constraints on sequence homogeneity [50,51].

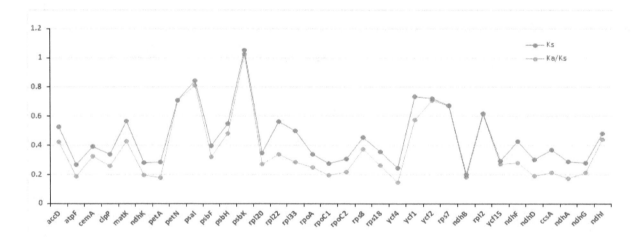

Figure 3. Non-synonymous (Ka) and synonymous (Ks) substitution rates and Ka/Ks ratio between *I. pinfanensis* and *H. triflora*. One gene *psbK* had Ka/Ks ratio greater than 1.0, whereas all the other genes were less than 1.0.

2.5. IR Expansion and Contraction

Despite of the highly conserved nature of the angiosperms inverted repeat (IRa/b) regions, the contraction or expansion at the IR junction are the usual evolutionary events resulting in varying cp genome sizes [52,53]. In our study, the IR/SSC and IR/LSC borders of *I. pinfanensis* and *H. triflora* were compared to those of the other six Ericales representatives (*P. persimilis, P. campechiana, D. lotus, B. fusicarpa, A. kolomikta* and *A. polysticta*) to identify the IR expansion or contraction (Figure 4). The IRb/SSC boundary expansions in all the eight species extended into the *ycf1* genes creating long *φycf1* pseudogene fragments with varying length. The *ycf1* pseudogene length in *I. pinfanensis* is 1101 bp, 1095 bp in *H. triflora*, 394 bp in *A. kolomikta*, 974 bp in *A. polysticta*, 1058 bp in *B. fusicarpa*, 1203 bp in *D. lotus*, 1078 bp in *P. campechiana* and 1018 bp in *P. persimilis*. Additionally, the *ndhF* gene is situated in the SSC region in *I. pinfanensis, H. triflora, A. kolomikta, D. lotus,* and *P. persimilis,* and it ranges from 32 bp, 9 bp, 71 bp, 10 bp and 44 bp away from the IRb/SSC boundary region respectively, but this gene formed an overlap with the *ycf1* pseudogene in *A. polystica*, *B. fusicarpa* and *P. campechiana* cp genomes sharing some nucleotides of 3 bp, 1 bp and 1 bp in that order. The *rps19* gene is located at the /IRb/LSC junction, of *I. pinfanensis*, *H. triflora* and of the other five cp genomes, apart from *A. kolomikta* in which this gene is found in the LSC region, 151 bp gap from the LSC/IRb junction. Moreover, the occurrence of *rps19* gene at the LSC/IRb junction resulted in partial duplication of this gene at the corresponding region (IRa/LSC border) in *I. pinfanensis*, *H. triflora*, and *A. polysticta* cp genomes. The *trnH* gene is detected in the LSC region in *I. pinfanensis* and *H. triflora*. However, complete gene rearrangement of this *trnH* gene was observed resulting in complete duplication in the IR in the *A. kolomikta* chloroplast genome, 630 bp apart from the IR/LSC junction with *psbA* gene extending towards LSC/IRa border, however this gene is found in the LSC regions of the other five chloroplast genomes.

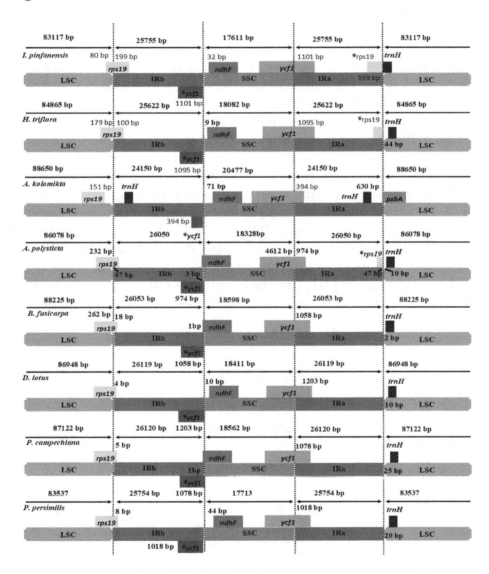

Figure 4. Comparison of IR, LSC and SSC border regions among eight Ericales cp genomes. The IRb/SSC junction extended into the *ycf1* genes creating various lengths of *ycf1* pseudogenes among the eight cp genomes. The numbers above, below or adjacent to genes shows the distance between the ends of genes and the boundary sites. The figure features are not to scale. $^\varphi$ indicates a pseudogene.

The border regions of the Ericales revealed that the *I. pinfanensis* and *H. triflora* cp genomes varied a little compared to other analyzed cp genomes. As shown in Figure 4, our analyses confirmed the

IR evolution as revealed by the incomplete *rps19* gene, which was duplicated in the IR region in *I. pinfanensis*, *H. triflora*, and *A. polysticta*. Conversely, this *rps19* gene was not duplicated among the remaining representatives of Ericales cp genomes. In a recent study [36,54] found that the *trnH* gene duplication occurs in Actinidiaceae, and Ericaceae. This duplication of genes in the LSC/IRb junction and the IRa/LSC junction would be of great importance in systematic studies. Furthermore, the *rps19* gene at the LSC/IRb in *I. pinfanensis* and *H. triflora* is largely extended into the IRb region (199 bp and 100 bp) respectively. The SSC region of *I. pinfanensis* is 471 bp smaller than that of sister species *H. triflora*, but also smallest among the other species used in this study. Additionally, the *I. pinfanensis* LSC region is smaller than that of other species. Previous studies have shown that there is expansion of single copy (SC) and IR regions of angiosperms cp genomes during evolution [50,55], the *I. pinfanensis* and *H. triflora* cp genomes revealed that the border areas were highly conserved despite of slight genome size differences between the two species.

2.6. Phylogenetic Analysis

Phylogenetic relationships within the order Ericales have been resolved in recent published reports but the position of Balsaminaceae still remains controversial [33,35–40]. In our study, the phylogenetic relationship of *I. pinfanensis*, and *H. triflora* and 38 other species of Ericales downloaded from GenBank (Table S3) was determined, with four cp genomes sequences belonging to Cornales being used as Outgroup species. Fifty-one common protein-coding sequences in all the selected cp genomes employed a single alignment data matrix of a total 35,548 characters (Supplementary Materials File S4). Almost all the nodes in the phylogenetic tree showed a strong bootstrap support. Though, Sapotaceae and Ebenaceae had low support (bootstrap < 70), this could be as a result of fewer samples in these families (Figure 5). *I. pinfanensis* and *H. triflora* as sister taxa (Balsaminaceae) formed the basal family of Ericales with intensive support. In general, all the 38 species together with the two Balsaminaceae family species formed a lineage (Ericales) recognizably discrete from the four outgroup species (Cornales). All the species grouped together into 10 clades corresponding to the 10 families in order Ericales according to APGIV system [31]. This study will provide resources for species identification and resolution of deeper phylogenetic branches among *Impatiens* and *Hydrocera* genera.

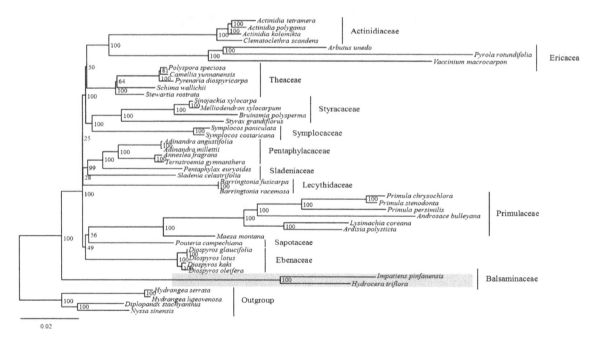

Figure 5. Phylogenetic relationships based on 51 common protein-coding genes of 38 representative species from order Ericales and four Cornales as Outgroup species with maximum likelihood. The numbers associated with the nodes indicate bootstrap values tested with 1000 replicates.

3. Materials and Methods

3.1. Plant Materials and DNA Extraction

Total genomic DNA was extracted from fresh leaves of the *I. pinfanensis* and *H. triflora* collected from Hubei province (108°42′19″ E, 30°12′33″ N) and Hainan province (110°18′57″ E, 19°23′10″ N) in China using a modified cetyltrimethylammonium bromide (CTAB) method [56]. The DNA quality was checked using spectrophotometry and their integrity examined by electrophoresis in 2% agarose gel. The voucher specimens (HIB-lzz07, HIB-lzz18) were deposited at the Wuhan Botanical Garden herbarium (HIB).

3.2. Chloroplast Genome Sequence Assembly and Annotation

The pair-end libraries were constructed using the Illumina Hiseq 2500 platform at NOVOgene Company (Beijing, China) with an average insert size of approximately 150 bp for each genome. The high-quality reads were filtered from Illumina raw reads using the PRINSEQ lite v0.20.4 (San Diego State University, San Diego, CA, USA) [57] (phredQ $\geq$ 20, Length $\geq$ 50), then assembled with closely related species cp genome using a BLASTn (with E value of 10^{-6}) with *Primula chrysochlora* (NC_034678) and *Diospyros lotus* (NC_030786) as reference species. In addition, the software Velvet v1.2.10 (Wellcome Trust Genome Campus, Hinxton, Cambridge, UK) [58] was used to assemble the obtained reads with K-mer length of 99–119. Then, consensus sequences with reference chloroplast genome was mapped using GENEIOUS 8.0.2 (Biomatters Ltd., Auckland, New Zealand) [59]. We used the online software local blast to verify the single copy (SC) and inverted repeat (IR) boundary regions of the assembled sequences.

The annotations of the complete cp genomes were performed using DOGMA (Dual Organellar GenoMe Annotator, University of Texas at Austin, Austin, TX, USA) [60]. The start and stop codons positions were further checked by local blast searches. Further, the tRNAs locations were confirmed with tRNAscan-SE v1.23 (http://lowelab.ucsc.edu/tRNAscan-SE/) [61]. The circular cp genome maps were generated using an online program (OGDrawV1.2, Max planck Institute of Molecular Plant Physiology, Potsdam, Germany) OrganellarGenomeDraw [62] with default settings plus manual corrections. Putative tRNAs, rRNAs and protein-coding genes were corrected by comparing them with the more similar reference species *Primula chrysochlora* (NC_034678) and *Diospyros lotus* (NC_030786) resulting from BLASTN and BLASTX searches against the nucleotide database NCBI (https://blast.ncbi.nlm.nih.gov/). The cp genome sequences were submitted to GenBank database, accession numbers *I. pinfanensis* (MG162586) and *H. triflora* (MG162585).

3.3. Genome Comparison and Structure Analyses

The IR and SC boundary regions of *I. pinfanensis* and *H. triflora*, and the other six Ericales species were compared and examined. For synonymous codon usage analysis, about 52 protein-coding genes of length > 300 bp were chosen. Online program CodonW1.4.2 (http://downloads.fyxm.net/CodonW-76666.html) was used to investigate the Nc and RSCU parameters. The simple sequence repeats (SSRs) of the two study species and other Ericales representatives were detected using MISA software [63] with SSR search parameters set same as Gichira et al. [48].

3.4. Substitution Rate Analysis—Synonymous (Ks) and Non-Synonymous (Ka)

We examined substitution rates synonymous (Ks) and non-synonymous (Ka) using Model Averaging in the KaKs_Cal-culator program (Institute of Genomics, Chinese Academy of Sciences, Beijing, China) [64]. Eighty common protein-coding genes shared by the *I. pinfanensis* and *H. triflora* were aligned separately using Geneious software v5.6.4 (Biomatters Ltd., Auckland, New Zealand) [59].

3.5. Phylogenetic Analyses

To locate the phylogenetic positions of *I. pinfanensis* and *H. triflora* (Balsaminaceae) within order Ericales, the chloroplast genome sequences of 38 species belonging to order Ericales and four Cornales species as outgroups, were used to reconstruct a phylogenetic relationships tree. The Phylogenetic tree was performed based on maximum likelihood (ML) analysis using RAxMLversion 8.0.20 (Scientific Computing Group, Heidelberg Institute for Theoretical Studies, Institute of Theoretical Informatics, Karlsruhe Institute of Technology, Karlsruhe, Germany) [65]. Consequently, based on the Akaike information criterion (AIC), the best-fitting substitution models (GTR + I + G) were selected (p-inv = 0.47, and gamma shape = 0.93) from jModelTest v2.1.7 [66]. The bootstrap test was performed in algorithm of RAxML with 1000 replicates.

4. Conclusions

The cp genomes of *I. pinfanensis*, and *H. triflora* from the family Balsaminaceae provide novel genome sequences and will be of benefit as a reference for further complete chloroplast genome sequencing within the family. The genome organization and gene content are well conserved typical of most angiosperms. Fifty protein-coding sequences, shared by selected species from Ericales as well as our study species, were used to construct the phylogenetic tree using the maximum likelihood (ML). Majority of the nodes showed strong bootstrap support values, and the few nodes with low support, should be solved using other methods (e.g., restriction-site-associated DNA sequencing). The two species (*I. pinfanensis*, and *H. triflora*) were placed close to each other. These findings strongly support Balsaminaceae as a basal family of the order Ericales. Lastly, the Balsaminaceae (*I. pinfanensis*, and *H. triflora*) has a relationship with the other 38 species, which are all grouped into one Clade (Ericales). This study will be of value in determining genome evolution and understanding phylogenomic relationships within Ericales and give precious resources for the evolutionary study of Balsaminaceae.

Acknowledgments: This study was supported by the Special Funds for the Young Scholars of Taxonomy of Chinese Academy of Sciences grants to Liao Kuo (Grant number; ZSBR-013), the Special Foundation for State Basic Working Program of China (2013FY112300) and the National Natural Science Foundation of China (grant no. 31570220).

Author Contributions: Qing-Feng Wang and Jin-Ming Chen conceived and designed the experiment; Zhi-Zhong Li, Josphat K. Saina, Andrew W. Gichira and Cornelius M. Kyalo assembled sequences and revised the manuscript; Zhi-Zhong Li and Josphat K. Saina performed the experiments, analyzed the data and wrote the paper; Jin-Ming Chen and Zhi-Zhong Li collected the plant materials. All authors have read and approved the final version of the manuscript.

Abbreviations

IR	Inverted repeat
LSC	Large single copy
SSC	Small single copy
SSR	Simple sequence repeats
RSCU	Relative synonymous codon usage

References

1.   Grey-Wilson, C. *Impatiens of Africa; Morphology, Pollination and Pollinators, Ecology, Phytogeography, Hybridization, Keys and a Systematics of All African Species with a Note on Collecting and Cultivation*; AA Balkema: Rotterdam, The Netherlands, 1980.

2.   Janssens, S.B.; Smets, E.F.; Vrijdaghs, A. Floral development of *Hydrocera* and *Impatiens* reveals evolutionary trends in the most early diverged lineages of the Balsaminaceae. *Ann. Bot.* **2012**, *109*, 1285–1296. [CrossRef] [PubMed]

3.   Fischer, E.; Rahelivololona, M.E. New taxa of *Impatiens* (Balsaminaceae) from Madagascar iii. *Adansonia* **2004**, *26*, 37–52.

4. Janssens, S.; Geuten, K.; Yuan, Y.-M.; Song, Y.; Küpfer, P.; Smets, E. Phylogenetics of *Impatiens* and *Hydrocera* (Balsaminaceae) using chloroplast atpb-rbcl spacer sequences. *Syst. Bot.* **2006**, *31*, 171–180. [CrossRef]
5. Janssens, S.B.; Knox, E.B.; Huysmans, S.; Smets, E.F.; Merckx, V.S. Rapid radiation of *Impatiens* (Balsaminaceae) during pliocene and pleistocene: Result of a global climate change. *Mol. Phylogenet. Evol.* **2009**, *52*, 806–824. [CrossRef] [PubMed]
6. Janssens, S.B.; Viaene, T.; Huysmans, S.; Smets, E.F.; Geuten, K.P. Selection on length mutations after frameshift can explain the origin and retention of the AP3/DEF-like paralogues in *Impatiens*. *J. Mol. Evol.* **2008**, *66*, 424–435. [CrossRef] [PubMed]
7. Yuan, Y.-M.; Song, Y.; Geuten, K.; Rahelivololona, E.; Wohlhauser, S.; Fischer, E.; Smets, E.; Küpfer, P. Phylogeny and biogeography of Balsaminaceae inferred from its sequences. *Taxon* **2004**, *53*, 391. [CrossRef]
8. Song, Y.; Yuan, Y.-M.; Küpfer, P. Chromosomal evolution in Balsaminaceae, with cytological observations on 45 species from Southeast Asia. *Caryologia* **2003**, *56*, 463–481. [CrossRef]
9. Tan, Y.-H.; Liu, Y.-N.; Jiang, H.; Zhu, X.-X.; Zhang, W.; Yu, S.-X. Impatiens pandurata (Balsaminaceae), a new species from Yunnan, China. *Bot. Stud.* **2015**, *56*, 29. [CrossRef] [PubMed]
10. Zeng, L.; Liu, Y.-N.; Gogoi, R.; Zhang, L.-J.; Yu, S.-X. *Impatiens tianlinensis* (Balsaminaceae), a new species from Guangxi, China. *Phytotaxa* **2015**, *227*, 253–260. [CrossRef]
11. Raju, R.; Dhanraj, F.I.; Arumugam, M.; Pandurangan, A. *Impatiens matthewiana*, a new scapigerous balsam from Western Ghats, India. *Phytotaxa* **2015**, *227*, 268–274. [CrossRef]
12. Guo, H.; Wei, L.; Hao, J.-C.; Du, Y.-F.; Zhang, L.-J.; Yu, S.-X. *Impatiens occultans* (Balsaminaceae), a newly recorded species from Xizang, China, and its phylogenetic position. *Phytotaxa* **2016**, *275*, 62–68. [CrossRef]
13. Cho, S.-H.; Kim, B.-Y.; Park, H.-S.; Phourin, C.; Kim, Y.-D. *Impatiens bokorensis* (Balsaminaceae), a new species from Cambodia. *PhytoKeys* **2017**, *77*, 33. [CrossRef] [PubMed]
14. Yang, B.; Zhou, S.-S.; Maung, K.W.; Tan, Y.-H. Two new species of *Impatiens* (Balsaminaceae) from Putao, Kachin state, northern Myanmar. *Phytotaxa* **2017**, *321*, 103–113. [CrossRef]
15. Hooker, J.D. Les Espèces du Genre "*Impatiens*" dans l'herbier du Museum de Paris. *Nov. Arch. Mus. Nat. Hist. Paris Ser.* **1908**, *10*, 233–272.
16. Bhaskar, V. *Taxonomic Monograph on 'Impatiens' ('Balsaminaceae') of Western Ghats, South India: The Key Genus for Endemism*; Centre for Plant Taxonomic Studies: Bengaluru, India, 2012.
17. Baskar, N.; Devi, B.P.; Jayakar, B. Anticancer studies on ethanol extract of *Impatiens balsamina*. *Int. J. Res. Ayurveda Pharm.* **2012**, *3*, 631–633.
18. Cimmino, A.; Mathieu, V.; Evidente, M.; Ferderin, M.; Banuls, L.M.Y.; Masi, M.; De Carvalho, A.; Kiss, R.; Evidente, A. Glanduliferins A and B, two new glucosylated steroids from *Impatiens glandulifera*, with in vitro growth inhibitory activity in human cancer cells. *Fitoterapia* **2016**, *109*, 138–145. [CrossRef] [PubMed]
19. Szewczyk, K.; Zidorn, C.; Biernasiuk, A.; Komsta, Ł.; Granica, S. Polyphenols from *Impatiens* (Balsaminaceae) and their antioxidant and antimicrobial activities. *Ind. Crops Prod.* **2016**, *86*, 262–272. [CrossRef]
20. Sugiura, M. The chloroplast genome. In *10 Years Plant Molecular Biology*; Springer: Berlin, Germany, 1992; pp. 149–168.
21. Chumley, T.W.; Palmer, J.D.; Mower, J.P.; Fourcade, H.M.; Calie, P.J.; Boore, J.L.; Jansen, R.K. The complete chloroplast genome sequence of *Pelargonium* × *hortorum*: Organization and evolution of the largest and most highly rearranged chloroplast genome of land plants. *Mol. Biol. Evol.* **2006**, *23*, 2175–2190. [CrossRef] [PubMed]
22. Tangphatsornruang, S.; Sangsrakru, D.; Chanprasert, J.; Uthaipaisanwong, P.; Yoocha, T.; Jomchai, N.; Tragoonrung, S. The chloroplast genome sequence of mungbean (*Vigna radiata*) determined by high-throughput pyrosequencing: Structural organization and phylogenetic relationships. *DNA Res.* **2009**, *17*, 11–22. [CrossRef] [PubMed]
23. Wicke, S.; Schneeweiss, G.M.; Müller, K.F.; Quandt, D. The evolution of the plastid chromosome in land plants: Gene content, gene order, gene function. *Plant Mol. Biol.* **2011**, *76*, 273–297. [CrossRef] [PubMed]
24. Jansen, R.K.; Cai, Z.; Raubeson, L.A.; Daniell, H.; Leebens-Mack, J.; Müller, K.F.; Guisinger-Bellian, M.; Haberle, R.C.; Hansen, A.K.; Chumley, T.W. Analysis of 81 genes from 64 plastid genomes resolves relationships in angiosperms and identifies genome-scale evolutionary patterns. *Proc. Natl. Acad. Sci. USA* **2007**, *104*, 19369–19374. [CrossRef] [PubMed]
25. Parks, M.; Cronn, R.; Liston, A. Increasing phylogenetic resolution at low taxonomic levels using massively parallel sequencing of chloroplast genomes. *BMC Biol.* **2009**, *7*, 84. [CrossRef] [PubMed]

26. Moore, M.J.; Soltis, P.S.; Bell, C.D.; Burleigh, J.G.; Soltis, D.E. Phylogenetic analysis of 83 plastid genes further resolves the early diversification of eudicots. *Proc. Natl. Acad. Sci. USA* **2010**, *107*, 4623–4628. [CrossRef] [PubMed]

27. Zhu, A.; Guo, W.; Gupta, S.; Fan, W.; Mower, J.P. Evolutionary dynamics of the plastid inverted repeat: The effects of expansion, contraction, and loss on substitution rates. *New Phytol.* **2016**, *209*, 1747–1756. [CrossRef] [PubMed]

28. Maliga, P. Engineering the plastid genome of higher plants. *Curr. Opin. Plant Biol.* **2002**, *5*, 164–172. [CrossRef]

29. Shaw, J.; Shafer, H.L.; Leonard, O.R.; Kovach, M.J.; Schorr, M.; Morris, A.B. Chloroplast DNA sequence utility for the lowest phylogenetic and phylogeographic inferences in angiosperms: The tortoise and the hare iv. *Am. J. Bot.* **2014**, *101*, 1987–2004. [CrossRef] [PubMed]

30. Yu, X.Q.; Gao, L.M.; Soltis, D.E.; Soltis, P.S.; Yang, J.B.; Fang, L.; Yang, S.X.; Li, D.Z. Insights into the historical assembly of East Asian subtropical evergreen broadleaved forests revealed by the temporal history of the tea family. *New Phytol.* **2017**, *215*, 1235–1248. [CrossRef] [PubMed]

31. Allantospermum, A.; Apodanthaceae, A.; Boraginales, B.; Buxaceae, C.; Centrolepidaceae, C.; Cynomoriaceae, D.; Dilleniales, D.; Dipterocarpaceae, E.; Forchhammeria, F.; Gesneriaceae, H. An update of the angiosperm phylogeny group classification for the orders and families of flowering plants: APG IV. *Bot. J. Linn. Soc.* **2016**, *181*, 1–20. [CrossRef]

32. Lan, Y.; Cheng, L.; Huang, W.; Cao, Q.; Zhou, Z.; Luo, A.; Hu, G. The complete chloroplast genome sequence of *Actinidia kolomikta* from north China. *Conserv. Genet. Resour.* **2017**, 1–3. [CrossRef]

33. Wang, W.-C.; Chen, S.-Y.; Zhang, X.-Z. Chloroplast genome evolution in Actinidiaceae: Clpp loss, heterogenous divergence and phylogenomic practice. *PLoS ONE* **2016**, *11*, e0162324. [CrossRef] [PubMed]

34. Logacheva, M.D.; Schelkunov, M.I.; Shtratnikova, V.Y.; Matveeva, M.V.; Penin, A.A. Comparative analysis of plastid genomes of non-photosynthetic Ericaceae and their photosynthetic relatives. *Sci. Rep.* **2016**, *6*, 30042. [CrossRef] [PubMed]

35. Fajardo, D.; Senalik, D.; Ames, M.; Zhu, H.; Steffan, S.A.; Harbut, R.; Polashock, J.; Vorsa, N.; Gillespie, E.; Kron, K. Complete plastid genome sequence of *Vaccinium macrocarpon*: Structure, gene content, and rearrangements revealed by next generation sequencing. *Tree Genet. Genomes* **2013**, *9*, 489–498. [CrossRef]

36. Fu, J.; Liu, H.; Hu, J.; Liang, Y.; Liang, J.; Wuyun, T.; Tan, X. Five complete chloroplast genome sequences from *Diospyros*: Genome organization and comparative analysis. *PLoS ONE* **2016**, *11*, e0159566. [CrossRef] [PubMed]

37. Jo, S.; Kim, H.-W.; Kim, Y.-K.; Cheon, S.-H.; Kim, K.-J. The first complete plastome sequence from the family Sapotaceae, *Pouteria campechiana* (kunth) baehni. *Mitochondr. DNA Part B* **2016**, *1*, 734–736. [CrossRef]

38. Ku, C.; Hu, J.-M.; Kuo, C.-H. Complete plastid genome sequence of the basal Asterid *Ardisia polysticta* miq. and comparative analyses of Asterid plastid genomes. *PLoS ONE* **2013**, *8*, e62548. [CrossRef]

39. Zhang, C.-Y.; Liu, T.-J.; Yan, H.-F.; Ge, X.-J.; Hao, G. The complete chloroplast genome of a rare candelabra primrose *Primula stenodonta* (Primulaceae). *Conserv. Genet. Resour.* **2017**, *9*, 123–125. [CrossRef]

40. Wang, L.-L.; Zhang, Y.; Yang, Y.-C.; Du, X.-M.; Ren, X.-L.; Liu, W.-Z. The complete chloroplast genome of *Sinojackia xylocarpa* (Ericales: Styracaceae), an endangered plant species endemic to China. *Conserv. Genet. Resour.* **2017**. [CrossRef]

41. Yao, X.; Tang, P.; Li, Z.; Li, D.; Liu, Y.; Huang, H. The first complete chloroplast genome sequences in Actinidiaceae: Genome structure and comparative analysis. *PLoS ONE* **2015**, *10*, e0129347. [CrossRef] [PubMed]

42. Kuroda, H.; Suzuki, H.; Kusumegi, T.; Hirose, T.; Yukawa, Y.; Sugiura, M. Translation of *psbC* mRNAs starts from the downstream GUG, not the upstream AUG, and requires the extended shine–dalgarno sequence in tobacco chloroplasts. *Plant Cell Physiol.* **2007**, *48*, 1374–1378. [CrossRef] [PubMed]

43. Takenaka, M.; Zehrmann, A.; Verbitskiy, D.; Härtel, B.; Brennicke, A. RNA editing in plants and its evolution. *Annu. Rev. Genet.* **2013**, *47*, 335–352. [CrossRef] [PubMed]

44. Zhao, J.; Qi, B.; Ding, L.; Tang, X. Based on RSCU and QRSCU research codon bias of F/10 and G/11 xylanase. *J. Food Sci. Biotechnol.* **2010**, *29*, 755–764.

45. Zuo, L.-H.; Shang, A.-Q.; Zhang, S.; Yu, X.-Y.; Ren, Y.-C.; Yang, M.-S.; Wang, J.-M. The first complete chloroplast genome sequences of *Ulmus* species by de novo sequencing: Genome comparative and taxonomic position analysis. *PLoS ONE* **2017**, *12*, e0171264. [CrossRef] [PubMed]

46. Zhou, J.; Chen, X.; Cui, Y.; Sun, W.; Li, Y.; Wang, Y.; Song, J.; Yao, H. Molecular structure and phylogenetic

analyses of complete chloroplast genomes of two *Aristolochia* medicinal species. *Int. J. Mol. Sci.* **2017**, *18*, 1839. [CrossRef] [PubMed]

47. Wang, W.; Yu, H.; Wang, J.; Lei, W.; Gao, J.; Qiu, X.; Wang, J. The complete chloroplast genome sequences of the medicinal plant *Forsythia suspensa* (Oleaceae). *Int. J. Mol. Sci.* **2017**, *18*, 2288. [CrossRef] [PubMed]

48. Gichira, A.W.; Li, Z.; Saina, J.K.; Long, Z.; Hu, G.; Gituru, R.W.; Wang, Q.; Chen, J. The complete chloroplast genome sequence of an endemic monotypic genus *Hagenia* (Rosaceae): Structural comparative analysis, gene content and microsatellite detection. *PeerJ* **2017**, *5*, e2846. [CrossRef] [PubMed]

49. Makałowski, W.; Boguski, M.S. Evolutionary parameters of the transcribed mammalian genome: An analysis of 2,820 orthologous rodent and human sequences. *Proc. Natl. Acad. Sci. USA* **1998**, *95*, 9407–9412. [CrossRef] [PubMed]

50. Hong, S.-Y.; Cheon, K.-S.; Yoo, K.-O.; Lee, H.-O.; Cho, K.-S.; Suh, J.-T.; Kim, S.-J.; Nam, J.-H.; Sohn, H.-B.; Kim, Y.-H. Complete chloroplast genome sequences and comparative analysis of *Chenopodium quinoa* and *C. album. Front. Plant Sci.* **2017**, *8*, 1696. [CrossRef] [PubMed]

51. Saina, J.K.; Gichira, A.W.; Li, Z.-Z.; Hu, G.-W.; Wang, Q.-F.; Liao, K. The complete chloroplast genome sequence of *Dodonaea viscosa*: Comparative and phylogenetic analyses. *Genetica* **2017**, 1–13. [CrossRef] [PubMed]

52. Raubeson, L.A.; Peery, R.; Chumley, T.W.; Dziubek, C.; Fourcade, H.M.; Boore, J.L.; Jansen, R.K. Comparative chloroplast genomics: Analyses including new sequences from the angiosperms *Nuphar advena* and *Ranunculus macranthus. BMC Genom.* **2007**, *8*, 174. [CrossRef] [PubMed]

53. Wang, R.-J.; Cheng, C.-L.; Chang, C.-C.; Wu, C.-L.; Su, T.-M.; Chaw, S.-M. Dynamics and evolution of the inverted repeat-large single copy junctions in the chloroplast genomes of monocots. *BMC Evol. Biol.* **2008**, *8*, 36. [CrossRef] [PubMed]

54. Huotari, T.; Korpelainen, H. Complete chloroplast genome sequence of *Elodea canadensis* and comparative analyses with other monocot plastid genomes. *Gene* **2012**, *508*, 96–105. [CrossRef] [PubMed]

55. Choi, K.S.; Chung, M.G.; Park, S. The complete chloroplast genome sequences of three Veroniceae species (Plantaginaceae): Comparative analysis and highly divergent regions. *Front. Plant Sci.* **2016**, *7*, 355. [CrossRef] [PubMed]

56. Doyle, J. DNA protocols for plants. In *Molecular Techniques in Taxonomy*; Springer: Berlin, Germany, 1991; pp. 283–293.

57. Schmieder, R.; Edwards, R. Quality control and preprocessing of metagenomic datasets. *Bioinformatics* **2011**, *27*, 863–864. [CrossRef] [PubMed]

58. Zerbino, D.R.; Birney, E. Velvet: Algorithms for de novo short read assembly using de bruijn graphs. *Genome Res.* **2008**, *18*, 821–829. [CrossRef] [PubMed]

59. Kearse, M.; Moir, R.; Wilson, A.; Stones-Havas, S.; Cheung, M.; Sturrock, S.; Buxton, S.; Cooper, A.; Markowitz, S.; Duran, C. Geneious basic: An integrated and extendable desktop software platform for the organization and analysis of sequence data. *Bioinformatics* **2012**, *28*, 1647–1649. [CrossRef] [PubMed]

60. Wyman, S.K.; Jansen, R.K.; Boore, J.L. Automatic annotation of organellar genomes with DOGMA. *Bioinformatics* **2004**, *20*, 3252–3255. [CrossRef] [PubMed]

61. Schattner, P.; Brooks, A.N.; Lowe, T.M. The tRNAscan-SE, snoscan and snoGPS web servers for the detection of tRNAs and snoRNAs. *Nucleic Acids Res.* **2005**, *33*, W686–W689. [CrossRef] [PubMed]

62. Lohse, M.; Drechsel, O.; Bock, R. OrganellarGenomeDRAW (OGDRAW): A tool for the easy generation of high-quality custom graphical maps of plastid and mitochondrial genomes. *Curr. Genet.* **2007**, *52*, 267–274. [CrossRef] [PubMed]

63. Thiel, T.; Michalek, W.; Varshney, R.; Graner, A. Exploiting EST databases for the development and characterization of gene-derived SSR-markers in barley (*Hordeum vulgare* L.). *TAG Theor. Appl. Genet.* **2003**, *106*, 411–422. [CrossRef] [PubMed]

64. Wang, D.; Liu, F.; Wang, L.; Huang, S.; Yu, J. Nonsynonymous substitution rate (Ka) is a relatively consistent parameter for defining fast-evolving and slow-evolving protein-coding genes. *Biol. Direct* **2011**, *6*, 13. [CrossRef] [PubMed]

65. Stamatakis, A. RAxML version 8: A tool for phylogenetic analysis and post-analysis of large phylogenies. *Bioinformatics* **2014**, *30*, 1312–1313. [CrossRef] [PubMed]

66. Posada, D. Jmodeltest: Phylogenetic model averaging. *Mol. Biol. Evol.* **2008**, *25*, 1253–1256. [CrossRef] [PubMed]

4

Different Natural Selection Pressures on the *atpF* Gene in Evergreen Sclerophyllous and Deciduous Oak Species: Evidence from Comparative Analysis of the Complete Chloroplast Genome of *Quercus aquifolioides* with Other Oak Species

Kangquan Yin, Yue Zhang, Yuejuan Li and Fang K. Du *

College of Forestry, Beijing Forestry University, Beijing 100083, China; yinkq@im.ac.cn (K.Y.);
zhangyue2016@bjfu.edu.cn (Y.Z.); liyuejuan@bjfu.edu.cn (Y.L.)
* Correspondence: dufang325@bjfu.edu.cn

Abstract: *Quercus* is an economically important and phylogenetically complex genus in the family Fagaceae. Due to extensive hybridization and introgression, it is considered to be one of the most challenging plant taxa, both taxonomically and phylogenetically. *Quercus aquifolioides* is an evergreen sclerophyllous oak species that is endemic to, but widely distributed across, the Hengduanshan Biodiversity Hotspot in the Eastern Himalayas. Here, we compared the fully assembled chloroplast (cp) genome of *Q. aquifolioides* with those of three closely related species. The analysis revealed a cp genome ranging in size from 160,415 to 161,304 bp and with a typical quadripartite structure, composed of two inverted repeats (IRs) separated by a small single copy (SSC) and a large single copy (LSC) region. The genome organization, gene number, gene order, and GC content of these four *Quercus* cp genomes are similar to those of many angiosperm cp genomes. We also analyzed the *Q. aquifolioides* repeats and microsatellites. Investigating the effects of selection events on shared protein-coding genes using the Ka/Ks ratio showed that significant positive selection had acted on the *atpF* gene of *Q. aquifolioides* compared to two deciduous oak species, and that there had been significant purifying selection on the *atpF* gene in the chloroplast of evergreen sclerophyllous oak trees. In addition, site-specific selection analysis identified positively selected sites in 12 genes. Phylogenetic analysis based on shared protein-coding genes from 14 species defined *Q. aquifolioides* as belonging to sect. *Heterobalanus* and being closely related to *Q. rubra* and *Q. aliena*. Our findings provide valuable genetic information for use in accurately identifying species, resolving taxonomy, and reconstructing the phylogeny of the genus *Quercus*.

Keywords: cp genome; repeat analysis; sequence divergence; non-synonymous substitution; electron transport chain; phylogeny

1. Introduction

The chloroplast (cp) is an organelle which plays an important role in photosynthesis and carbon fixation in plant cells. In angiosperms, the cp is a uniparentally inherited organelle, and it has its own circular, haploid, evolutionarily conserved genome. The cp genome is therefore considered to be a useful and informative genetic resource for studies on evolutionary relationships in the plant kingdom at various taxonomic levels [1]. In most cases, the cp genome is between 120 and 160 kb in size and has a structure composed of two copies of a large inverted repeat (IR) region, a large single copy (LSC) region, and a small single copy (SSC) region [2].

Oaks (*Quercus* L.), which comprise approximately 500 shrub and tree species, form a phylogenetically complex and economically important genus of the beech family, Fagaceae [3]. Distributed throughout much of the Northern Hemisphere, oaks are located in the northern temperate region, and they also occur in the Andes of South America and subtropical and tropical Asia [4]. Oaks are dominant in various habitats, such as temperate deciduous forest, oak-pine forest and temperate and subtropical evergreen forest [5]. They are intimately associated with many other organisms, including fungi, ferns, birds, mammals, and insects [4]. For this reason, their interactions have been the subject of a large number of ecological studies. Human beings have a close connection with oak, as throughout history it has been a common symbol of strength and courage and has been chosen as the national tree in many countries. Moreover, oaks are of great economic value, being used in, for example, the construction of fine furniture and the wine industry.

Oak species are notoriously difficult to classify taxonomically, due to morphological variation caused in part by hybridization [6–14]. Some studies stated that *Quercus* contained two subgenera, *Cyclobalanopsis* and *Quercus*, the latter including three sections: *Quercus* (white oaks), *Lobatae* (red oaks), and *Protobalanus* (golden cup or intermediate oaks) [3,15]. Because previous classifications of oaks have been based solely on morphological characters which are often homoplastic in oaks, these classifications have always been subject to debate [3,15]. With advances in molecular phylogenetics and techniques based on pollen morphology, views on oak classification are changing [15–19]. Recently, Denk et al. proposed an updated classification for *Quercus* with two subgenera: subgenus *Quercus*, the 'New World clade' or 'high-latitude clade', and subgenus *Cerris*, the exclusively Eurasian 'Old World clade' or 'mid-latitude clade' [19]. There are five sections (*Protobalanus, Ponticae, Virentes, Quercus*, and *Lobatae*) in subgenus *Quercus* and three sections (*Cyclobalanopsis, Ilex* and *Cerris*) in subgenus *Cerris*.

China, which is a center of *Quercus* diversity, has 35–51 species [20]. Based on morphological characters, including 25 qualitative and 18 quantitative characters, oaks in China were divided into five sections, namely *Aegilops, Quercus, Brachylepides, Engleriana*, and *Echinolepides*. Recently, we studied the phylogeography of *Quercus aquifolioides*, which is endemic to the Hengduanshan Biodiversity Hotspot, based on 58 populations distributed throughout the species range, using four chloroplast DNA fragments and 11 nuclear microsatellite loci [21]. Up till now, to our knowledge, very few studies have focused on the phylogenetic relationships and population genetics of oaks in China [22], in part due to the challenges arising from introgressive hybridization, lineage sorting, and molecular markers failing to give sufficient phylogenetic signals.

In this study, we produced the first cp genome sequence for *Q. aquifolioides* using next-generation sequencing technology. This complete cp genome, combined with previously reported cp genome sequences for other members of the genus, will enhance our understanding of the systematic evolution of *Quercus*. We analyzed the completely assembled cp genome of *Q. aquifolioides* and compared it to those of three other oak species to investigate common structural patterns and hotspot regions of sequence divergence in these four *Quercus* cp genomes, examined whether selection pressure had acted on protein coding genes, and reconstructed the phylogenetic relationships of the four *Quercus* species. Our findings will not only enrich the complete cp genome resources available for the genus *Quercus* but also provide abundant genetic information for use in subsequent taxonomic and phylogenetic identification of members of the genus, and assist geneticists and breeders in improving commercially-grown oak trees.

2. Results and Discussion

2.1. Chloroplast Genome Organization in Q. aquifolioides

The *Q. aquifolioides* cp genome is a typical circular double-stranded DNA molecule with a length of 160,415 bp, which falls within the normal angiosperm length range [23,24]. The cp genome has the usual quadripartite structure, featuring a LSC region (large single copy region, 89,493 bp), a SSC region (small single copy region, 16,594 bp), and a pair of IRs (inverted repeats, 25,857 bp) (Figure 1; GenBank accession No. KP340971). The GC contents of the LSC, SSC, and IR regions individually, and of the cp genome as a whole, are 34.8%, 31.2%, 42.7%, and 36.9%, respectively. These GC contents are within the range previously reported for other plant species. Approximately 48.0% of the cp genome encodes proteins, 5.6% encodes rRNAs and 1.3% encodes tRNAs. Noncoding regions (intergenic regions, introns and pseudogenes) constitute the remaining 45.1% of the genome. The *Q. aquifolioides* cp genome encodes 127 genes: 80 protein-coding genes, eight ribosomal RNA genes, and 39 tRNA genes. *ycf2* is the largest gene, having a length of 6834 bp. We found that 18 genes have one intron (10 protein coding genes and 8 tRNA genes) and two genes (*clpP* and *ycf3*) have two introns each. Two identical rRNA gene clusters (16S-23S-4.5S-5S) were found in the IR regions. There are two tRNA genes, *trnI* and *trnA*, in the 16S~23S spacer region of each cluster. The sequence of the rRNA coding region is highly conserved: sequence identities of four rRNA genes with those of *Arabidopsis thaliana* (L.) Heynh were over 98%.

Figure 1. Gene map of the *Q. aquifolioides* chloroplast genome. The annotated chloroplast (cp) genome of *Q. aquifolioides* is represented as concentric circles. Genes shown outside the outer circle are transcribed counter-clockwise and genes indicated inside the outer circle are transcribed clockwise. Two inverted repeats (IRs), the large single copy (LSC) and the small single copy (SSC) are shown in the inner circle.

2.2. Repeat Sequence Analysis and Simple Sequence Repeats (SSR)

Repeat sequences have been used extensively for phylogeny, population genetics, genetic mapping, and forensic studies [25]. In the cp genome of *Q. aquifolioides*, 38 pairs of repeats longer than 30 bp were detected; they consisted of 24 palindromic repeats and 14 forward repeats (Figure 2). Among these repeats, 36 are 30–40 bp long, one is 44 bp long, and one is 64 bp long (Figure 2). A large proportion of the repeats (73.7%) are present in non-coding regions, but some repeats are embedded in coding regions, such as the *trnS-GCU*, *trnS-GGA*, *psaB*, *psaA*, *ycf1*, *ycf2*, and *accD* genes

(Table S1). As previous studies reported, many repeats were found in the *ycf2* gene [26–29]. Apart from the IR region, the longest repeats, which were 64 bp in length, were present in the *ndhD/psaC* intergenic region.

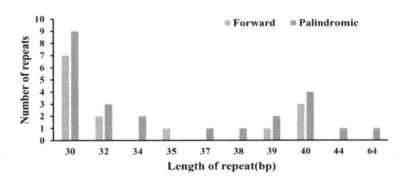

Figure 2. Analysis of repeated sequences in *Q. aquifolioides*.

SSR, also known as microsatellites, are highly polymorphic and thus widely used as molecular markers. A total of 78 perfect microsatellites were identified in the *Q. aquifolioides* cp genome. Among them, 70.51% were present in the LSC regions, whereas 10.26% and 19.23% were identified in the IR and SSC regions respectively (Figure 3A). This result is consistent with previous reports that SSRs are not evenly distributed in cp genomes [30]. Twelve of the SSRs were present in protein-coding regions, six were in introns, and 60 were located in intergenic spacers of the *Q. aquifolioides* cp genome (Figure 3B). Of the motifs forming these SSRs, 58 are mononucleotides, six are dinucleotides, five are trinucleotides, six are tetranucleotides, and three are pentanucleotides (Figure 3C). Most of the mononucleotides (98.28%) and dinucleotides (100%) are composed of A and T. (Figure 3C). These results are consistent with previous reports that SSRs in cp genomes generally consist of short polyA or polyT repeats [31]. The high AT content of cp SSRs contributes to the AT richness of the *Q. aquifolioides* cp genome, which is similar in this respect to other cp genomes [31].

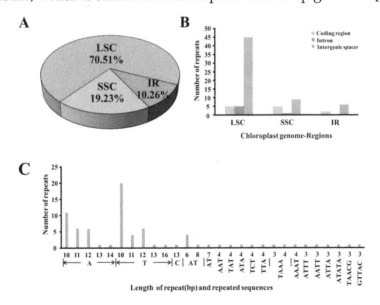

Figure 3. Analysis of simple sequence repeats (SSRs) in the *Q. aquifolioides* cp genome. (**A**) Frequency of SSRs identified in the LSC, SSC, and IR regions; (**B**) Frequency of SSRs identified in the coding regions, intergenic spacers and introns of the LSC, SSC and IR regions; (**C**) Frequency distribution of different classes of polymer in the cp genome of *Q. aquifolioides*.

2.3. Comparison of the cp Genomes of Q. aquifolioides and Three Related Quercus Species

Three complete cp genomes, those of *Q. aliena* (GenBank accession number: KP301144), *Q. rubra* (GenBank accession number: JX970937), and *Q. spinosa* (GenBank accession number: KM841421), belonging to three different sections within the *Quercus* genus, were selected for comparison with *Q. aquifolioides* (Table 1). *Q. rubra* has the largest cp genome; this is mostly attributable to variations in the lengths of the LSC and SSC regions. The GC content of these four cp genomes is very similar, at ~37%. *Q. aquifolioides* has the same number of protein coding genes and rRNA genes as the other three *Quercus* species. Although *Q. spinosa* has one tRNA fewer than the other three *Quercus* species, the total length of its tRNA genes is greater than that in any of the other three species. We found that *Q. aquifolioides* shared 80 protein-coding genes with the cp genomes of all three of the other *Quercus* species.

Table 1. Summary of the features of four complete *Quercus* plastomes.

Genome Features	Sect. *Heterobalanus*		Sect. *Lobatae*	Sect. *Quercus*
	Q. aquifolioides	*Q. spinosa*	*Q. rubra*	*Q. aliena*
Genome size/GC content	160,415/37.0	160,825/36.9	161,304/36.8	160,921/36.9
Coding genes: number/size	80(7)/80,270	80(7)/80,812	80(7)/80,946	80(7)/80,052
tRNA: number/size	39/10,625	38/11,402	39/10,756	39/10,753
rRNA: number/size	8/9048	8/9050	8/9050	8/9048
LSC: size/percent/GC content	89,807/56/34.8	90,371/56.2/34.7	90,541/56.1/34.7	90,258/56.1/34.7
SSC: size/percent/GC content	18,894/11.8/31.2	18,732/11.6/31.2	19,023/56.1/30.9	18,972/11.8/31.0
IR: size/percent/GC content	51,754/32.2/42.7	51,722/32.2/47.2	52,740/32.7/42.7	51,682/32.1/42.7
Introns: size/percent	20,473/12.8	19,757/12.3	20,217/12.5	20,014/12.4
Intergenic spacer: size/percent	49,548/31.0	50,207/31.2	47,473/29.4	47,304/29.3

Numbers in brackets denote the numbers of genes duplicated in the IR regions.

We compared the other three complete cp genomes with that of *Q. aquifolioides* (Figure 4). The sequence identity between these four *Quercus* cp genomes was analyzed. Our results revealed perfect conservation of gene order along the cp genomes of the four species and very high similarity between them.

Although the overall quadripartite structure, including the gene number and order, is usually well conserved, the IR region often undergoes expansion or contraction, a phenomenon called ebb and flow in cp genomes [32]. Generally, the expansion or contraction involves no more than a few hundred nucleotides. Kim and Lee proposed that length variation in angiosperm cp genomes was primarily caused by expansion and contraction of the IR region and the single-copy (SC) boundary regions [33]. The IR/SC boundary regions of these four complete *Quercus* cp genomes were compared, and found to exhibit clear differences in junction positions (Figure 5). The inverted repeat b (IRb)/SSC borders are located in the coding region of the *ycf1* gene with a region of 4590–4611 bp located in the SSC regions. The shortened *ycf1* gene crossed the inverted repeat a (IRa)/SSC borders, with 25–28 bp falling within the SSC regions, and the *ndhF* gene was located in the SSC region with its distance to the IRa/SSC borders ranging from 8 to 22 bp. At the LSC/IRa junction, the distances between *rps19* and the border ranged from 12 to 35 bp, while the distances between *rpl2* and the border were from 39 to 63 bp. At the LSC/IRb junction, the distances between *rpl2* and the border ranged from 54 to 226 bp and the distances between *trnH* and the border were the same, at 16 bp. Thus, variations at the IR/SC borders in these four cp genomes contribute to the differences in length of the cp genome sequence as a whole.

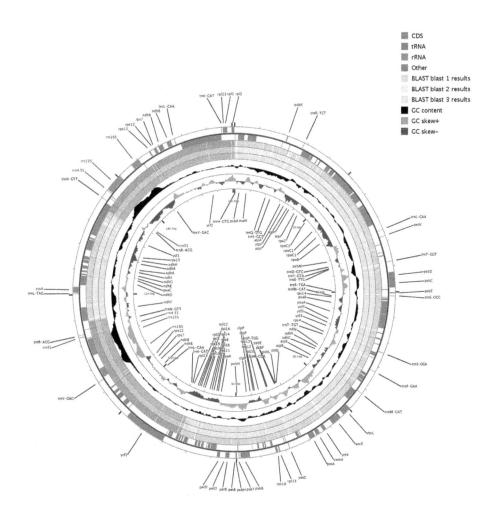

Figure 4. Comparison of four *Quercus* cp genome sequences. The outer four rings show the coding sequences, tRNA genes, rRNA genes, and other genes in the forward and reverse strands. The next three rings show the blast results between the cp genomes of *Q. aquifolioides* and three other *Quercus* species based on BlastN (blast 1–3 results: *Q. aquifolioides* Vs *Q. aliena*, *Q. rubra*, and *Q. spinosa*, respectively). The following black ring is the GC content curve for the *Q. aquifolioides* cp genome. The innermost ring is a GC skew curve for the *Q. aquifolioides* cp genome. GC skew+ (green) indicates G > C, GC skew− (purple) indicates G < C.

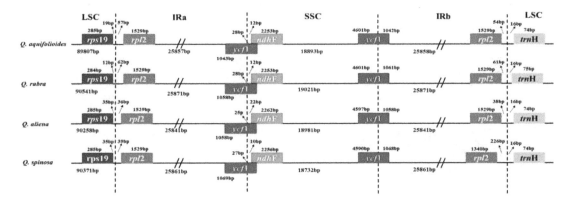

Figure 5. Comparisons of borders between neighboring genes and junctions of the LSC, SSC, and IR regions among the four *Quercus* cp genomes. Boxes above or below the main line indicate genes adjacent to borders. The figure is not to scale with regard to sequence length and shows only relative changes at or near (inverted repeats/single copy) IR/SC borders.

The whole-genome alignment revealed high sequence similarity across these four cp genomes, suggesting that *Quercus* cp genomes are well conserved (Figure 6). As observed in other angiosperms [34–36], we also found that among these four cp genomes the SC regions are more divergent than the IR regions, possibly due to error correction occurring via gene conversion between IRs [37]. Our results also showed that coding regions are more conserved than non-coding regions, as seen in other plants [38,39]. The most divergent coding region in these four *Quercus* cp genomes was *rpl22*. Non-coding regions showed various degrees of sequence divergence among these four *Quercus* cp genomes, with the *trnH-GUG/psbA* regions having the highest level of divergence. These hotspot regions furnish valuable information as a basis for developing molecular markers for phylogenetic studies and identification of *Quercus* species.

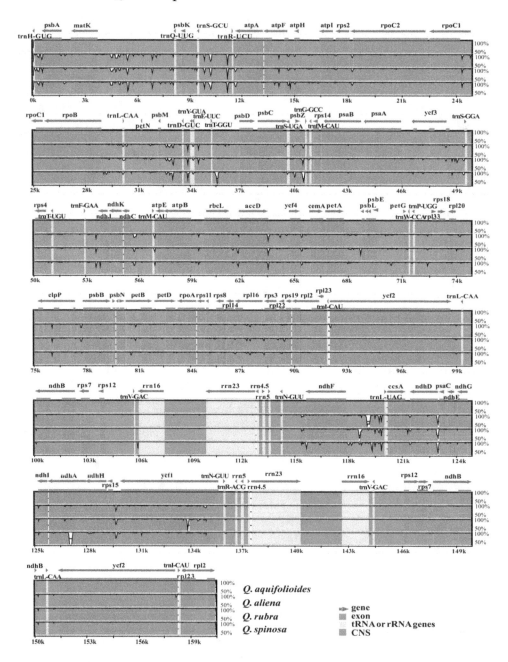

Figure 6. Alignment of four *Quercus* cp genome sequences. Sequence identity plot for four *Quercus* species, with *Q. aquifolioides* as a reference. The X-axis corresponds to coordinates within the cp genome. The Y-axis shows the percentage identity in the range 50% to 100%.

2.4. Genome Sequence Divergence among Quercus Species

To investigate the extent of sequence divergence among these four *Quercus* cp genomes, the nucleotide variability (*Pi*) values within 600 bp windows (200 bp stepwise moving) in the LSC, SSC, and IR regions of the genomes were calculated (Figure 7). In the LSC region, the values varied from 0 to 0.02389 with a mean of 0.00603, while the SSC regions were from 0 to 0.02 with a mean of 0.00863, and the IR regions were from 0 to 0.00417, with a mean of 0.00098. These results suggest that the differences between these genomic regions are very small. However, we also found certain highly variable regions in the LSC, SSC, and IRs. In the LSC, the highly variable regions were *trnH/psbA* and *petA/psbJ*, with *Pi* > 0.02. In the SSC, highly variable regions included *ndhF/rpl32*, *ndhA/ndhH* and *ycf1* (*Pi* > 0.015). In the IRs, two regions, *trnR/trnN* and *ndhB*, with *Pi* > 0.004 were identified (Figure 8). Four of these regions, *trnH/psbA*, *petA/psbJ*, *ndhF/rpl32*, and *ycf1*, have also been identified as highly variable in other plants [33,40–42]. On the basis of our results, five of these variable regions (*trnH/psbA*, *petA/psbJ*, *ndhF/rpl32*, *ndhA/ndhH*, and *ycf1*) show great potential as sources of useful phylogenetic markers for *Quercus*.

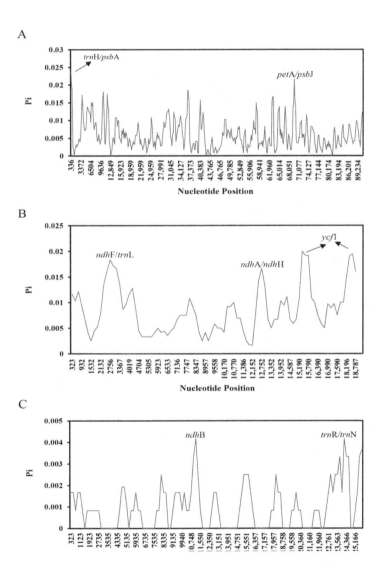

Figure 7. Comparative analysis of nucleotide variability (*Pi*) values among the four *Quercus* cp genome sequences. (**A**) Analysis of the LSC regions; (**B**) Analysis of the SSC regions; (**C**) Analysis of the IR regions. (Window length: 600 bp, step size: 200 bp). X-axis: position of the midpoint of a window, Y-axis: nucleotide diversity of each window.

2.5. Selection Events in Protein Coding Genes

The non-synonymous (Ka) to synonymous (Ks) nucleotide substitution rate ratio (denoted by Ka/Ks) is a very important tool used in studying in protein coding gene evolution. The Ka/Ks ratio is used to evaluate the rate of gene divergence and determine whether positive, purifying or neutral selection has been in operation. A Ka/Ks ratio of >1 indicates positive selection, while Ka/Ks < 1 (especially if it is less than 0.5) indicates purifying selection. A value of close to 1 indicates neutral selection [43].

In this study, we compared Ka/Ks ratios for 73 shared unique protein coding genes in the *Q. aquifolioides* cp genome and the cp genomes of three other related *Quercus* species: *Q. rubra*, *Q. spinosa*, and *Q. aliena* (Table S2). The results are shown in Figure 8. Interestingly, we found that the Ka/Ks ratios were both region specific and gene specific. The average Ka/Ks ratio of the 73 protein coding genes analyzed across the four cp genomes was 0.1653. The most conserved genes with average Ka/Ks values of 0, suggesting very strong purifying selective pressure, were *rpl23*, *rps7*, *psaC*, *rps12*, *psbA*, *psbK*, *psbI*, *atpH*, *atpI*, *rps2*, *petN*, *psbM*, *psbD*, *rps14*, *ycf3*, *ndhJ*, *ndhK*, *ndhC*, *psbJ*, *psbL*, *psbF*, *psbE*, *petL*, *rpl33*, *clpP*, *psbT*, *psbN*, *psbH*, *rpl36*, *rps8*, *rpl14*, *rpl16*, and *rps19* (Table S2). The averaging Ka/Ks method showed no gene with Ka/Ks > 1, which suggests that no gene had been under positive selection in the *Q. aquifolioides* cp genome. Average Ka/Ks value within the range 0.5 to 1, indicating relaxed selection, were observed for *accD*, *petA*, *rpl20*, *rpl22*, *ycf2*, *ndhF*, *ccsA*, *matK*, *atpF*, and *rpoC2*. The remaining genes showed average Ka/Ks values of between 0 and 0.49, which suggested that most genes in the *Q. aquifolioides* cp genome were under purifying selection.

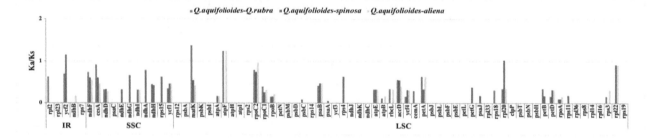

Figure 8. Ka/Ks ratios for protein-coding genes from *Q. rubra*, *Q. spinosa*, and *Q. aliena* chloroplast genome in comparison with *Q. aquifolioides*.

Although no genes were observed with average Ka/Ks > 1, there were four genes (*ycf2*, *matK*, *atpF*, and *rpl20*) with Ka/Ks > 1 in at least one pairwise comparison (Figure 8). Of these, only *atpF* had Ka/Ks > 1 in two pairwise comparisons. More interestingly, the Ka/Ks ratio for the *atpF* gene was more than 1 in the comparisons with the two deciduous *Quercus* species. In contrast, the Ka/Ks ratio for this gene was 0 in the comparison between *Q. aquifolioides* and *Q. spinosa*, which are both evergreen sclerophyllous oak trees. Deciduous oak trees completely lose their foliage during the winter or the dry season, usually as an adaptation to cold and/or drought, whereas evergreen sclerophyllous oak trees retain their leaves throughout of the year. For leaves to tolerate cold and drought stresses requires energy. The *atpF* gene encodes one of the subunits of the H$^+$-ATP synthase, which is essential for electron transport and photophosphorylation during photosynthesis [44]. This finding suggests that differential selection acting on the *atpF* gene may indicate that it has played a role in deciduous-evergreen oak tree divergence.

Alongside the Ka/Ks analysis, we also investigated site-specific selection events. We found a total of 12 genes exhibiting site-specific selection (Table 2). Of these, *rpoC2* was found to have 26 sites under positive selection. This gene encodes one of the four subunits of RNA polymerase type I (plastid-encoded polymerase, PEP), which is a key enzyme required for transcription of photosynthesis-related genes in the chloroplast [45]. Our identification of the positively-selected sites in this analysis could lead to a better understanding of the evolution of *Quercus* species.

Table 2. Positive selection sites identified by Selecton.

Gene	NULL (M8a)	POSITIVE (M8)	Putative Sites under Positive Selection
rpl2	−1177.05	−1174.54	1 (131 S)
ycf2	−9154.8	−9154.87	5 (96 K, 932 W, 1174 P, 1291 W, 2007 R)
rps7	−615.83	−615.616	1 (130 E)
ndhD	−2145.31	−2141.84	8 (170 T, 188 G, 200 L, 206 A, 362 R, 375 P, 413 Q, 504 F)
ycf1	−8057.78	−8051.12	8 (426 F, 529 L, 757 L, 761 L, 1007 I, 1490 Q, 1491 G, 1492 F)
rpoC2	−5814.68	−5814.8	26 (33 H, 131 P, 280 L, 364 I, 505 H, 542 E, 587 E, 595 P, 598 V, 626 N, 643 K, 691 G, 697 T, 815 Y, 849 G, 856 H, 898 D, 947 S, 1013 K, 1074 I, 1081 A, 1132 E, 1176 I, 1273 C, 1374 D, 1394 N)
rpoC1	−2857.37	−2855.15	1 (145 Y)
psaB	−2999.97	−2998.62	3 (145 L, 238 E, 239 K)
ndhJ	−633.793	−633.569	1 (107 A)
ndhC	−828.789	−828.761	2 (68 V, 86 F)
rpl36	−145.376	−145.135	1 (20 R)

Likelihood ratio test (LRT) analysis of models comparison M8 vs. M8a. M8 represents a model with positive selection; M8a represents null model without positive selection. Degree of freedom (df) = 1.

2.6. Phylogenetic Analysis of the cp Genomes of Q. aquifolioides and Related Quercus Species

The phylogeny of oak trees is complex due to extensive introgression, hybridization, incomplete lineage sorting, and convergent evolution [46]. However, phylogenetic issues in many angiosperms have been addressed successfully with the help of cp genome sequences [47–49]. Maximum parsimony (MP) analysis with 73 protein-coding genes from 12 Fragaceae with two tobacco species as outgroup revealed 10 out of 11 nodes with bootstrap values ≥ 95%, which is very high for an MP tree (Figure 9). The MP phylogenetic tree was even more strongly supported by eight 100% bootstrap values, showing that *Q. aquifolioides* was grouped with *Q. spinosa* within *Quercus*. Both of these are members of sect. *Heterobalanus*. The MP tree also revealed that *Q. rubra* and *Q. aliena* were the closest relatives of *Q. aquifolioides* and *Q. spinosa* (Figure 9). However, this phylogenetic tree is solely based on cp DNA. To fully understand their phylogenetic relationships, nuclear DNA is required to be investigated to assess the effect of introgression and hybridization on phylogeny.

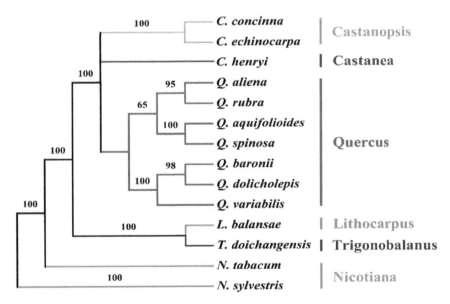

Figure 9. Phylogenetic relationship between *Q. aquifolioides* and related species, inferred from 73 protein-coding genes shared by all cp genomes. The phylogenetic tree was constructed by the maximum parsimony method, with two *Nicotiana* species as outgroups.

3. Materials and Methods

3.1. Plant Material

We collected a *Q. aquifolioides* tree less than 3 years old from Lijiang Alpine Botanic Garden, China and transplanted it to Beijing Forestry University. *Q. aquifolioides* is a common, non-endangered tree species in China. No specific protective policy was implemented in this area. The plants were grown in a growth chamber under 150 mmol·m^{-2}·s^{-1} light, with 16 h light/8 h dark cycles, at 24 °C with a constant humidity of 65%. Voucher specimens were deposited in the herbarium of Beijing Forestry University, Beijing, China.

3.2. Chloroplast Isolation, DNA Extraction, and Sequencing

A 0.3–0.5 g sample of the fresh young leaves was collected after the plant had grown in the dark for 24–36 h to promote starch degradation in chloroplasts (Nobel 1974). The chloroplast DNA (cpDNA) extraction and enrichment method followed the protocol developed by our group [50]. After amplifying the cpDNA by rolling circle amplification (RCA), we purified the RCA product and used 5 μg for library preparation. A 101-bp paired-end run was performed on an Illumina-HiSeq 1500 (Illumina, San Diego, CA, USA) at the gene sequencing platform of the School of Life Sciences, Tsinghua University, China. Briefly, library preparation was carried out following the manufacturer's instructions with an insert size of up to 500 bp. Base calling was performed with RTA v.1.6 (Illumina, San Diego, CA, USA).

3.3. Chloroplast Genome Assembly

We assembled the *Q. aquifolioides* chloroplast genome using a pipeline developed in our lab [21]. Briefly, we used an in-house Perl script to eliminate low quality (probability of error > 1%) nucleotides in each read. SOAPdenovo 2 [51] was used for de novo assembly with default parameters, except that an insert size of 500 bp was set. Next, the primary contigs were assembled using the *Quercus rubra* chloroplast genome (GenBank accession number: JX970937) as the reference sequence. Gaps between two neighbor contigs were filled with N. These gaps were resolved as previously described [50].

3.4. Genome Annotation

We used CpGAVAS [52] for chloroplast genome annotation then manually corrected the output. This program uses a chloroplast genome sequence in FASTA format to identify protein-coding genes by performing BLASTX searches against a custom database of known chloroplast genomes. The program also produces a circular map of the chloroplast genome, displaying the protein-coding genes, transfer RNAs (tRNAs), and ribosomal RNAs (rRNAs) based on the annotations.

3.5. Repeat Analysis

Simple sequence repeats (SSRs) in the cp genomes were detected using the Perl script MISA [53]. The thresholds set for the SSRs were 10, 6, 4, 3, 3, and 3 for mono-, di-, tri-, tetra-, penta-, and hexa-nucleotides, respectively. Tandem repeat sequences (>10 bp in length) were detected using the online program Tandem Repeats Finder [54]. The minimum alignment score and maximum period size were 90 and 500 respectively. The online REPuter software tool (Available online: https://bibiserv.cebitec.uni-bielefeld.de/reputer/) was used to identify forward, palindrome, reverse, and complement sequences with a minimum repeat size of 30 bp, and sequence identity greater than 90% (Hamming distance equal to 3) [55].

3.6. CCT Map

Comparative genome maps of *Q. aquifolioides* and the other three *Quercus* cp genomes were constructed by BLAST using CCT software [56] and the results were displayed as a circular map.

Additional features such as the Clusters of Orthologous Groups of proteins (COG) and GC Skew in the reference genome were also included.

3.7. Sequence Divergence Analysis

The alignments of the cp genomes of *Q. aquifolioides* and the other three *Quercus* cp genome were visualized using mVISTA [57] (Available online: http://genome.lbl.gov/vista/mvista/submit.shtml) in Shuffle-LAGAN mode [34] in order to show interspecific variation. The sequence divergences of four *Quercus* protein coding genes were evaluated using MEGA 7 [58]. A sliding window analysis was conducted to generate nucleotide diversity (*Pi*) values for the three data sets (the aligned LSC, SSC, and IR regions of the four complete *Quercus* cp genomes) using DnaSP 5 [59]. The step size was set to 200 bp, with a 600 bp window length. The Tamura 3-parameter (T92) model was selected to calculate pairwise sequence divergences [60].

3.8. Selection Pressure Analysis

To estimate selection pressures, non-synonymous (Ka) and synonymous (Ks) substitution rates of 73 protein coding genes between the cp genomes of *Q. aquifolioides* and the other three *Quercus* species were calculated using DnaSP 5. For identification of site-specific selection, protein coding gene alignments were analyzed using Selecton [61], with *Q. aquifolioides* as a reference sequence. Two models, M8 (allows for positive selection operating on the protein) and M8a (does not allow for positive selection), were used and likelihood scores estimated by models were evaluated using a log-likelihood ratio test (LRT) with degree of freedom (df) = 1. Only sites with posterior probabilities > 0.8 were selected.

3.9. Phylogenetic Analysis

The sequences were aligned using MAFFT 7 [62]. Maximum parsimony (MP) analysis was executed using PAUP 4 [63]. A total of 73 protein-coding genes shared by all cp genomes were used for this phylogenetic analysis, which included 12 Fagaceae species (*Q. aquifolioides* KP340971; *Q. aliena* KP301144; *Q. rubra* JX970937; *Q. spinosa* KM841421; *Q. variabilis* KU240009; *Q. dolicholepis* KU240010; *Q. baronii* KT963087; *Castanopsis concinna* KT793041; *C. echinocarpa* KJ001129; *Castanea henryi* KX954615; *Lithocarpus balansae* KP299291; *Trigonobalanus doichangensis* KF990556), with two *Nicotiana* species (*N. sylvestris* AB237912; *N. tabacum* Z00044) as outgroups.

Acknowledgments: This research was supported by the National Key Research and Development Plan "Research on protection and restoration of typical small populations of wild plants" (Grant No. 2016YFC0503106), Fundamental Research Funds for the Central Universities (No. 2015ZCQ-LX-03), the National Science Foundation of China (grant 41671039) and the Beijing Nova Program (grant Z151100000315056) to FKD.

Author Contributions: Fang K. Du and Kangquan Yin designed the research; Fang K. Du and Kangquan Yin collected the samples; Yue Zhang, Yuejuan Li and Kangquan Yin performed the experiments and analysis; Fang K. Du and Kangquan Yin wrote the manuscript; all authors revised the manuscript.

References

1. McCauley, D.E.; Stevens, J.E.; Peroni, P.A.; Raveill, J.A. The spatial distribution of chloroplast DNA and allozyme polymorphisms within a population of *Silene alba* (Caryophyllaceae). *Am. J. Bot.* **1996**, *83*, 727–731. [CrossRef]

2. Yurina, N.P.; Odintsova, M.S. Comparative structural organization of plant chloroplast and mitochondrial genomes. *Genetika* **1998**, *34*, 5–22.

3. Nixon, K.C. The genus *Quercus* in Mexico. In *Biological Diversity of Mexico: Origins and Distribution*; Oxford University Press: New York, NY, USA, 1993; pp. 447–458, ISBN 019506674X.

4. Keator, G.; Bazel, S. *The Life of an Oak: An Intimate Portrait*; Heyday Books: Berkeley, CA, USA; California Oak Foundation: Oakland, CA, USA, 1998; p. 256, ISBN 9780930588984.

5. Nixon, K.C. Global and neotropical distribution and diversity of oak (genus *Quercus*) and oak forests. In *Ecology and Conservation of Neotropical Montane Oak Forests*; Springer: Berlin/Heidelberg, Germany, 2006; pp. 3–13, ISBN 364206695X.

6. Trelease, W. The American oaks. *Mem. Natl. Acad. Sci.* **1924**, *20*, 1–255.

7. Palmer, E.J. Hybrid oaks of North America. *J. Arnold Arbor.* **1948**, *29*, 1–48.

8. Muller, C.H. Ecological control of hybridization in *Quercus*: A factor in the mechanism of evolution. *Evolution* **1952**, *6*, 147–161.

9. Tucker, J.M. Studies in the *Quercus undulata* complex. I. A preliminary statement. *Am. J. Bot.* **1961**, *48*, 202–208. [CrossRef]

10. Hardin, J.W. Hybridization and introgression in *Quercus alba*. *J. Arnold Arbor.* **1975**, *56*, 336–363.

11. Rushton, B.S. Natural hybridization within the genus *Quercus*. *Ann. For. Sci.* **1993**, *50*, 73s–90s. [CrossRef]

12. Spellenberg, R. On the hybrid nature of *Quercus basaseachicensis* (Fagaceae, sect. Quercus). *SIDA Contrib. Bot.* **1995**, *16*, 427–437.

13. Bacilieri, R.; Ducousso, A.; Petit, R.J.; Kremer, A. Mating system and asymmetric hybridization in a mixed stand of European oaks. *Evolution* **1996**, *50*, 900–908. [CrossRef] [PubMed]

14. Howard, D.J.; Preszler, R.W.; Williams, J.; Fenchel, S.; Boecklen, W.J. How discrete are oak species? Insights from a hybrid zone between *Quercus grisea* and *Quercus gambelii*. *Evolution* **1997**, *51*, 747–755. [CrossRef] [PubMed]

15. Manos, P.S.; Doyle, J.J.; Nixon, K.C. Phylogeny, biogeography, and processes of molecular differentiation in *Quercus* subgenus *Quercus* (Fagaceae). *Mol. Phylogenet. Evol.* **1999**, *12*, 333–349. [CrossRef] [PubMed]

16. Manos, P.S.; Stanford, A.M. The historical biogeography of Fagaceae: Tracking the tertiary history of temperate and subtropical forests of the Northern Hemisphere. *Int. J. Plant Sci.* **2001**, *162*, S77–S93. [CrossRef]

17. Grímsson, F.; Zetter, R.; Grimm, G.W.; Pedersen, G.K.; Pedersen, A.K.; Denk, T. Fagaceae pollen from the early Cenozoic of West Greenland: Revisiting Engler's and Chaney's Arcto-Tertiary hypotheses. *Plant Syst. Evol.* **2015**, *301*, 809–832. [CrossRef] [PubMed]

18. Simeone, M.C.; Grimm, G.W.; Papini, A.; Vessella, F.; Cardoni, S.; Tordoni, E.; Piredda, R.; Franc, A.; Denk, T. Plastome data reveal multiple geographic origins of *Quercus* Group Ilex. *PeerJ* **2016**, *4*, e1897. [CrossRef] [PubMed]

19. Denk, T.; Grimm, G.W.; Manos, P.S.; Deng, M.; Hipp, A.L. An updated infrageneric classification of the oaks: Review of previous taxonomic schemes and synthesis of evolutionary patterns. In *Oaks Physiological Ecology. Exploring the Functional Diversity of Genus Quercus L.*; Gil-Pelegrín, E., Peguero-Pina, J., Sancho-Knapik, D., Eds.; Springer: Cham, Switzerland, 2017; pp. 13–38. ISBN 978-3-319-69098-8.

20. Huang, C.J.; Zhang, Y.T.; Bartholomew, B. Fagaeeae. In *Flora of China*; Wu, Z.Y., Raven, P.H., Eds.; Science Press: Beijing, China, 1999; pp. 370–380, ISBN 0915279703.

21. Du, F.K.; Hou, M.; Wang, W.; Mao, K.S.; Hampe, A. Phylogeography of *Quercus aquifolioides* provides novel insights into the Neogene history of a major global hotspot of plant diversity in southwest China. *J. Biogeogr.* **2017**, *44*, 294–307. [CrossRef]

22. Yang, Y.; Zhou, T.; Duan, D.; Yang, J.; Feng, L.; Zhao, G. Comparative analysis of the complete chloroplast genomes of five *Quercus* species. *Front. Plant Sci.* **2016**, *7*, 959. [CrossRef] [PubMed]

23. Jansen, R.K.; Raubeson, L.A.; Boore, J.L.; dePamphilis, C.W.; Chumley, T.W.; Haberle, R.C.; Wyman, S.K.; Alverson, A.J.; Peery, R.; Herman, S.J.; et al. Methods for obtaining and analyzing whole chloroplast genome sequences. *Method Enzymol.* **2005**, *395*, 348–384.

24. Chumley, T.W.; Palmer, J.D.; Mower, J.P.; Fourcade, H.M.; Calie, P.J.; Boore, J.L.; Jansen, R.K. The complete chloroplast genome sequence of *Pelargonium* × *hortorum*: Organization and evolution of the largest and most highly rearranged chloroplast genome of land plants. *Mol. Biol. Evol.* **2006**, *23*, 2175–2190. [CrossRef] [PubMed]

25. Bull, L.N.; Pabón-Peña, C.R.; Freimer, N.B. Compound microsatellite repeats: Practical and theoretical features. *Genome Res.* **1999**, *9*, 830–838. [CrossRef] [PubMed]

26. Jansen, R.K.; Kaittanis, C.; Saski, C.; Lee, S.B.; Tomkins, J.; Alverson, A.J.; Daniell, H. Phylogenetic analyses of *Vitis* (Vitaceae) based on complete chloroplast genome sequences: Effects of taxon sampling and phylogenetic methods on resolving relationships among rosids. *BMC Evol. Biol.* **2006**, *6*, 32. [CrossRef] [PubMed]

27. Ruhlman, T.; Lee, S.B.; Jansen, R.K.; Hostetler, J.B.; Tallon, L.J.; Town, C.D.; Daniell, H. Complete plastid genome sequence of *Daucus carota*: Implications for biotechnology and phylogeny of angiosperms. *BMC Genom.* **2006**, *7*, 222. [CrossRef] [PubMed]

28. Silva, S.R.; Diaz, Y.C.; Penha, H.A.; Pinheiro, D.G.; Fernandes, C.C.; Miranda, V.F.; Todd, P.; Michael, T.P.; Varani, A.M. The Chloroplast Genome of *Utricularia reniformis* Sheds Light on the Evolution of the *ndh* Gene Complex of Terrestrial Carnivorous Plants from the Lentibulariaceae Family. *PLoS ONE* **2016**, *11*, e0165176. [CrossRef] [PubMed]

29. Liu, L.X.; Li, R.; Worth, J.R.; Li, X.; Li, P.; Cameron, K.M.; Fu, C.X. The Complete Chloroplast Genome of Chinese Bayberry (*Morella rubra*, Myricaceae): Implications for Understanding the Evolution of Fagales. *Front. Plant Sci.* **2017**, *8*, 968. [CrossRef] [PubMed]

30. Qian, J.; Song, J.; Gao, H.; Zhu, Y.; Xu, J.; Pang, X.; Yao, H.; Sun, C.; Li, X.; Li, C.; et al. The complete chloroplast genome sequence of the medicinal plant *Salvia miltiorrhiza*. *PLoS ONE* **2013**, *8*, e57607. [CrossRef] [PubMed]

31. Kuang, D.Y.; Wu, H.; Wang, Y.L.; Gao, L.M.; Zhang, S.Z.; Lu, L. Complete chloroplast genome sequence of *Magnolia kwangsiensis* (Magnoliaceae): Implication for DNA barcoding and population genetics. *Genome* **2011**, *54*, 663–673. [CrossRef] [PubMed]

32. Goulding, S.E.; Wolfe, K.H.; Olmstead, R.G.; Morden, C.W. Ebb and flow of the chloroplast inverted repeat. *Mol. Gen. Genet.* **1996**, *252*, 195–206. [CrossRef] [PubMed]

33. Kim, K.J.; Lee, H.L. Complete chloroplast genome sequences from Korean ginseng (*Panax schinseng* Nees) and comparative analysis of sequence evolution among 17 vascular plants. *DNA Res.* **2004**, *11*, 247–261. [CrossRef] [PubMed]

34. Dong, W.; Xu, C.; Cheng, T.; Zhou, S. Complete chloroplast genome of *Sedum sarmentosum* and chloroplast genome evolution in Saxifragales. *PLoS ONE* **2013**, *8*, e77965. [CrossRef] [PubMed]

35. Lu, R.; Li, P.; Qiu, Y. The complete chloroplast genomes of three *Cardiocrinum* (Liliaceae) species: Comparative genomic and phylogenetic analyses. *Front. Plant Sci.* **2016**, *7*, 2054. [CrossRef] [PubMed]

36. Zhang, Y.; Du, L.; Liu, A.; Chen, J.; Wu, L.; Hu, W.; Zhang, W.; Kim, K.; Lee, S.C.; Tae-Jin Yang, T.J.; et al. The Complete Chloroplast Genome Sequences of Five *Epimedium* Species: Lights into Phylogenetic and Taxonomic Analyses. *Front. Plant Sci.* **2016**, *7*, 306. [CrossRef] [PubMed]

37. Khakhlova, O.; Bock, R. Elimination of deleterious mutations in plastid genomes by gene conversion. *Plant J.* **2006**, *46*, 85–94. [CrossRef] [PubMed]

38. Liu, Y.; Huo, N.; Dong, L.; Wang, Y.; Zhang, S.; Young, H.A.; Feng, X.; Gu, Y.Q. Complete Chloroplast Genome Sequences of Mongolia Medicine *Artemisia frigida* and Phylogenetic Relationships with Other Plants. *PLoS ONE* **2013**, *8*, e57533. [CrossRef] [PubMed]

39. Nazareno, A.G.; Carlsen, M.; Lohmann, L.G. Complete Chloroplast Genome of *Tanaecium tetragonolobum*: The First Bignoniaceae Plastome. *PLoS ONE* **2015**, *10*, e0129930. [CrossRef] [PubMed]

40. Dong, W.; Liu, J.; Yu, J.; Wang, L.; Zhou, S. Highly variable chloroplast markers for evaluating plant phylogeny at low taxonomic levels and for DNA barcoding. *PLoS ONE* **2012**, *7*, e35071. [CrossRef] [PubMed]

41. Awad, M.; Fahmy, R.M.; Mosa, K.A.; Helmy, M.; El-Feky, F.A. Identification of effective DNA barcodes for *Triticum* plants through chloroplast genome-wide analysis. *Comput. Biol. Chem.* **2017**, *71*, 20–31. [CrossRef] [PubMed]

42. Shaw, J.; Lickey, E.B.; Schilling, E.E.; Small, R.L. Comparison of whole chloroplast genome sequences to choose noncoding regions for phylogenetic studies in angiosperms: The tortoise and the hare III. *Am. J. Bot.* **2007**, *94*, 275–288. [CrossRef] [PubMed]

43. Kimura, M. The neutral theory of molecular evolution and the world view of the neutralists. *Genome* **1983**, *31*, 24–31. [CrossRef]

44. Hudson, G.S.; Mason, J.G. The chloroplast genes encoding subunits of the H^+-ATP synthase. In *Molecular Biology of Photosynthesis*; Govindjee, Ed.; Springer: Dordrecht, The Netherlands, 1988; pp. 565–582, ISBN 978-94-010-7517-6.

45. Cummings, M.P.; King, L.M.; Kellogg, E.A. Slipped-strand mispairing in a plastid gene: *RpoC2* in grasses (*Poaceae*). *Mol. Biol. Evol.* **1994**, *11*, 1–8. [PubMed]

46. Aldrich, P.R.; Cavender-Bares, J. Quercus. In *Wild Crop Relatives: Genomic and Breeding Resources*; Kole, C., Ed.; Springer: Berlin/Heidelberg, Germany, 2011; pp. 89–129, ISBN 978-3-642-21249-9.

47. Jansen, R.K.; Cai, Z.; Raubeson, L.A.; Daniell, H.; dePamphilis, C.W.; Leebens-Mack, J.; Müller, K.F.; Guisinger-Bellian, M.; Haberle, R.C.; Hansen, A.K.; et al. Analysis of 81 genes from 64 plastid genomes resolves relationships in angiosperms and identifies genome-scale evolutionary patterns. *Proc. Natl. Acad. Sci. USA* **2007**, *104*, 19369–19374. [CrossRef] [PubMed]

48. Moore, M.J.; Bell, C.D.; Soltis, P.S.; Soltis, D.E. Using plastid genome-scale data to resolve enigmatic relationships among basal angiosperms. *Proc. Natl. Acad. Sci. USA* **2007**, *104*, 19363–19368. [CrossRef] [PubMed]

49. Yang, J.B.; Li, D.Z.; Li, H.T. Highly effective sequencing whole chloroplast genomes of angiosperms by nine novel universal primer pairs. *Mol. Ecol. Res.* **2014**, *14*, 1024–1031. [CrossRef] [PubMed]

50. Du, F.K.; Lang, T.; Lu, S.; Wang, Y.; Li, J.; Yin, K. An improved method for chloroplast genome sequencing in non-model forest tree species. *Tree Genet. Genomes* **2015**, *11*, 114. [CrossRef]

51. Luo, R.; Liu, B.; Xie, Y.; Li, Z.; Huang, W.; Yuan, J.; He, G.; Chen, Y.; Pan, Q.; Liu, Y.; et al. SOAPdenovo2: An empirically improved memory-efficient short-read de novo assembler. *Gigascience* **2012**, *1*, 18. [CrossRef] [PubMed]

52. Liu, C.; Shi, L.; Zhu, Y.; Chen, H.; Zhang, J.; Lin, X.; Guan, X. CpGAVAS, an integrated web server for the annotation, visualization, analysis, and GenBank submission of completely sequenced chloroplast genome sequences. *BMC Genom.* **2012**, *13*, 715. [CrossRef] [PubMed]

53. Thiel, T.; Michalek, W.; Varshney, R.; Graner, A. Exploiting EST databases for the development and characterization of gene-derived SSR-markers in barley (*Hordeum vulgare* L.). *Theor. Appl. Genet.* **2003**, *106*, 411–422. [CrossRef] [PubMed]

54. Benson, G. Tandem repeats finder: A program to analyze DNA sequences. *Nucleic Acids Res.* **1999**, *27*, 573. [CrossRef] [PubMed]

55. Kurtz, S.; Schleiermacher, C. REPuter: Fast computation of maximal repeats in complete genomes. *Bioinformatics* **1999**, *15*, 426–427. [CrossRef] [PubMed]

56. Grant, J.R.; Stothard, P. The CGView Server: A comparative genomics tool for circular genomes. *Nucleic Acids Res.* **2008**, *36*, W181–W184. [CrossRef] [PubMed]

57. Frazer, K.A.; Pachter, L.; Poliakov, A.; Rubin, E.M.; Dubchak, I. VISTA: Computational tools for comparative genomics. *Nucleic Acids Res.* **2004**, *32*, W273–W279. [CrossRef] [PubMed]

58. Kumar, S.; Stecher, G.; Tamura, K. MEGA7: Molecular evolutionary genetics analysis version 7.0 for bigger datasets. *Mol. Biol. Evol.* **2016**, *33*, 1870–1874. [CrossRef] [PubMed]

59. Librado, P.; Rozas, J. DnaSP v5: A software for comprehensive analysis of DNA polymorphism data. *Bioinformatics* **2009**, *25*, 1451–1452. [CrossRef] [PubMed]

60. Tamura, K. Estimation of the number of nucleotide substitutions when there are strong transition-transversion and G+C-content biases. *Mol. Biol. Evol.* **1992**, *9*, 678–687. [PubMed]

61. Stern, A.; Doron-Faigenboim, A.; Erez, E.; Martz, E.; Bacharach, E.; Pupko, T. Selecton 2007: Advanced models for detecting positive and purifying selection using a Bayesian inference approach. *Nucleic Acids Res.* **2007**, *35*, W506–W511. [CrossRef] [PubMed]

62. Katoh, K.; Standley, D.M. MAFFT multiple sequence alignment software version 7: Improvements in performance and usability. *Mol. Biol. Evol.* **2013**, *30*, 772–780. [CrossRef] [PubMed]

63. Swofford, D.L. *PAUP*. *Phylogenetic Analysis Using Parsimony (and Other Methods).* *Version 4*; Sinauer Associates: Sunderland, MA, USA, 2003.

5

Insights into the Mechanisms of Chloroplast Division

Yamato Yoshida

Department of Science, College of Science, Ibaraki University, Ibaraki 310-8512, Japan;
yamato.yoshida.sci@vc.ibaraki.ac.jp

Abstract: The endosymbiosis of a free-living cyanobacterium into an ancestral eukaryote led to the evolution of the chloroplast (plastid) more than one billion years ago. Given their independent origins, plastid proliferation is restricted to the binary fission of pre-existing plastids within a cell. In the last 25 years, the structure of the supramolecular machinery regulating plastid division has been discovered, and some of its component proteins identified. More recently, isolated plastid-division machineries have been examined to elucidate their structural and mechanistic details. Furthermore, complex studies have revealed how the plastid-division machinery morphologically transforms during plastid division, and which of its component proteins play a critical role in generating the contractile force. Identifying the three-dimensional structures and putative functional domains of the component proteins has given us hints about the mechanisms driving the machinery. Surprisingly, the mechanisms driving plastid division resemble those of mitochondrial division, indicating that these division machineries likely developed from the same evolutionary origin, providing a key insight into how endosymbiotic organelles were established. These findings have opened new avenues of research into organelle proliferation mechanisms and the evolution of organelles.

Keywords: chloroplast division; mitochondrial division; endosymbiotic organelle; FtsZ; dynamin-related protein; glycosyltransferase protein

---

1. Introduction

Chloroplasts (plastids) produce organic molecules and oxygen via photosynthesis, directly and indirectly providing a diverse array of living organisms with the materials they need to grow and develop. The activity of plastids over the past billion years has resulted in the dramatic greening of the Earth. Due to their endosymbiotic origin, plastids contain their own genomes and multiply by the binary fission of pre-existing plastids [1–4]. Although the mechanisms driving this division were long unclear, the discovery of a specialized ring structure at the division site of plastids in primitive unicellular alga provided groundbreaking insights into this process [5], and the rise of genomics and proteomics has further accelerated studies in this field. Consequently, some components for plastid division have been identified in the last 25 years [1–4,6]. It is now known that plastid division is carried out by a ring-shaped supermolecule termed the plastid-division machinery, which contains three or more types of ring structures: the plastid-dividing (PD) ring, which forms the main framework of the division machinery and comprises a ring-shaped bundle of nanofilaments on the cytosolic side of the outer envelope membrane of the corresponding organelle [4,5,7]; the FtsZ ring, a single ring constructed from homologs of the bacterial fission protein FtsZ located beneath the inner envelope membrane at the division site [8]; and the dynamin ring, a disconnected ring-like structure formed of a dynamin superfamily member on the cytosolic side of the outer envelope membrane at the site of organelle division (Figure 1) [9,10]. The importance of each component for plastid division has been well studied and summarized in detail elsewhere [3,11]; however, the functional mechanical details of the plastid-division machinery remain unclear. Various types of functional domains have been identified in the component proteins, including GTPase and glycosyltransferase domains, some of which are conserved not only within the plant kingdom but also in bacteria and non-photosynthetic eukaryotes; therefore, considerations and comparisons of these component proteins in other species will support our understanding of their fundamental functions during plastid division. Furthermore, it is now known that the mitochondrial- and peroxisome-division machineries carry out similar division processes to those of the plastids, suggesting that the elucidation of the plastid-division system will also provide insights into the proliferation of other membranous organelles within eukaryotic cells. This review summarizes and considers the domain architectures of the major proteins involved in plastid division, enabling the further exploration of the proliferation mechanisms in plastids as well as the proliferation mechanisms in mitochondria.

2. Structure and Assembly of the Plastid-Division Machinery

As mentioned above, plastids divide under the regulation of their division machinery, supramolecular complexes comprising a dynamic trio of rings, the PD ring, FtsZ ring, and dynamin ring, which span the plastid double membrane. Although the molecular function of each ring in the division machinery has not been fully revealed, studies using the alga *Cyanidioschyzon merolae* and the model dicot *Arabidopsis thaliana* have begun to uncover the molecular mechanisms by which this machinery functions. The formation of the plastid-division machinery is executed in a specific order during plastid division (Figure 1). As in bacteria, the assembly of the FtsZ ring in the stromal region is the first known event to occur at the plastid-division site, followed by the appearance of the inner PD ring beneath the inner envelope membrane at the division site. After the formation of the inner rings, the interaction between FtsZ and certain membrane proteins might transfer positional information about the FtsZ ring to the outside of the plastid, resulting in the binding of the glycosyltransferase protein PLASTID-DIVIDING RING 1 (PDR1) to the outer membrane, where it assembles the outer PD ring [12]. Another possibility was suggested following a series of electron microscopy (EM) observations in *C. caldarium*; small electron-dense vesicles were visualized along the putative division site before the assembly of the outer PD ring. Interestingly, the boundary of the vesicles appeared to coincide with one end of the PD ring filament, suggesting that these filaments might be biosynthesized on the surface of these vesicles using their components [5].

These inner and outer rings appeared to be linked to each other through nano-scale holes that appear on the groove of the division site, as revealed by scanning EM [13]. Recent yeast two-hybrid studies using several plastid-division proteins in *A. thaliana* showed that the FtsZ ring interacts with the inner membrane proteins ACCUMULATION AND REPLICATION OF CHLOROPLAST6 (ARC6) and PALAROG OF ARC6 (PARC6) in the stromal region [14,15]. ARC6 and PARC6 then further interact with the outer membrane proteins PDV2 and PDV1 in the intermembrane space to form the ARC6-PDV2 and PARC6-PDV1 complexes [15–17]. The intermembrane structure enables the translation of positional information regarding the FtsZ ring from the stromal region to the outer envelope membrane. In the final assembly step of the plastid-division machinery, dynamin-related protein (Dnm2/DRP5B) molecules cross-link the outer PD ring filaments. The sequence of assembly was confirmed in a range of knockdown/knockout experiments involving the plastid-division genes. When the expression of *PDR1* was downregulated, the FtsZ ring was assembled, but the Dnm2 proteins were not recruited to the division site [12]. Meanwhile, expression of a gene encoding the Dnm2 K135A mutant protein, which corresponds to a GTP-binding deficient mutant (K44A) of human dynamin 1, disturbed the formation of the dynamin ring and inhibited plastid division in *C. merolae*, despite the normal assembly of the FtsZ and PD rings [18–20]. Thus, neither FtsZ ring formation nor PD ring formation depends on the subsequent assembly of the plastid-division machinery, while the localization of Dnm2 relies on formation of the outer PD ring.

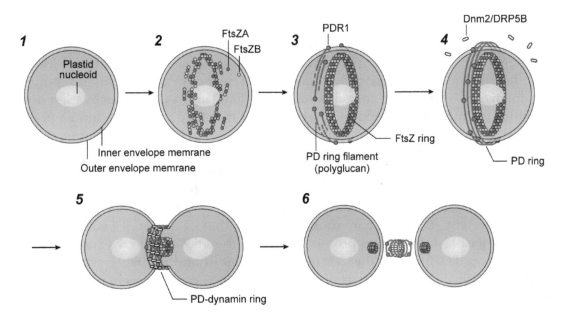

Figure 1. Representation of the plastid-division process. Plastid division occurs as follows: (**1,2**) Two types of FtsZ protein assemble in a heterodimer in the stromal region, then polymerize to form the FtsZ ring in the center of the plastid. To tether to the inner envelope membrane, FtsZ proteins bind to several membrane proteins. (**3,4**) PDR1 proteins attach to the outer envelope membrane above the site of the FtsZ ring, and it is hypothesized that PDR1 biosynthesizes polyglucan nanofilaments to form the PD ring from UDP-glucose molecules. (**5**) The GTPase protein Dnm2 (also known as DRP5B) binds to the PD ring filaments and is likely to generate the motive force for constriction. (**6**) Dnm2 proteins accumulate at the contracting bridge of two daughter plastids and pinch off the membranes. After the abscission of the plastids, the division machinery is disassembled. The inner PD ring and membrane proteins such as ARC6, and PDV2 are omitted from this representation. Modified from Yoshida et al. (2016) [21].

A series of ultrastructural studies provided structural insights into the contractile mechanism of the plastid-division machinery. Sequential EM observations of cells during plastid division revealed that the thickness of the inner PD ring does not change, but its volume decreases at a constant rate during contraction [22]. By contrast, the width and thickness of the outer PD ring monotonically increase during contraction in *C. merolae*, retaining its density and volume [22,23]. Similar EM observations of the outer PD ring changes have also been made in the green alga *Nannochloris bacillaris* [24], and in the land plants [25]. These morphological transitions of the outer PD ring during plastid division led to the idea that the outer PD ring filaments can slide and squeeze the plastid membranes. The establishment of a technique for isolating the intact plastid-division machinery from *C. merolae* cells provided a major breakthrough on this issue [13]; the isolated plastid-division machinery not only formed a circular structure, but also formed super-twisted and spiral structures featuring both clockwise and anticlockwise spirals. The existence of plastid-division machinery in these twisted states indicates the motive force involved in the contraction of plastid membranes; indeed, the plastid-division machinery autonomously contracted when plastid membranes were dissolved by detergent. In addition, reconstituted plastid FtsZ rings also displayed contractile ability. Although the detailed molecular mechanism involved is still unclear, the contraction process of the FtsZ ring accompanies the transition of the FtsZ protofilament from a less dynamic to a more dynamic state; therefore, this dynamic transition of protofilament states is assumed to induce a decrease in the average protofilament length in the FtsZ ring and thus cause contraction [21]. This led to the undertaking of complex studies to reveal how the plastid-division machinery is transformed morphologically during contraction, and which of its component proteins play a key role in generating this force.

3. The FtsZ Ring

Plastid FtsZ is a homolog of the bacterial fission protein FtsZ [8,26,27], and was the first factor found to be responsible for plastid division, assembling into a ring beneath the inner envelope membrane at the division site [22,28,29]. Both plastid and bacterial FtsZ proteins are composed of two functional domains, a GTP-binding domain at the N-terminal and a GTPase-activating domain at the C-terminal, which interact with the opposing terminal of other FtsZ proteins to form a polymer strand (Figure 2) [30,31]. Interestingly, whereas bacteria have one *FtsZ* gene in their genome, the plastid *FtsZ* gene underwent duplication and these duplicated loci, now present in the nuclear genomes, are widely conserved throughout photosynthetic eukaryotes [8,32,33]. Thus, the *C. merolae* genome contains two *FtsZ* genes, *FtsZ2-1* (*FtsZ2* in *A. thaliana*; FtsZA group) and *FtsZ2-2* (*FtsZ1* in *A. thaliana*; FtsZB group), for plastid division [27,32,34]. Phylogenetic studies have found that FtsZA is more ancestral [32], containing a conserved C-terminal core motif similar to that of bacterial FtsZ, which is assumed to interact with other plastid-division proteins to tether the FtsZ ring to the inner membrane [6,8]. The other FtsZ group, FtsZB, lacks the C-terminal core motif and probably evolved following the duplication of the original *FtsZ* gene or its precursor during the establishment of the plastids after the endosymbiotic event.

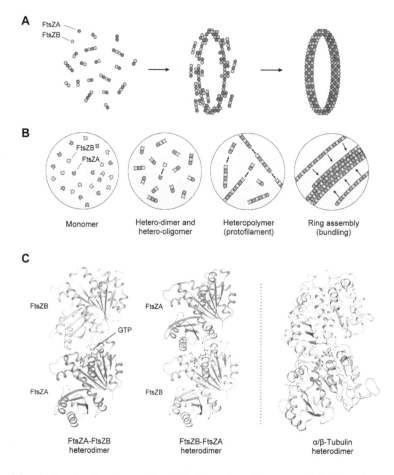

Figure 2. Assembly of the FtsZ ring. (**A,B**) FtsZ molecules assemble into hetero-oligomers, then these FtsZ protofilaments bundle and assemble into a ring structure in the stroma region. (**C**) The two types of FtsZs can assemble into heteropolymer structures via FtsZA-FtsZB and FtsZB-FtsZA hetero-interactions. The protein structure of the tubulin heterodimer (PDB: 1TUB) is also shown on the right. The protein structures of *A. thaliana* FtsZ2 (shown as FtsZA in the Figure) and FtsZ1 (FtsZB in the Figure) were obtained using homology modeling in the Modeller program [35], and structural data for each protein molecule were visualized using CueMol: Molecular Visualization Framework software (http://www.cuemol.org/). Reproduced and modified from Yoshida et al. (2016) [21].

Although the mechanistic details of FtsZ ring assembly and dynamics remain unclear, recent studies using a heterologous yeast system revealed that the two types of plastid FtsZs spontaneously formed heteropolymers then assembled into a single ring that could generate contractile force in the absence of any other related proteins (Figure 2A,B) [21,36]. Interestingly, the plastid FtsZ heteropolymer had higher kinetic dynamics for protofilament assembly, mobility, and flexibility than either homopolymer, suggesting that the gene duplication of plastid *FtsZ* led to innovation in the kinetic functions of the FtsZ ring for plastid division [21]. The FtsZ heteropolymers are structurally similar to the microtubules comprised of α- and β-tubulin, suggesting the convergent evolution of functions in the plastid FtsZs and eukaryotic tubulins (Figure 2C).

Consistent with molecular genetic studies using *A. thaliana*, the assembly of the plastid FtsZ ring in vivo is further promoted by several regulatory factors. ARC6 positively regulates the assembly of the FtsZ protofilament through interactions with the FtsZA molecules [6,8]. PARC6 also interacts with FtsZA, inducing the remodeling of the FtsZ protofilaments for ring formation with support from ARC3 [6,8,15]. Moreover, ARC3 and GIANT CHLOROPLAST1 (GC1) negatively regulate FtsZ ring assembly [37,38]; however, as these FtsZ-associated genes are not well conserved in the plant kingdom, another regulatory system with unknown factors might also coordinate the assembly of the FtsZ ring.

4. The Dynamin Ring

A member of the dynamin superfamily, Dnm2 (also known as DRP5B or ARC5 in *A. thaliana*), is also observed in a ring structure at the plastid-division site (Figure 1) [9,10]. Originally, the classical dynamin protein was identified as a 100-kDa mechanochemical GTPase required for the scission of clathrin-coated vesicles from the plasma membrane [39,40]. Dynamin-related proteins are now known to be involved in diverse membrane remodeling events; for example, Dnm1 (also known as DRP1 in animals) is involved in mitochondrial and peroxisome divisions [40,41].

The dynamin superfamily proteins possess several functional domain subunits, including a large GTPase domain (approximately 300 amino acids), a middle domain, a pleckstrin-homology (PH) domain, a GTPase effector domain (GED), and a carboxy-terminal proline-rich domain (PRD) (Figure 3A,B, upper) [40,42,43]. Although the molecular mass of Dnm2 is similar to that of classical dynamin, a conserved domain search identified only the GTPase domain at its amino-terminal region (Figure 3A). The three uncharacterized regions occupying the rest of the sequence are highly conserved between Dnm2 proteins in plant species, implying that these regions might be responsible for key roles during plastid division (Figure 3A,B, bottom).

The functional importance of Dnm2 proteins during plastid division has been demonstrated in *C. merolae*, the moss *Physcomitrella patens*, and *A. thaliana* [9,10,44]. Dnm2 proteins formed a discontinuous ring structure at the plastid-division site in the early phases of plastid division, after which they appeared to link and assemble into a single ring [9,10]. Surprisingly, the plastid-division dynamin-related proteins are phylogenetically related to a group of dynamin-related proteins involved in cytokinesis [45]. Considering that the primitive algal genome encodes only two dynamin-related proteins, Dnm2 for plastid division and Dnm1 for mitochondrial/peroxisome division [34,46,47], these findings raise the question of the original function of the ancestral dynamin protein.

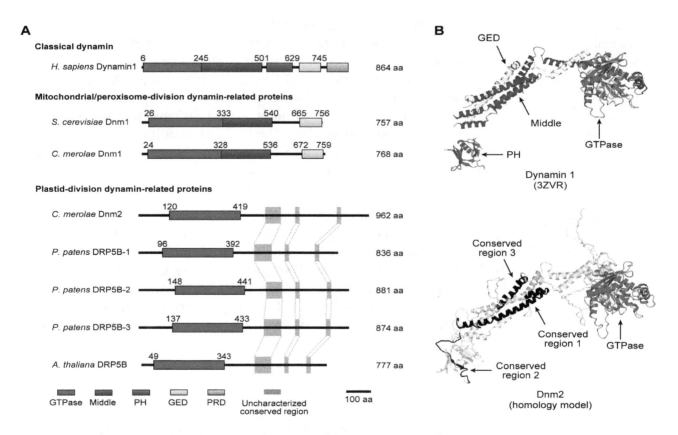

Figure 3. Structures of the dynamin superfamily. (**A**) Domain architectures of the dynamin superfamily. Dynamin 1 catalyzes clathrin-coated vesicle scission at the plasma membrane. Dnm1/DRP1 is involved in the division of mitochondria and peroxisomes. Dnm2 (also known as DRP5B/ARC5 in *A. thaliana* and moss) is involved in plastid division. GTPase domain (red); middle domain (purple); pleckstrin homology domain (PH, blue); GTPase effector domain (GED, yellow); and proline-rich domain (PRD, light brown). Domain architectures were identified using a conserved-domain search program [48]. (**B**) Protein structures of human dynamin 1 (classical dynamin) for vesicle scission and *C. merolae* Dnm2 for plastid division. The structure of dynamin 1 is represented with crystal structure data from an assembly-deficient dynamin 1 mutant, G397D ΔPRD (PDB: 3ZVR) [43], while the structure of Dnm2 is visualized using homology modeling based on the dynamin 1 structure. The functional domains in dynamin 1 are shown in red (GTPase domain), purple (middle domain), blue (PH domain), and yellow (GED domain); the proline-rich domain (PRD) is not shown. Uncharacterized conserved regions in Dnm2 are shown in black. The protein structure of Dnm2 was modeled as described in Figure 2.

5. The PD Ring

In the 1980s and 1990s, EM observations revealed the presence of the PD ring at the division site of plastids in numerous photosynthetic eukaryotes [5,49]. Two (or three) types of electron-dense specialized ring structures comprise the PD ring: the outer PD ring is the main skeletal structure of the plastid-division machinery, and is composed of a ring-shaped bundle of nanofilaments, 5 to 7 nm in width, on the cytosolic side of the outer envelope membrane (Figure 1) [13,23]. In addition, an inner 5-nm-thick belt-like PD ring forms on the stromal side of the inner envelope membrane [22,50]. The outer and inner PD rings have been widely observed in the plant kingdom [5,49], while a third, intermediate, PD ring has only been identified in the intermembrane space of *C. merolae* and the green alga *N. bacillaris* [24,51]. Many angiosperms also form double rings at the constricted isthmi of dividing plastids, including proplastids, amyloplasts, and chloroplasts [5,49]. The PD ring structure can be observed only during the late phases of plastid division in land plants [49,52], suggesting that the number and electron density of the PD rings in these organisms might be too low to detect during the early phases of plastid division using EM, making their morphological dynamics more difficult to study in these species. Interestingly, a phylogenetic study of the PD ring identified a clear trend, in which the PD rings of the primitive unicellular eukaryotes with smaller genomes are larger than those of the multiplastidic cells of land plants [5,49]; therefore, *C. merolae* is one of the most suitable organisms for studying the involvement of the PD ring in plastid division. Based on these previous studies, the PD rings, especially the outer rings, were concluded to be universal across the plant kingdom, where they play an important role in plastidokinesis.

Although the molecular components of the outer PD ring have not yet been elucidated, despite more than 25 years passing since its discovery, a chemical-staining screen shed light on this issue [12]. Periodic acid-horseradish peroxidase staining indicated that the outer PD ring is likely to contain saccharic components, which led us to perform a proteomic analysis of the isolated plastid-division machinery fraction and identify a novel glycosyltransferase protein, PDR1 [12]. Interestingly, the expression of *PDR1* was specifically detected in the plastid-division phase, and PDR1 proteins assembled a single-ring structure at the plastid-division site.

Ultrastructural studies clearly showed that PDR1 proteins localized on the outer PD ring filaments. Furthermore, analyses of the components of the purified PD ring filaments revealed that glucan molecules were components of the outer PD ring. PDR1 has sequence similarity to glycogenin, which acts as a priming protein for glycogen biosynthesis (Figure 4A) [53]; therefore, it is now hypothesized that PDR1 can elongate the glucan chain to biosynthesize PD ring filaments from UDP-glucose molecules, analogous to the biosynthesis of glycogen (Figure 4B). Taking these findings together, although the biosynthesis mechanism is still unclear, PDR1 is probably involved in the biosynthesis of the PD ring polyglucan filaments. Potential orthologs of *PDR1* have also been identified in land plant genomes.

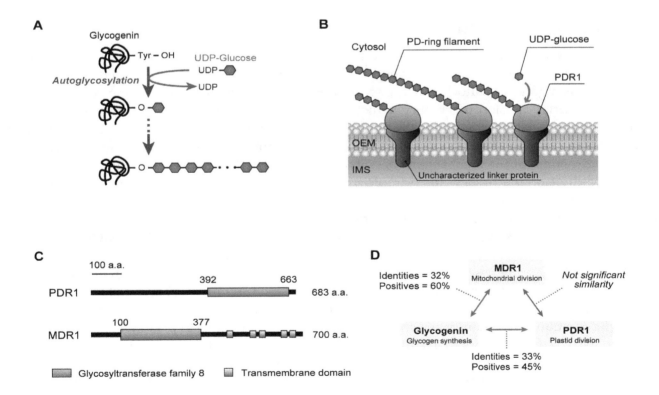

Figure 4. *Cont.*

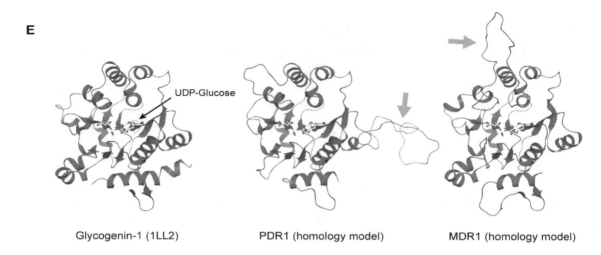

E

Figure 4. Working models of the glycosyltransferases glycogenin and PDR1. (**A**) Glycogenin is required for the initiation of glycogen biosynthesis, and can be autoglycosylated at a specific tyrosine residue to form a short oligosaccharide chain of glucose molecules to act as a priming chain for the subsequent biosynthesis of glycogen. (**B**) A schematic representation of PD ring filament biosynthesis by PDR1. A series of results suggested that PD ring filaments are composed of both PDR1 and glucose molecules. Considering the sequence similarity with glycogenin, the glycosyltransferase domain of PDR1 may biosynthesize the polyglucan nanofilaments from UDP-glucose residues to form the PD ring filaments. OEM, outer envelope membrane; IMS, intermembrane space. (**C**) Schematic of *C. merolae* PDR1 and MDR1 domain structures. The glycosyltransferase domains of PDR1 and MDR1 identified them as type-8 subgroup members of the glycosyltransferase family. (**D**) Protein sequence similarities between PDR1, MDR1 and glycogenin-1. (**E**) Comparisons of the protein structure of glycogenin-1 (PDB: 1LL2) and the putative structures of the glycosyltransferase domains of PDR1 and MDR1. Orange arrows indicate specific insertion regions in the glycosyltransferase domains of PDR1 and MDR1. The protein structures of the PDR1 and MDR1 glycosyltransferase domains were modeled as described in Figure 2.

6. The Homology between Plastid- and Mitochondrial-Division Machinery

Over the past 25 years, the mode of plastid division has been unveiled and several key components of the division machinery have been identified. Furthermore, studies have revealed that the other endosymbiotic organelles, mitochondria, also proliferate via the activity of a supramolecular complex, the mitochondrial-division machinery, which, like the plastid-division machinery, comprises an electron-dense specialized ring called the mitochondrion-dividing (MD) ring which is the counterpart of the PD ring in mitochondrial division, the FtsZ ring, and a dynamin ring (Figure 5) [1,54,55]. Interestingly, some of the mechanisms of mitochondrial division are also very similar to those of the plastid-division machinery. In addition, a recent multi-omics analysis of isolated plastid- and mitochondrial-division machineries showed that 185 proteins, including 54 uncharacterized proteins, were present in the fraction, indicating that many unknown components may be involved in these machineries [54]. Indeed, a glycosyltransferase homologous to PDR1, MITOCHONDRION-DIVIDING RING1 (MDR1) (Figure 4C), was identified from these candidates, and a series of analyses showed that MDR1 is required for the assembly of the MD ring, which also consists of polyglucan filaments and is required for mitochondrial division [54]. Despite the low sequence similarity between PDR1 and MDR1 (Figure 4D,E), these proteins both have a glycosyltransferase domain belongs to the type-8 subgroup of the glycosyltransferase family and they have homologous functions in plastid and mitochondrial division [54]. Given that both plastids and mitochondria evolved from free-living bacteria, the compelling structural and mechanical similarity between the plastid- and mitochondrial-division machineries indicates that they were established in host cells to dominate and control the proliferation of these endosymbiotic organelles during their early evolution.

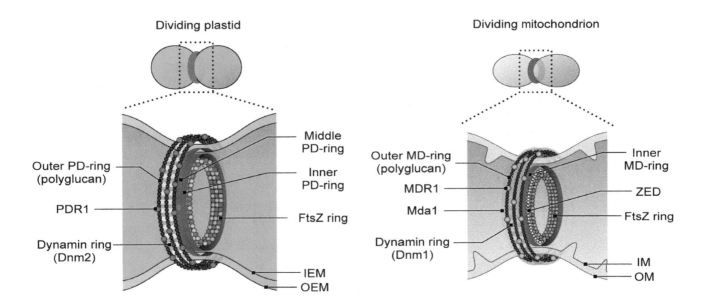

Figure 5. Schematic representations of the division machinery in plastids and mitochondria. IEM, inner envelope membrane; OEM, outer envelope membrane; IM, inner membrane; OM, outer membrane. For details on the mitochondrial-division machinery, see Refs. [1,54–58]. Modified from Yoshida et al. (2016) [21] and Yoshida et al. (2017) [54].

7. Conclusions and Perspectives

One of the current hot topics in this field is the elucidation of the system coordinating the cell-division cycle and the plastid/mitochondrial-division cycle [59–61]. Combinational analyses using genetic engineering and synchronized *C. merolae* cells has revealed the existence of a plastid-division checkpoint in the cell cycle, which is very likely to contribute to the permanent possession of plastids [19]; however, the fundamental mechanisms driving this process are still unclear. To identify the coordination system, the selection of suitable organisms for each analysis will be very important. The unicellular alga *C. merolae* and the land plant *A. thaliana* have enabled the molecular study of plastid division via their species-specific advantages.

As a model organism, *A. thaliana* enabled many types of genetic studies to be conducted, leading to the identification of some of the genes responsible for plastid division (see for more details Chen et al. (2018) [6]). Studies in this area also benefit from using *C. merolae* cells, which can be synchronized to enable the isolation of intact plastid/mitochondrial-division machineries [13,54,62]. Furthermore, many genetic-engineering techniques have been recently established in *C. merolae* [18,63–66]. These innovations will enable dramatic advances in the investigation of the molecular mechanisms driving plastid division over the next decade; for example, an impressive recent study elucidated the crystal structure of the ARC6-PDV2 complex to reveal how protein-protein interactions translate information regarding plastid division across the double membrane, from the stromal region to the cytosol [17]. It is expected that the further identification of three-dimensional protein structures involved in the plastid-division machinery will open up a completely new avenue in the field.

Acknowledgments: This work was supported by Human Frontier Science Program Long-Term Fellowship LT000356/2011-L and a Japan Society for the Promotion of Science Postdoctoral Research Fellowship for Research Abroad.

References

1.   Osteryoung, K.W.; Nunnari, J. The division of endosymbiotic organelles. *Science* **2003**, *302*, 1698–1704. [CrossRef] [PubMed]

2.   Kuroiwa, T.; Misumi, O.; Nishida, K.; Yagisawa, F.; Yoshida, Y.; Fujiwara, T.; Kuroiwa, H. Vesicle, mitochondrial, and plastid division machineries with emphasis on dynamin and electron-dense rings. *Int. Rev. Cell Mol. Biol.* **2008**, *271*, 97–152. [CrossRef] [PubMed]

3.   Miyagishima, S.Y.; Nakanishi, H.; Kabeya, Y. Structure, regulation, and evolution of the plastid division machinery. *Int. Rev. Cell Mol. Biol.* **2011**, *291*, 115–153. [CrossRef] [PubMed]

4.   Yoshida, Y.; Miyagishima, S.Y.; Kuroiwa, H.; Kuroiwa, T. The plastid-dividing machinery: Formation, constriction and fission. *Curr. Opin. Plant Biol.* **2012**, *15*, 714–721. [CrossRef] [PubMed]

5.   Kuroiwa, T.; Kuroiwa, H.; Sakai, A.; Takahashi, H.; Toda, K.; Itoh, R. The division apparatus of plastids and mitochondria. *Int. Rev. Cytol.* **1998**, *181*, 1–41. [PubMed]

6.   Chen, C.; MacCready, J.S.; Ducat, D.C.; Osteryoung, K.W. The molecular machinery of chloroplast division. *Plant Physiol.* **2018**, *176*, 138–151. [CrossRef] [PubMed]

7.   Miyagishima, S.Y.; Nishida, K.; Kuroiwa, T. An evolutionary puzzle: Chloroplast and mitochondrial division rings. *Trends Plant Sci.* **2003**, *8*, 432–438. [CrossRef]

8.   TerBush, A.D.; Yoshida, Y.; Osteryoung, K.W. FtsZ in chloroplast division: Structure, function and evolution. *Curr. Opin. Cell Biol.* **2013**, *25*, 461–470. [CrossRef] [PubMed]

9.   Miyagishima, S.Y.; Nishida, K.; Mori, T.; Matsuzaki, M.; Higashiyama, T.; Kuroiwa, H.; Kuroiwa, T. A plant-specific dynamin-related protein forms a ring at the chloroplast division site. *Plant Cell Online* **2003**, *15*, 655–665. [CrossRef]

10. Gao, H.; Kadirjan-Kalbach, D.; Froehlich, J.E.; Osteryoung, K.W. ARC5, a cytosolic dynamin-like protein from plants, is part of the chloroplast division machinery. *Proc. Natl. Acad. Sci. USA* **2003**, *100*, 4328–4333. [CrossRef] [PubMed]

11. Osteryoung, K.W.; Pyke, K.A. Division and dynamic morphology of plastids. *Annu. Rev. Plant Biol.* **2014**, *65*, 443–472. [CrossRef] [PubMed]

12. Yoshida, Y.; Kuroiwa, H.; Misumi, O.; Yoshida, M.; Ohnuma, M.; Fujiwara, T.; Yagisawa, F.; Hirooka, S.; Imoto, Y.; Matsushita, K.; et al. Chloroplasts divide by contraction of a bundle of nanofilaments consisting of polyglucan. *Science* **2010**, *329*, 949–953. [CrossRef] [PubMed]

13. Yoshida, Y.; Kuroiwa, H.; Misumi, O.; Nishida, K.; Yagisawa, F.; Fujiwara, T.; Nanamiya, H.; Kawamura, F.; Kuroiwa, T. Isolated chloroplast division machinery can actively constrict after stretching. *Science* **2006**, *313*, 1435–1438. [CrossRef] [PubMed]

14. Glynn, J.M.; Yang, Y.; Vitha, S.; Schmitz, A.J.; Hemmes, M.; Miyagishima, S.-Y.; Osteryoung, K.W. PARC6, a novel chloroplast division factor, influences FtsZ assembly and is required for recruitment of PDV1 during chloroplast division in Arabidopsis. *Plant J.* **2009**, *59*, 700–711. [CrossRef] [PubMed]

15. Zhang, M.; Chen, C.; Froehlich, J.E.; Terbush, A.D.; Osteryoung, K.W. Roles of Arabidopsis PARC6 in coordination of the chloroplast division complex and negative regulation of FtsZ assembly. *Plant Physiol.* **2016**, *170*, 250–262. [CrossRef] [PubMed]

16. Glynn, J.M.; Froehlich, J.E.; Osteryoung, K.W. Arabidopsis ARC6 coordinates the division machineries of the inner and outer chloroplast membranes through interaction with PDV2 in the intermembrane space. *Plant Cell* **2008**, *20*, 2460–2470. [CrossRef] [PubMed]

17. Wang, W.; Li, J.; Sun, Q.; Yu, X.; Zhang, W.; Jia, N.; An, C.; Li, Y.; Dong, Y.; Han, F.; et al. Structural insights into the coordination of plastid division by the ARC6-PDV2 complex. *Nat. Plants* **2017**, *3*, 1–9. [CrossRef] [PubMed]

18. Sumiya, N.; Fujiwara, T.; Kobayashi, Y.; Misumi, O.; Miyagishima, S.Y. Development of a heat-shock inducible gene expression system in the red alga *Cyanidioschyzon merolae*. *PLoS ONE* **2014**, *9*, 1–11. [CrossRef] [PubMed]

19. Sumiya, N.; Fujiwara, T.; Era, A.; Miyagishima, S.Y. Chloroplast division checkpoint in eukaryotic algae. *Proc. Natl. Acad. Sci. USA* **2016**, *113*, E7629–E7638. [CrossRef] [PubMed]

20. Sumiya, N.; Miyagishim, S.Y. Hierarchal order in the formation of chloroplast division machinery in the red alga *Cyanidioschyzon merolae*. *Commun. Integr. Biol.* **2017**, *10*, 1–5. [CrossRef] [PubMed]

21. Yoshida, Y.; Mogi, Y.; TerBush, A.D.; Osteryoung, K.W. Chloroplast FtsZ assembles into a contractible ring via tubulin-like heteropolymerization. *Nat. Plants* **2016**, *2*, 16095. [CrossRef] [PubMed]

22. Miyagishima, S.Y.; Takahara, M.; Mori, T.; Kuroiwa, H.; Higashiyama, T.; Kuroiwa, T. Plastid division is driven by a complex mechanism that involves differential transition of the bacterial and eukaryotic division rings. *Plant Cell* **2001**, *13*, 2257–2268. [CrossRef]

23. Miyagishima, S.Y.; Takahara, M.; Kuroiwa, T. Novel filaments 5 nm in diameter constitute the cytosolic ring of the plastid division apparatus. *Plant Cell* **2001**, *13*, 707–721. [CrossRef] [PubMed]

24. Sumiya, N.; Hirata, A.; Kawano, S. Multiple FtsZ ring formation and reduplicated chloroplast DNA in *Nannochloris bacillaris* (Chlorophyta, Trebouxiophyceae) under phosphate-enriched culture. *J. Phycol.* **2008**, *44*, 1476–1489. [CrossRef] [PubMed]

25. Kuroiwa, H.; Mori, T.; Takahara, M.; Miyagishima, S.Y.; Kuroiwa, T. Chloroplast division machinery as revealed by immunofluorescence and electron microscopy. *Planta* **2002**, *215*, 185–190. [CrossRef] [PubMed]

26. Osteryoung, K.W.; Vierling, E. Conserved cell and organelle division. *Nature* **1995**, *376*, 473–474. [CrossRef] [PubMed]

27. Takahara, M.; Takahashi, H.; Matsunaga, S.Y.; Miyagishima, S.; Takano, H.; Sakai, A.; Kawano, S.; Kuroiwa, T. A putative mitochondrial FtsZ gene is present in the unicellular primitive red alga *Cyanidioschyzon merolae*. *Mol. Gen. Genet.* **2000**, *264*, 452–460. [CrossRef] [PubMed]

28. Mori, T.; Kuroiwa, H.; Takahara, M.; Miyagishima, S.Y.; Kuroiwa, T. Visualization of an FtsZ ring in chloroplasts of *Lilium longiflorum* leaves. *Plant Cell Physiol.* **2001**, *42*, 555–559. [CrossRef] [PubMed]

29. Vitha, S.; McAndrew, R.S.; Osteryoung, K.W. FtsZ ring formation at the chloroplast division site in plants. *J. Cell Biol.* **2001**, *153*, 111–120. [CrossRef] [PubMed]

30. Löwe, J.; Amos, L.A. Crystal structure of the bacterial cell-division protein FtsZ. *Nature* **1998**, *391*, 203–206. [CrossRef]

31. Oliva, M.A.; Cordell, S.C.; Löwe, J. Structural insights into FtsZ protofilament formation. *Nat. Struct. Mol. Biol.* **2004**, *11*, 1243–1250. [CrossRef] [PubMed]

32. Miyagishima, S.Y.; Nozaki, H.; Nishida, K.; Nishida, K.; Matsuzaki, M.; Kuroiwa, T. Two types of FtsZ proteins in mitochondria and red-lineage chloroplasts: The duplication of FtsZ is implicated in endosymbiosis. *J. Mol. Evol.* **2004**, *58*, 291–303. [CrossRef] [PubMed]

33. Schmitz, A.J.; Glynn, J.M.; Olson, B.J.S.C.; Stokes, K.D.; Osteryoung, K.W. Arabidopsis FtsZ2-1 and FtsZ2-2 are functionally redundant, but FtsZ-based plastid division is not essential for chloroplast partitioning or plant growth and development. *Mol. Plant* **2009**, *2*, 1211–1222. [CrossRef] [PubMed]

34. Matsuzaki, M.; Misumi, O.; Shin-I, T.; Maruyama, S.; Takahara, M.; Miyagishima, S.Y.; Mori, T.; Nishida, K.; Yagisawa, F.; Nishida, K.; et al. Genome sequence of the ultrasmall unicellular red alga *Cyanidioschyzon merolae* 10D. *Nature* **2004**, *428*, 653–657. [CrossRef] [PubMed]

35. Eswar, N.; John, B.; Mirkovic, N.; Fiser, A.; Ilyin, V.A.; Pieper, U.; Stuart, A.C.; Marti-Renom, M.A.; Madhusudhan, M.S.; Yerkovich, B.; et al. Tools for comparative protein structure modeling and analysis. *Nucleic Acids Res.* **2003**, *31*, 3375–3380. [CrossRef] [PubMed]

36. TerBush, A.D.; Osteryoung, K.W. Distinct functions of chloroplast FtsZ1 and FtsZ2 in Z-ring structure and remodeling. *J. Cell Biol.* **2012**, *199*, 623–637. [CrossRef] [PubMed]

37. Shimada, H.; Koizumi, M.; Kuroki, K.; Mochizuki, M.; Fujimoto, H.; Ohta, H.; Masuda, T.; Takamiya, K. ARC3, a chloroplast division factor, is a chimera of prokaryotic FtsZ and part of eukaryotic phosphatidylinositol-4-phosphate 5-kinase. *Plant Cell Physiol.* **2004**, *45*, 960–967. [CrossRef] [PubMed]

38. Maple, J.; Fujiwara, M.T.; Kitahata, N.; Lawson, T.; Baker, N.R.; Yoshida, S.; Møller, S.G. GIANT CHLOROPLAST 1 is essential for correct plastid division in *Arabidopsis*. *Curr. Biol.* **2004**, *14*, 776–781. [CrossRef] [PubMed]

39. Shpetner, H.S.; Vallee, R.B. Identification of dynamin, a novel mechanochemical enzyme that mediates interactions between microtubules. *Cell* **1989**, *59*, 421–432. [CrossRef]

40. Praefcke, G.J.K.; McMahon, H.T. The dynamin superfamily: Universal membrane tubulation and fission molecules? *Nat. Rev. Mol. Cell Biol.* **2004**, *5*, 133–147. [CrossRef] [PubMed]

41. Bleazard, W.; McCaffery, J.M.; King, E.J.; Bale, S.; Mozdy, A.; Tieu, Q.; Nunnari, J.; Shaw, J.M. The dynamin-related GTPase Dnm1 regulates mitochondrial fission in yeast. *Nat. Cell Biol.* **1999**, *1*, 298–304. [CrossRef] [PubMed]

42. Faelber, K.; Posor, Y.; Gao, S.; Held, M.; Roske, Y.; Schulze, D.; Haucke, V.; Noé, F.; Daumke, O. Crystal structure of nucleotide-free dynamin. *Nature* **2011**, *477*, 556–560. [CrossRef] [PubMed]

43. Ford, M.G.J.; Jenni, S.; Nunnari, J. The crystal structure of dynamin. *Nature* **2011**, *477*, 561–566. [CrossRef] [PubMed]

44. Sakaguchi, E.; Takechi, K.; Sato, H.; Yamada, T.; Takio, S.; Takano, H. Three dynamin-related protein 5B genes are related to plastid division in *Physcomitrella patens*. *Plant Sci.* **2011**, *180*, 789–795. [CrossRef] [PubMed]

45. Miyagishima, S.Y.; Kuwayama, H.; Urushihara, H.; Nakanishi, H. Evolutionary linkage between eukaryotic cytokinesis and chloroplast division by dynamin proteins. *Proc. Natl. Acad. Sci. USA* **2008**, *105*, 15202–15207. [CrossRef] [PubMed]

46. Nishida, K.; Takahara, M.; Miyagishima, S.Y.; Kuroiwa, H.; Matsuzaki, M.; Kuroiwa, T. Dynamic recruitment of dynamin for final mitochondrial severance in a primitive red alga. *Proc. Natl. Acad. Sci. USA* **2003**, *100*, 2146–2151. [CrossRef] [PubMed]

47. Imoto, Y.; Kuroiwa, H.; Yoshida, Y.; Ohnuma, M.; Fujiwara, T.; Yoshida, M.; Nishida, K.; Yagisawa, F.; Hirooka, S.; Miyagishima, S.; Misumi, O.; Kawano, S.; Kuroiwa, T. Single-membrane-bounded peroxisome division revealed by isolation of dynamin-based machinery. *Proc. Natl. Acad. Sci. USA* **2013**, *110*, 9583–9588. [CrossRef] [PubMed]

48. Marchler-Bauer, A.; Derbyshire, M.K.; Gonzales, N.R.; Lu, S.; Chitsaz, F.; Geer, L.Y.; Geer, R.C.; He, J.; Gwadz, M.; Hurwitz, D.I.; et al. CDD: NCBI's conserved domain database. *Nucleic Acids Res.* **2015**, *43*, D222–D226. [CrossRef] [PubMed]

49. Kuroiwa, T. The primitive red algae *Cyanidium caldarium* and *Cyanidioschyzon merolae* as model system for investigating the dividing apparatus of mitochondria and plastids. *BioEssays* **1998**, *20*, 344–354. [CrossRef]

50. Hashimoto, H. Double ring structure around the constricting neck of dividing plastids of *Avena sativa*. *Protoplasma* **1986**, *135*, 166–172. [CrossRef]

51. Miyagishima, S.Y.; Itoh, R.; Toda, K.; Takahashi, H.; Kuroiwa, H.; Kuroiwa, T. Identification of a triple ring structure involved in plastid division in the primitive red alga *Cyanidioschyzon merolae*. *J. Electron Microsc.* **1998**, *47*, 269–272. [CrossRef]

52. Kuroiwa, H.; Mori, T.; Takahara, M.; Miyagishima, S.Y.; Kuroiwa, T. Multiple FtsZ rings in a pleomorphic chloroplast in embryonic cap cells of *Pelargonium zonale*. *Cytologia* **2001**, *66*, 227–233. [CrossRef]

53. Lomako, J.; Lomako, W.M.; Whelan, W.J. Glycogenin: The primer for mammalian and yeast glycogen synthesis. *Biochim. Biophys. Acta* **2004**, *1673*, 45–55. [CrossRef] [PubMed]

54. Yoshida, Y.; Kuroiwa, H.; Shimada, T.; Yoshida, M.; Ohnuma, M.; Fujiwara, T.; Imoto, Y.; Yagisawa, F.; Nishida, K.; Hirooka, S.; et al. Glycosyltransferase MDR1 assembles a dividing ring for mitochondrial proliferation comprising polyglucan nanofilaments. *Proc. Natl. Acad. Sci. USA* **2017**, *114*, 13284–13289. [CrossRef] [PubMed]

55. Kuroiwa, T.; Nishida, K.; Yoshida, Y.; Fujiwara, T.; Mori, T.; Kuroiwa, H.; Misumi, O. Structure, function and evolution of the mitochondrial division apparatus. *Biochim. Biophys. Acta* **2006**, *1763*, 510–521. [CrossRef] [PubMed]

56. Van der Bliek, A.M.; Shen, Q.; Kawajiri, S. Mechanisms of mitochondrial fission and fusion. *Cold Spring Harb. Perspect. Biol.* **2013**, *5*, 1–16. [CrossRef] [PubMed]

57. Friedman, J.R.; Nunnari, J. Mitochondrial form and function. *Nature* **2014**, *505*, 335–343. [CrossRef] [PubMed]

58. Roy, M.; Reddy, P.H.; Iijima, M.; Sesaki, H. Mitochondrial division and fusion in metabolism. *Curr. Opin. Cell Biol.* **2015**, *33*, 111–118. [CrossRef] [PubMed]

59. Kobayashi, Y.; Kanesaki, Y.; Tanaka, A.; Kuroiwa, H.; Kuroiwa, T.; Tanaka, K. Tetrapyrrole signal as a cell-cycle coordinator from organelle to nuclear DNA replication in plant cells. *Proc. Natl. Acad. Sci. USA* **2009**, *106*, 803–807. [CrossRef] [PubMed]

60. Kobayashi, Y.; Imamura, S.; Hanaoka, M.; Tanaka, K. A tetrapyrrole-regulated ubiquitin ligase controls algal nuclear DNA replication. *Nat. Cell Biol.* **2011**, *13*, 483–487. [CrossRef] [PubMed]

61. Miyagishima, S.Y.; Fujiwara, T.; Sumiya, N.; Hirooka, S.; Nakano, A.; Kabeya, Y.; Nakamura, M. Translation-independent circadian control of the cell cycle in a unicellular photosynthetic eukaryote. *Nat. Commun.* **2014**, *5*, 3807. [CrossRef] [PubMed]

62. Suzuki, K.; Ehara, T.; Osafune, T.; Kuroiwa, H.; Kawano, S.; Kuroiwa, T. Behavior of mitochondria, chloroplasts and their nuclei during the mitotic cycle in the ultramicroalga *Cyanidioschyzon merolae*. *Eur. J. Cell Biol.* **1994**, *63*, 280–288. [PubMed]

63. Ohnuma, M.; Yokoyama, T.; Inouye, T.; Sekine, Y.; Tanaka, K. Polyethylene glycol (PEG)-mediated transient gene expression in a red alga, *Cyanidioschyzon merolae* 10D. *Plant Cell Physiol.* **2008**, *49*, 117–120. [CrossRef] [PubMed]

64. Ohnuma, M.; Misumi, O.; Fujiwara, T.; Watanabe, S.; Tanaka, K.; Kuroiwa, T. Transient gene suppression in a red alga, *Cyanidioschyzon merolae* 10D. *Protoplasma* **2009**, *236*, 107–112. [CrossRef] [PubMed]

65. Fujiwara, T.; Kanesaki, Y.; Hirooka, S.; Era, A.; Sumiya, N.; Yoshikawa, H.; Tanaka, K.; Miyagishima, S.Y. A nitrogen source-dependent inducible and repressible gene expression system in the red alga *Cyanidioschyzon merolae*. *Front. Plant Sci.* **2015**, *6*, 1–10. [CrossRef] [PubMed]

66. Fujiwara, T.; Ohnuma, M.; Kuroiwa, T.; Ohbayashi, R.; Hirooka, S.; Miyagishima, S.Y. Development of a double nuclear gene-targeting method by two-step transformation based on a newly established chloramphenicol-selection system in the red alga *Cyanidioschyzon merolae*. *Front. Plant Sci.* **2017**, *8*, 1–10. [CrossRef] [PubMed]

6

Molecular Structure and Phylogenetic Analyses of Complete Chloroplast Genomes of Two *Aristolochia* Medicinal Species

Jianguo Zhou [1], Xinlian Chen [1], Yingxian Cui [1], Wei Sun [2], Yonghua Li [3], Yu Wang [1], Jingyuan Song [1] and Hui Yao [1,*]

[1] Key Lab of Chinese Medicine Resources Conservation, State Administration of Traditional Chinese Medicine of the People's Republic of China, Institute of Medicinal Plant Development, Chinese Academy of Medical Sciences & Peking Union Medical College, Beijing 100193, China; jgzhou1316@163.com (J.Z.); chenxinlian1053@163.com (X.C.); yxcui2017@163.com (Y.C.); ywang@implad.ac.cn (Y.W.); jysong@implad.ac.cn (J.S.)

[2] Institute of Chinese Materia Medica, China Academy of Chinese Medicinal Sciences, Beijing 100700, China; wsun@icmm.ac.cn

[3] Department of Pharmacy, Guangxi Traditional Chinese Medicine University, Nanning 530200, China; liyonghua185@126.com

* Correspondence: scauyaoh@sina.com

Abstract: The family Aristolochiaceae, comprising about 600 species of eight genera, is a unique plant family containing aristolochic acids (AAs). The complete chloroplast genome sequences of *Aristolochia debilis* and *Aristolochia contorta* are reported here. The results show that the complete chloroplast genomes of *A. debilis* and *A. contorta* comprise circular 159,793 and 160,576 bp-long molecules, respectively and have typical quadripartite structures. The GC contents of both species were 38.3% each. A total of 131 genes were identified in each genome including 85 protein-coding genes, 37 tRNA genes, eight rRNA genes and one pseudogene (*ycf1*). The simple-sequence repeat sequences mainly comprise A/T mononucletide repeats. Phylogenetic analyses using maximum parsimony (MP) revealed that *A. debilis* and *A. contorta* had a close phylogenetic relationship with species of the family Piperaceae, as well as Laurales and Magnoliales. The data obtained in this study will be beneficial for further investigations on *A. debilis* and *A. contorta* from the aspect of evolution, and chloroplast genetic engineering.

Keywords: *Aristolochia debilis*; *Aristolochia contorta*; chloroplast genome; molecular structure; phylogenetic analyses

1. Introduction

The traditional Chinese medicine plants, *Aristolochia debilis* and *Aristolochia contorta*, are herbaceous climbers in the family Aristolochiaceae. *Aristolochiae fructus* originates from the mellow fruit of the two species, while *Aristolochiae herba* originates from their dried aerial parts. *Aristolochiae fructus* and *Aristolochiae herba* have been recorded as traditional herbal medicines which can clear lung-heat to stop coughing and activate meridians to stop pain, respectively [1]. Modern pharmacology studies have shown that the primary chemical constituents of the two species are aristolochic acid analogues including aristolochic acids (AAs) and aristolactams (ALs) [2,3]. AAs and ALs have been found among species from the family Aristolochiaceae [4]. Previous researches have revealed that AAs are able to react with DNA to form covalent dA-aristolactam (dA-AL) and dG-aristolactam (dG-AL) adducts [5,6]. With further research, current evidence from studies of AAs has demonstrated that AAs can cause nephrotoxicity, carcinogenicity, and mutagenicity [7–10], especially after prolonged low-dose or shortdated high-dose intake [11,12]. Some nephropathy and malignant tumours including renal interstitial fibrosis, Balkan endemic nephropathy, and upper tract urothelial carcinomas are caused by AAs [13–15]. Currently, there are different degrees of restrictions on the sale and use of AAs-containing herbal preparations in many countries.

Chloroplasts are key and semi-autonomous organelles for photosynthesis and biosynthesis in plant cells [16–18]. The chloroplast genome, one of three major genetic systems (the other two are nuclear and mitochondrial genomes), is a circular molecule with a typical quadripartite structure of 115 to 165 kb in length [19,20]. All chloroplast genomes of land plants, apart from several rare exceptions, are highly conserved in terms of size, structure, gene content, and gene [21–23]. Due to its self-replication mechanism and relatively independent evolution, the genetic information from the chloroplast genome has been used in studies of molecular markers, barcoding identification, plant evolution and phylogenetic [24–26]. In 1976, Bedbrook and Bogorad produced the first chloroplast physical mapping of *Zea mays* by digestion with multiple restriction enzymes [27]. Subsequently, the first complete chloroplast genome sequence of *Nicotiana tabacum* was determined [28]. With the development of sequencing technology and bioinformatics, research into the chloroplast genome has increased rapidly. By now, the number of chloroplast genome sequence recorded in the National Center for Biotechnology Information (NCBI) has reached more than 1,500 plant species [29].

About eight genera and 600 species are classified within Aristolochiaceae, and are primarily distributed in tropical and subtropical regions. Of these plants, there are four genera (one endemic) and 86 species (69 endemic) distributed widely in China. The genus *Aristolochia L.*, comprising about 400 species (45 species in China), is the largest and most representative genus of Aristolochiaceae [30]. However, there are no reports on the chloroplast genomes of the family Aristolochiaceae at present, and this has hindered our understanding and progress in the research of the evolution, phylogeny, species identification, and genetic engineering of Aristolochiaceae.

In this study, we determined the complete chloroplast genome sequences of *A. debilis* and *A. contorta*, which are the first two sequenced members of the family Aristolochiaceae. Furthermore, to reveal the phylogenetic positions of the two species, we conducted a phylogenetic tree using the maximum parsimony (MP) method based on common protein-coding genes from 37 species. Overall, the results provide basic genetic information on the chloroplast of *A. debilis* and *A. contorta*, and the role of the two species in plant systematics.

2. Results and Discussion

2.1. The Chloroplast Genome Structures of A. debilis and A. contorta

Both species displayed a typical quadripartite structure, and the corresponding regions were of similar lengths. The complete chloroplast genome of *A. debilis* is a circular molecule of 159,793 bp in length comprising a large single-copy (LSC) region of 89,609 bp and a small single-copy (SSC) region of 19,834 bp separated by a pair of inverted repeats (IRs), each 25,175 bp in length (Figure 1, Table 1). The complete chloroplast genome of *A. contorta* is 160,576 bp in length, which is divided into one LSC (89,781 bp), one SSC (19,877 bp) and two IRs, each 25,459 bp in length (Figure 2, Table 1).

Table 1. Base composition in the chloroplast genomes of *A. debilis* and *A. contorta*.

Species	Regions	Positions	T(U) (%)	C (%)	A (%)	G (%)	Length (bp)
A. debilis	LSC	-	32.2	18.7	31.2	17.9	89,609
	SSC	-	34.0	17.4	33.2	15.5	19,834
	IRa	-	28.4	22.4	28.3	21.0	25,175
	IRb	-	28.3	21.0	28.4	22.4	25,175
	Total	-	31.2	19.5	30.5	18.8	159,793
	CDS	-	30.9	18.1	30.2	20.8	78,717
	-	1st position	23.5	18.8	30.5	27.2	26,239
	-	2nd position	32.2	20.5	29.2	18.1	26,239
	-	3rd position	36.9	15.1	31.1	17.0	26,239

Table 1. *Cont.*

Species	Regions	Positions	T(U) (%)	C (%)	A (%)	G (%)	Length (bp)
	LSC	-	32.2	18.7	31.2	17.8	89,781
	SSC	-	33.9	17.4	33.3	15.4	19,877
	IRa	-	28.4	22.4	28.2	21.0	25,459
	IRb	-	28.2	21.0	28.4	22.4	25,459
A. contorta	Total	-	31.2	19.5	30.6	18.8	160,576
	CDS	-	30.9	18.1	30.3	20.7	78,765
	-	1st position	23.5	18.8	30.5	27.2	26,255
	-	2nd position	32.2	20.6	29.2	18.1	26,255
	-	3rd position	37.0	15.0	31.1	16.9	26,255

* CDS: protein-coding regions.

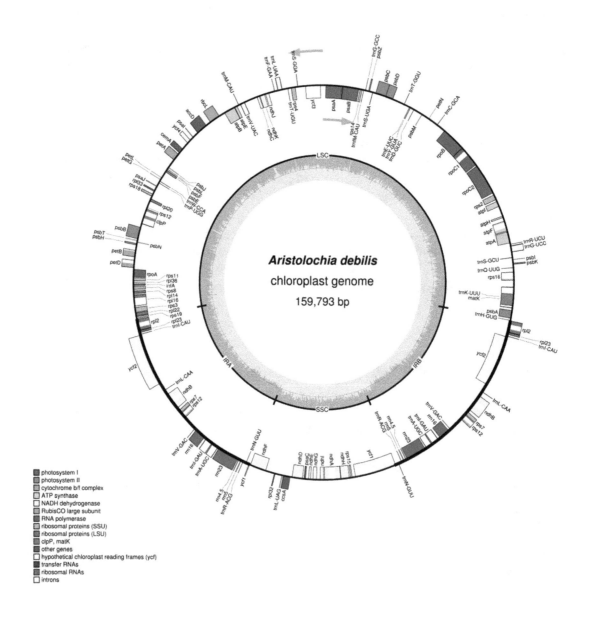

Figure 1. Gene map of the complete chloroplast genome of *A. debilis*. Genes on the inside of the circle are transcribed clockwise, while those outside are transcribed counter clockwise. The darker gray in the inner circle corresponds to GC content, whereas the lighter gray corresponds to AT content.

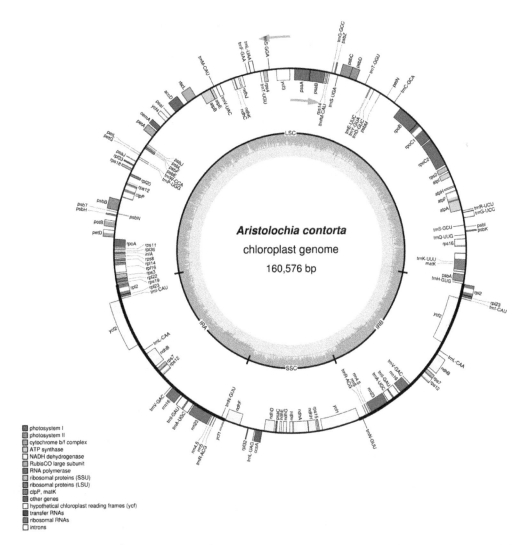

Figure 2. Gene map of the complete chloroplast genome of *A. contorta*. Genes on the inside of the circle are transcribed clockwise, while those outside are transcribed counter clockwise. The darker gray in the inner circle corresponds to GC content, whereas the lighter gray corresponds to AT content.

The analysis results revealed that both species had a GC content of 38.3%. However, this was unevenly distributed across the whole chloroplast genome. In both species, the GC content exhibited the highest values of the IR regions across the complete chloroplast genome, 43.4% both in *A. debilis* and *A. contorta*. The high GC content in the IR regions was the result of four rRNA genes (*rrn16*, *rrn23*, *rrn4.5* and *rrn5*) that occur in this region [31]. In addition, the LSC regions have GC contents of 36.6% and 35.5%, as well as the lowest values of 32.9% and 32.8% are seen in SSC regions in *A. debilis* and *A. contorta*, respectively. Within the protein-coding regions (CDS) of chloroplast genome of *A. debilis*, the percentage of AT content for the first, second and third codon positions were 54%, 61.4% and 68%, respectively (Table 1). A bias towards a higher AT representation at the third codon position has also been observed in other land plant chloroplast genomes [32–34].

A total of 131 genes were identified from each genome including 85 protein-coding genes, 37 tRNAs, eight rRNAs, and one pseudogene (*ycf1*) (Table 2). The functional *ycf1* copy existed encompassing IR-SSC boundary and the other pseudogene *ycf1* copy was on the other IR region. Six protein-coding genes, seven tRNA genes, and all rRNA genes were duplicated in the IR regions. Coding regions including protein-coding genes (CDS), tRNAs, and rRNAs constituted 56.7% and 56.4% in the chloroplast genomes of *A. debilis* and *A. contorta*, respectively; while the non-coding regions including introns, pseudogenes, and intergenic spacers constituted 43.3% and 43.6% of the genome, respectively.

Table 2. Gene contents in the chloroplast genomes of *A. debilis* and *A. contorta*.

No.	Group of Genes	Gene names	Amount
1	Photosystem I	*psaA, psaB, psaC, psaI, psaJ*	5
2	Photosystem II	*psbA, psbB, psbC, psbD, psbE, psbF, psbH, psbI, psbJ, psbK, psbL, psbM, psbN, psbT, psbZ*	15
3	Cytochrome b/f complex	*petA, petB *, petD *, petG, petL, petN*	6
4	ATP synthase	*atpA, atpB, atpE, atpF *, atpH, atpI*	6
5	NADH dehydrogenase	*ndhA *, ndhB *(×2)[1], ndhC, ndhD, ndhE, ndhF, ndhG, ndhH, ndhI, ndhJ, ndhK*	12(1)
6	RubisCO large subunit	*rbcL*	1
7	RNA polymerase	*rpoA, rpoB, rpoC1 *, rpoC2*	4
8	Ribosomal proteins (SSU)	*rps2, rps3, rps4, rps7(×2), rps8, rps11, rps12 **(×2), rps14, rps15, rps16 *, rps18, rps19*	14(2)
9	Ribosomal proteins (LSU)	*rpl2 *(×2), rpl14, rpl16 *, rpl20, rpl22, rpl23(×2), rpl32, rpl33, rpl36*	11(2)
10	Proteins of unknown function	*ycf1, ycf2(×2), ycf3 **, ycf4*	5(1)
11	Transfer RNAs	37 *tRNAs* (6 contain an intron, 7 in the IRs)	37(7)
12	Ribosomal RNAs	*rrn4.5(×2), rrn5(×2), rrn16(×2), rrn23(×2)*	8(4)
13	Other genes	*accD, clpP **, matK, ccsA, cemA, infA*	6

* Gene contains one intron; ** gene contains two introns; (×2) indicates the number of the repeat unit is 2.

Introns play an important role in the regulation of gene expression and can enhance the expression of exogenous genes at specific sites and specific times of the plant [35]. The intron content of genes reserved in the chloroplast genomes of *A. debilis* and *A. contorta* are maintained in other angiosperms [31,36]. Data revealed the presence of 18 genes containing introns in each chloroplast genome, including *atpF, rpoC1, ycf3, rps12, rpl2, rpl16, clpP, petB, petD, rps16, ndhA, ndhB,* and six tRNA genes (Table 3). In addition, the *ycf3* gene and *rps12* gene each contain two introns and three exons. The *ycf3* gene is located in LSC region and the *rps12* gene is a special trans-splicing gene, the 5′ exon is located in LSC, while the 3′ exon is located in IR, which is similar to that in *Aquilaria sinensis* [25], *Panax ginseng* [36] and *Cistanche deserticola* [37].

Table 3. Genes with introns in the chloroplast genomes of *A. debilis* and *A. contorta* as well as the lengths of the exons and introns.

Species	Gene	Location	Exon I (bp)	Intron I (bp)	Exon II (bp)	Intron II (bp)	Exon III (bp)
	atpF	LSC	145	805	410	-	-
	clpP	LSC	71	781	292	678	255
	ndhA	SSC	552	1090	540	-	-
	ndhB	IR	777	705	756	-	-
	petB	LSC	6	214	642	-	-
	petD	LSC	6	485	476	-	-
	rpl16	LSC	8	1065	403	-	-
	rpl2	IR	391	657	431	-	-
A. debilis	*rpoC1*	LSC	430	776	1622	-	-
	rps12	LSC	114	-	232	536	26
	rps16	LSC	46	853	191	-	-
	trnA-UGC	IR	38	809	35	-	-
	trnG-UCC	LSC	24	761	48	-	-
	trnI-GAU	IR	37	937	35	-	-
	trnK-UUU	LSC	37	2658	35	-	-
	trnL-UAA	LSC	35	521	50	-	-
	trnV-UAC	LSC	39	597	37	-	-
	ycf3	LSC	126	777	228	753	147

Table 3. *Cont.*

Species	Gene	Location	Exon I (bp)	Intron I (bp)	Exon II (bp)	Intron II (bp)	Exon III (bp)
	atpF	LSC	145	771	410	-	-
	clpP	LSC	71	821	292	664	255
	ndhA	SSC	552	1091	540	-	-
	ndhB	IR	777	716	756	-	-
	petB	LSC	6	214	642	-	-
	petD	LSC	7	485	476	-	-
	rpl16	LSC	8	1088	403	-	-
	rpl2	IR	391	657	431	-	-
A. contorta	*rpoC1*	LSC	430	776	1619	-	-
	rps12	LSC	114	-	232	536	26
	rps16	LSC	46	832	221	-	-
	trnA-UGC	IR	38	809	35	-	-
	trnG-UCC	LSC	24	751	48	-	-
	trnI-GAU	IR	37	938	35	-	-
	trnK-UUU	LSC	37	2648	35	-	-
	trnL-UAA	LSC	35	552	50	-	-
	trnV-UAC	LSC	39	605	37	-	-
	ycf3	LSC	126	764	228	760	147

2.2. IR Contraction and Expansion

Although genomic structure and size were highly conserved in Angiosperms chloroplast genomes, the IR/SC boundary regions still varied slightly (Figure 3). The contraction and expansion at the borders of the IR regions are common evolutionary events and represent the main reasons for size variation of the chloroplast genomes [33,38]. From Figure 3, the junctions of the IR and LSC regions of four species including *Arabidopsis thaliana* (accession number: NC_000932), *Nicotiana tabacum* (NC_001879), as well as two *Aristolochia* species were compared. The IRb/SSC border extended into the *ycf1* genes to cause long *ycf1* pseudogenes in all species; however, compared with *A. thaliana* and *N. tabacum*, the length of *ycf1* pseudogene of two *Aristolochia* species were only 171 and 169 bp, respectively. The IRa/SSC border was located in the CDS of the *ycf1* gene and expanded the same length into the 5′ portion of *ycf1* gene as IRb expanded in the four chloroplast genomes. The *trnH* genes were located in the LSC regions in *Nicotiana tabacum*, *Sesamum indicum*, *Arabidopsis thaliana*, and *Salvia miltiorrhiza* [31], while this gene was usually located in the IR region in the monocot chloroplast genomes [39]. Interestingly, the IRa/LSC borders were located in the coding region of *trnH* genes in the two *Aristolochia* species.

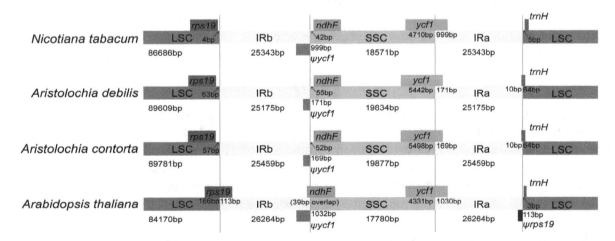

Figure 3. Comparison of the borders of LSC, SSC and IR regions among four chloroplast genomes. Number above the gene features means the distance between the ends of genes and the borders sites. The IRb/SSC border extended intothe *ycf1* genes to create various lengths of *ycf1* pseudogenes among four chloroplast genomes. These features are not to scale.

2.3. Codon Usage and RNA Editing Sites

All the protein-coding genes were composed of 26,239 and 26,255 codons in the chloroplast genomes of *A. debilis* and *A. contorta*, respectively. Among these codons, 2737 encode leucine and 315 encode cysteine, respectively, the most and least universal amino acids in the *A. debilis* chloroplast genome. The codon usages of protein-coding genes in the *A. debilis* and *A. contorta* chloroplast genomes are deduced and summarized in Figure 4 and Table S1. Figure 4 shows that the relative synonymous codon usage (RSCU) value increased with the quantity of codons that code for a specific amino acid. Most of the amino acid codons have preferences except for methionine and tryptophan. The results presented here are similar in codon usage with the chloroplast genomes of species within the genus *Ulmus* [40] and *Aq. sinensis* [25]. In addition, potential RNA editing sites were predicted for 35 genes of the chloroplast genomes of two species. A total of 92 RNA editing sites were identified (Table S2). The amino acid conversion S to L occurred most frequently, while P to S and R to W occurred least. Seventy-six common RNA editing sites were shared in genes of the two species.

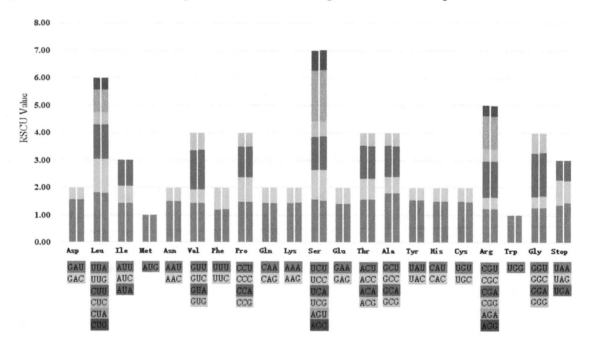

Figure 4. Codon content of 20 amino acid and stop codons in all protein-coding genes of the chloroplast genomes of two *Aristolochia* species. The histogram on the left-hand side of each amino acid shows codon usage within the *A. debilis* chloroplast genome, while the right-hand side illustrates the genome of *A. contorta*.

2.4. Repeat Structure and Simple Sequence Repeats Analyses

The repeats were mostly distributed in the intergenic spacer (IGS) and intron sequences. Figure 5 shows the repeat structure analyses of six species. The results revealed that the repeats of chloroplast genome of *A. contorta* had the greatest number, comprising of 41 forward, 43 palindromic, 29 reverse, and 25 complement repeats. Followed by *A. debilis*, contained 14 forward, 23 palindromic, 23 reverse, and six complement repeats. Simple sequence repeats (SSRs), which are ubiquitous throughout the genomes and are also known as microsatellites, are tandemly repeated DNA sequences that consist of 1–6 nucleotide repeat units [41]. SSRs are widely used for molecular markers in species identification, population genetics, and phylogenetic investigations based on their high level of polymorphism [42–44]. A total of 129 and 156 SSRs were identified using the microsatellite identification tool (MISA) in the chloroplast genomes of *A. debilis* and *A. contorta*, respectively (Table 4; Tables S3 and S4). In these SSRs, mononucletide repeats were largest in number, which were found 81 and 96 times in *A. debilis* and *A. contorta*, respectively. A/T mononucleotide repeats (96.3% and 94.8%, respectively) were the most common, while the majority of dinucleotide repeat sequences comprised of AT/TA repeats (100% and 92.8%, respectively). This result agreed with the previous studies where proportions of polyadenine (polyA) and polythymine (polyT) were higher than polycytosine (polyC) and polyguanine (polyG) within chloroplast SSRs in many plants [24].

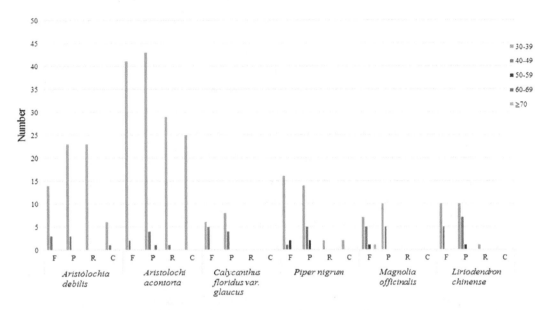

Figure 5. Repeat sequences in six chloroplast genomes. REPuter was used to identify repeat sequences with length ≥30 bp and sequence identified ≥90% in the chloroplast genomes. F, P, R, and C indicate the repeat types F (forward), P (palindrome), R (reverse), and C (complement), respectively. Repeats with different lengths are indicated in different colours.

Table 4. Types and amounts of SSRs in the *A. debilis* and *A. contorta* chloroplast genomes.

SSR Type	Repeat Unit	Amount		Ratio (%)	
		A. debilis	*A. contorta*	*A. debilis*	*A. contorta*
Mono	A/T	78	91	96.3	94.8
	C/G	3	5	3.7	5.2
Di	AC/GT	0	1	0	3.6
	AG/CT	0	1	0	3.6
	AT/TA	19	26	100	92.8
Tri	AAC/GTT	1	1	10	8.3
	AAG/CTT	1	1	10	8.3
	ATC/ATG	1	0	10	0
	AAT/ATT	7	10	70	83.4
Tetra	AAAC/GTTT	2	2	16.7	14.3
	AAAT/ATTT	4	5	33.3	35.7
	AATC/ATTG	1	1	8.3	7.1
	AGAT/ATCT	2	1	16.7	7.1
	AATT/AATT	0	1	0	7.1
	ACAT/ATGT	0	1	0	7.1
	AACT/AGTT	1	1	8.3	7.1
	AATG/ATTC	2	2	16.7	14.3
Penta	AATAT/ATATT	2	2	33.3	50
	AAATT/AATTT	1	0	16.7	0
	AAATC/ATTTG	1	0	16.7	0
	AACAT/ATGTT	0	1	0	25
	AAAAT/ATTTT	2	1	33.3	25
Hexa	AAATAG/ATTTCT	0	1	0	50
	ACATAT/ATATGT	0	1	0	50
	ACTGAT/AGTATC	1	0	100	0

2.5. Comparative Genomic Analysis

The whole chloroplast genome sequences of *A. debilis* and *A. contorta* were compared to those of *Calycanthus floridus* var. *glaucus* (accession number: NC_004993), *Magnolia officinalis* (NC_020316), and *Liriodendron chinense* (NC_030504) using the mVISTA program (Figure 6). The comparison showed that the two IR regions were less divergent than the LSC and SSC regions. The four rRNA genes were the most conserved, while the most divergent coding regions were *ndhF*, *rpl22*, *ycf1*, *rpoC2* and *ccsA*. Additionally, the results revealed that non-coding regions exhibited a higher divergence than coding regions, and the most divergent regions localized in the intergenic spacers among the five chloroplast genomes.

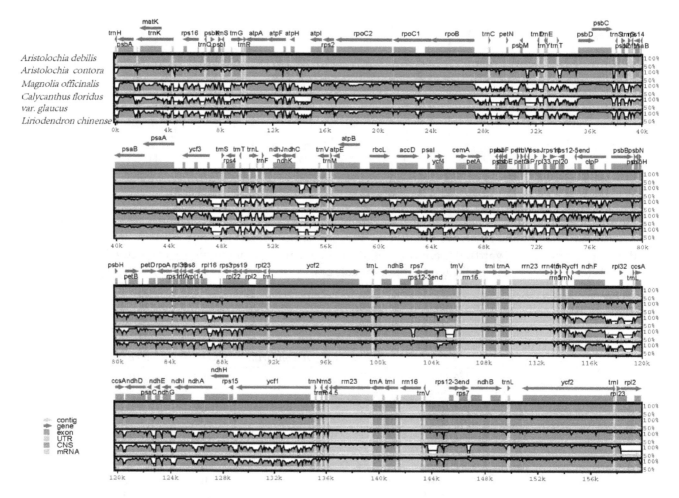

Figure 6. Sequence identity plot comparing the five chloroplast genomes with *A. debilis* as a reference by using mVISTA. Grey arrows and thick black lines above the alignment indicate genes with their orientation and the position of the IRs, respectively. A cut-off of 70% identity was used for the plots, and the Y-scale represents the percent identity ranging from 50% to 100%.

2.6. Phylogenetic Analyses

Chloroplast genomes provide abundant resources, which are significant for evolutionary, taxonomic, and phylogenetic studies [31,45,46]. The whole chloroplast genomes and protein-coding genes have been successfully used to resolve phylogenetic relationships at almost any taxonomic level during the past decade [31,37]. *Aristolochia*, consisting of nearly 400 species, is the largest genus in the family Aristolochiaceae [30]. Phylogenetic analyses employing one or several genes have been performed in previous studies [47–49]; however, these analyses were restricted to the species of Aristolochiaceae and included few species from other families. In this study, to identify the phylogenetic positions of *A. debilis* and *A. contorta* within Angiosperms, 60 protein-coding genes commonly present in 37 species from Piperales, Laurales, Magnoliales, Ranunculales, Fabales, Rosales, Chloranthales, as well as two *Aristolochia* species were used to construct the phylogenetic tree using the Maximum parsimony (MP) method (Figure 7). All the nodes in the MP trees have high bootstrap support values, and 30 out of 34 nodes with 100% bootstrap values were found. The result illustrated that two *Aristolochia* species were sister taxa with respect to four *Piper* species (Piperaceae), and these species were grouped with four species from Laurales and five species from Magnoliales. Additionally, all species are clustered within a lineage distinct from the outgroup. This result (inferred from the chloroplast genome data) obtained high support values, which suggested that the chloroplast genome could effectively resolve the phylogenetic positions and relationships of this family. Nevertheless, to accurately illustrate the evolution of the family Aristolochiaceae, it is necessary to use more species to analyze the phylogeny. This study will also provide a reference for species identification among *Aristolochia* and other genus using the chloroplast genome.

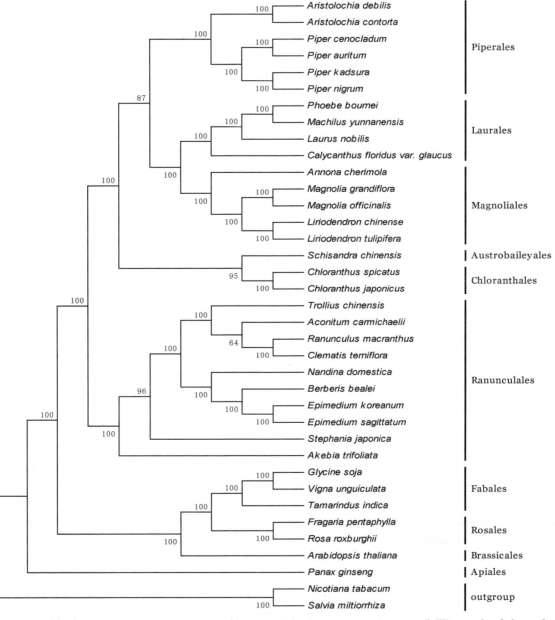

Figure 7. Phylogenetic tree constructed using Maximum parsimony (MP) method based on 60 protein-coding genes from different species. Numbers at nodes are values for bootstrap support.

3. Materials and Methods

3.1. Plant Material, DNA Extraction, and Sequencing

Fresh plants of *A. debilis* and *A. contorta* were collected from Lichuan City in Hubei Province and Tonghua City in Jilin Province, respectively. All samples were identified by Professor Yulin Lin, who is based at the Institute of Medicinal Plant Development (IMPLAD), Chinese Academy of Medical Sciences (CAMS) and Peking Union Medical College (PUMC). The voucher specimens were deposited in the herbarium of the IMPLAD. The leaves were cleansed and preserved in a $-80\ °C$ refrigerator. Total genomic DNA was extracted from approximately 100 mg of samples using DNeasy Plant Mini Kit with a standard protocol (Qiagen Co., Hilden, Germany). Final DNA quality was assessed based on spectrophotometry and their integrity was examined by electrophoresis in 1% (w/v) agarose gel. The DNA was used to construct shotgun libraries with insert sizes of 500 bp and sequenced according to the manufacturer's manual for the Illumina Hiseq X. Approximately 6.3 Gb of raw data from *A. debilis* and 5.8 Gb from *A. contorta* were produced with 150 bp pair-end read lengths.

3.2. Chloroplast Genome Assembly and Annotation

First, we used the software Trimmomatic (v0.36, Max Planck Institute of Molecular Plant Physiology, Potsdam, Germany) [50] to trim the low-quality reads. After quality control, the clean reads were used to assemble the chloroplast genome. All chloroplast genomes of plants recorded in the National Center for Biotechnology Information (NCBI) were used to construct a reference database. Next, the clean reads were mapped to the database on the basis of their coverage and similarity, and the mapped reads extracted. Extracted reads were assembled to contigs using SOAPdenovo (v2, BGI HK Research Institute, Hong Kong, China) [51], and the resulting contigs were combined and extended to obtain a complete chloroplast genome sequence. To verify the accuracy of assembly, four boundaries of single copy (SC) and inverted repeat (IR) regions of the assembled sequences were confirmed by PCR amplification and Sanger sequencing using the primers listed in Table S5.

We used the online program Dual Organellar GenoMe Annotator (DOGMA), (University of Texas at Austin, Austin, TX, USA) [52] and the software Chloroplast Genome Annotation, Visualization, Analysis, and GenBank Submission (CPGAVAS), (Institute of Medicinal Plant Development, Chinese Academy of Medical Sciences and Peking Union Medical College, Beijing, China) [53] coupled with manual corrections to perform the preliminarily gene annotation of chloroplast genomes of two species. The tRNA genes were identified using the software tRNAscan-SE (v2.0, University of California, Santa Cruz, CA, USA) [54] and DOGMA [52]. The gene map was drawn using the Organellar Genome DRAW (OGDRAW) (v1.2, Max Planck Institute of Molecular Plant Physiology, Potsdam, Germany) [55] with default settings and checked manually. The complete and correct chloroplast genome sequences of the two species were deposited in GenBank, accession numbers of *A. debilis* and *A. contorta* are MF539928 and MF539927, respectively.

3.3. Genome Structure Analyses and Genome Comparison

The distribution of codon usage was investigated using the software CodonW (University of Texas, Houston, TX, USA) with the RSCU ratio [56]. Thirty-five protein-coding genes of the chloroplast genomes of two species were used to predict potential RNA editing sites using the online program Predictive RNA Editor for Plants (PREP) suite [57] with a cutoff value of 0.8. GC content was analyzed using Molecular Evolutionary Genetics Analysis (MEGA v6.0, Tokyo Metropolitan University, Tokyo, Japan) [58]. REPuter (University of Bielefeld, Bielefeld, Germany) [59] to identify the size and location of repeat sequences, including forward, palindromic, reverse, and complement repeats in the chloroplast genomes of six species *C. floridus* var. *glaucus*, *M. officinalis*, *L. chinense* and *Piper nigrum* (NC_034692). For all repeat types, the minimal size was 30 bp and the two repeat copies had at least 90% similarity. Simple sequence repeats (SSRs) were detected using MISA software [60] with parameters set the same as Li et al. [61]. The whole-genome alignment for the chloroplast genomes of the five species including *A. debilis*, *A. contorta*, *C. floridus* var. *glaucus*, *M. officinalis*, and *L. chinense* were performed and plotted using the mVISTA program [62].

3.4. Phylogenetic Analyses

A total of 35 complete chloroplast genomes were downloaded from the NCBI Organelle Genome Resources database (Table S6). The 60 protein-coding gene sequences commonly present in 37 species, including the two species in this study, were aligned using the Clustal algorithm [63]. To determine the phylogenetic positions of *A. debilis*, and *A. contorta*, we analyzed the chloroplast genomes of these 60 protein-coding genes. Maximum parsimony (MP) analysis was performed with PAUP*4.0b10 [64], using a heuristic search performed with the MULPARS option, the random stepwise addition with 1000 replications and tree bisection-reconnection (TBR) branch swapping. Bootstrap analysis was also performed with 1,000 replicates with TBR branch swapping.

4. Conclusions

The complete chloroplast genomes of *A. debilis* and *A. contorta*, the first two sequenced members of the family Aristolochiaceae, were determined in this study. The genome structure and gene content were relatively conserved. The phylogenetic analyses illustrated that these two *Aristolochia* species were positioned close to four species from the family Piperaceae and had a close phylogenetic relationship with Laurales and Magnoliales. The results provided the basis for the study of the evolutionary history of *A. debilis* and *A. contorta*. All the data presented in this paper will facilitate the further investigation of these two medicinal plants.

Acknowledgments: This work was supported by Chinese Academy of Medical Sciences (CAMS) Innovation Fund for Medical Sciences (CIFMS) (No. 2016-I2M-3-016), Major Scientific and Technological Special Project for "Significant New Drugs Creation" (No. 2014ZX09304307001) and The Key Projects in the National Science and Technology Pillar Program (No. 2011BAI07B08).

Author Contributions: Jianguo Zhou, Xinlian Chen, and Yingxian Cui, performed the experiments; Jianguo Zhou, Wei Sun, and Jingyuan Song, assembled sequences and analyzed the data; Jianguo Zhou wrote the manuscript; Yonghua Li, and Yu Wang, collected plant material; Hui Yao conceived the research and revised the manuscript. All authors have read and approved the final manuscript.

Abbreviations

LSC	Large single copy
SSC	Small single copy
IR	Inverted repeat
MP	Maximum parsimony
SSR	Simple sequence repeats
ATP	Adenosine triphosphate
NADH	Nicotinamide adenine dinucleotide

References

1. Chinese Pharmacopoeia Commission. *The Chinese Pharmacopoeia*; Chemical Industry Press: Beijing, China, 2015; pp. 51–52.

2. Chen, C.X. Studies on the chemical constituents from the fruit of *Aristolochia debilis*. *J. Chin. Med. Mater.* **2010**, *33*, 1260–1261.

3. Xu, Y.; Shang, M.; Ge, Y.; Wang, X.; Cai, S. Chemical constituent from fruit of *Aristolochia contorta*. *Chin. J. Chin. Mater. Medica* **2010**, *35*, 2862.

4. Mix, D.B.; Guinaudeau, H.; Shamma, M. The aristolochic acids and aristolactams. *J. Nat. Prod.* **1982**, *45*, 657–666. [CrossRef]

5.   Arlt, V.M.; Stiborova, M.; Schmeiser, H.H. Aristolochic acid as a probable human cancer hazard in herbal remedies: A review. *Mutagenesis* **2002**, *17*, 265. [CrossRef] [PubMed]

6.   Schmeiser, H.H.; Janssen, J.W.; Lyons, J.; Scherf, H.R.; Pfau, W.; Buchmann, A.; Bartram, C.R.; Wiessler, M. Aristolochic acid activates *RAS* genes in rat tumors at deoxyadenosine residues. *Cancer Res.* **1990**, *50*, 5464–5469. [PubMed]

7.   Chen, L.; Mei, N.; Yao, L.; Chen, T. Mutations induced by carcinogenic doses of aristolochic acid in kidney of big blue transgenic rats. *Toxicol. Lett.* **2006**, *165*, 250–256. [CrossRef] [PubMed]

8.   Cheng, C.L.; Chen, K.J.; Shih, P.H.; Lu, L.Y.; Hung, C.F.; Lin, W.C.; Yesong, G.J. Chronic renal failure rats are highly sensitive to aristolochic acids, which are nephrotoxic and carcinogenic agents. *Cancer Lett.* **2006**, *232*, 236–242. [CrossRef] [PubMed]

9.   Cosyns, J.P.; Goebbels, R.M.; Liberton, V.; Schmeiser, H.H.; Bieler, C.A.; Bernard, A.M. Chinese herbs nephropathy-associated slimming regimen induces tumours in the forestomach but no interstitial nephropathy in rats. *Arch. Toxicol.* **1998**, *72*, 738–743. [CrossRef] [PubMed]

10. Hoang, M.L.; Chen, C.H.; Sidorenko, V.S.; He, J.; Dickman, K.G.; Yun, B.H.; Moriya, M.; Niknafs, N.; Douville, C.; Karchin, R. Mutational signature of aristolochic acid exposure as revealed by whole-exome sequencing. *Sci. Transl. Med.* **2013**, *5*. [CrossRef] [PubMed]

11. Balachandran, P.; Wei, F.; Lin, R.C.; Khan, I.A.; Pasco, D.S. Structure activity relationships of aristolochic acid analogues: Toxicity in cultured renal epithelial cells. *Kidney Int.* **2005**, *67*, 1797. [CrossRef] [PubMed]

12. Tsai, D.M.; Kang, J.J.; Lee, S.S.; Wang, S.Y.; Tsai, I.; Chen, G.Y.; Liao, H.W.; Li, W.C.; Kuo, C.H.; Tseng, Y.J. Metabolomic analysis of complex Chinese remedies: Examples of induced nephrotoxicity in the mouse from a series of remedies containing aristolochic acid. *Evid. Based Compl. Alt.* **2013**, *2013*, 263757. [CrossRef] [PubMed]

13. Grollman, A.P.; Shibutani, S.; Moriya, M.; Miller, F.; Wu, L.; Moll, U.; Suzuki, N.; Fernandes, A.; Rosenquist, T.; Medverec, Z.; et al. Aristolochic acid and the etiology of endemic (Balkan) nephropathy. *Proc. Natl. Acad. Sci. USA* **2007**, *104*, 12129–12134. [CrossRef] [PubMed]

14. Lord, G.M.; Tagore, R.; Cook, T.; Gower, P.; Pusey, C.D. Nephropathy caused by Chinese herbs in the UK. *Lancet* **1999**, *354*, 481–482. [CrossRef]

15. Vanherweghem, J.L.; Tielemans, C.; Abramowicz, D.; Depierreux, M.; Vanhaelen-Fastre, R.; Vanhaelen, M.; Dratwa, M.; Richard, C.; Vandervelde, D.; Verbeelen, D.; et al. Rapidly progressive interstitial renal fibrosis in young women: Association with slimming regimen including Chinese herbs. *Lancet* **1993**, *341*, 387–391. [CrossRef]

16. Dong, W.; Xu, C.; Cheng, T.; Lin, K.; Zhou, S. Sequencing angiosperm plastid genomes made easy: A complete set of universal primers and a case study on the phylogeny of Saxifragales. *Genome Biol. Evol.* **2013**, *5*, 989–997. [CrossRef] [PubMed]

17. Leliaert, F.; Smith, D.R.; Moreau, H.; Herron, M.D.; Verbruggen, H.; Delwiche, C.F.; Clerck, O.D. Phylogeny and molecular evolution of the green algae. *Crit. Rev. Plant Sci.* **2012**, *31*, 1–46. [CrossRef]

18. Raman, G.; Park, S. Analysis of the complete chloroplast genome of a medicinal plant, *Dianthus superbus* var. *longicalyncinus*, from a comparative genomics perspective. *PLoS ONE* **2015**, *10*. [CrossRef]

19. Jansen, R.K.; Raubeson, L.A.; Boore, J.L.; Depamphilis, C.W.; Chumley, T.W.; Haberle, R.C.; Wyman, S.K.; Alverson, A.J.; Peery, R.; Herman, S.J. Methods for obtaining and analyzing whole chloroplast genome sequences. *Methods Enzymol.* **2005**, *395*, 348–384. [PubMed]

20. Wolfe, K.H.; Li, W.H.; Sharp, P.M. Rates of nucleotide substitution vary greatly among plant mitochondrial, chloroplast and nuclear DNA. *Proc. Natl. Acad. Sci. USA* **1988**, *84*, 9054–9058. [CrossRef]

21. Smith, D.R.; Keeling, P.J. Mitochondrial and plastid genome architecture: Reoccurring themes, but significant differences at the extremes. *Proc. Natl. Acad. Sci. USA* **2015**, *112*, 10177–10184. [CrossRef] [PubMed]

22. Tonti-Filippini, J.; Nevill, P.G.; Dixon, K.; Small, I. What can we do with 1,000 plastid genomes? *Plant J.* **2017**, *90*, 808–818. [CrossRef] [PubMed]

23. Wicke, S.; Schneeweiss, G.M.; Müller, K.F.; Quandt, D. The evolution of the plastid chromosome in land plants: Gene content, gene order, gene function. *Plant. Mol. Boil.* **2011**, *76*, 273–297. [CrossRef] [PubMed]

24. Kuang, D.Y.; Wu, H.; Wang, Y.L.; Gao, L.M.; Zhang, S.Z.; Lu, L. Complete chloroplast genome sequence of *Magnolia kwangsiensis* (Magnoliaceae): Implication for DNA barcoding and population genetics. *Genome* **2011**, *54*, 663–673. [CrossRef] [PubMed]

25. Wang, Y.; Zhan, D.F.; Jia, X.; Mei, W.L.; Dai, H.F.; Chen, X.T.; Peng, S.Q. Complete chloroplast genome sequence of *Aquilaria sinensis* (lour.) gilg and evolution analysis within the Malvales order. *Front. Plant Sci.* **2016**, *7*, 280. [CrossRef] [PubMed]

26. Wu, F.H.; Chan, M.T.; Liao, D.C.; Chentran, H.; Yiwei, L.; Daniell, H.; Duvall, M.R.; Lin, C.S. Complete chloroplast genome of *Oncidium* Gower Ramsey and evaluation of molecular markers for identification and breeding in Oncidiinae. *BMC Plant Biol.* **2010**, *10*, 68. [CrossRef] [PubMed]

27. Bedbrook, J.R.; Bogorad, L. Endonuclease recognition sites mapped on *Zea mays* chloroplast DNA. *Proc. Natl. Acad. Sci. USA* **1976**, *73*, 4309–4313. [CrossRef] [PubMed]

28. Shinozaki, K.; Ohme, M.; Tanaka, M.; Wakasugi, T.; Hayshida, N.; Matsubayasha, T.; Zaita, N.; Chunwongse, J.; Obokata, J.; Yamaguchi-Shinozaki, K. The complete nucleotide sequence of the tobacco chloroplast genome. *EMBO J.* **1986**, *4*, 111–148. [CrossRef]

29. NCBI, Genome. Available online: https://www.ncbi.nlm.nih.gov/genome/browse/?report=5 (accessed on 30 June 2017).

30. The Editorial Committee of Flora of China. *Flora of China*; Science Press: Beijing, China; Missouri Botanical Garden Press: St. Louis, MO, USA, 2003; Volume 5, pp. 246–269.

31. Qian, J.; Song, J.; Gao, H.; Zhu, Y.; Xu, J.; Pang, X.; Yao, H.; Sun, C.; Li, X.; Li, C. The complete chloroplast genome sequence of the medicinal plant *Salvia miltiorrhiza*. *PLoS ONE* **2013**, *8*. [CrossRef] [PubMed]

32. Clegg, M.T.; Gaut, B.S.; Learn, G.H., Jr.; Morton, B.R. Rates and patterns of chloroplast DNA evolution. *Proc. Natl. Acad. Sci. USA* **1994**, *91*, 6795–6801. [CrossRef] [PubMed]

33. Yang, M.; Zhang, X.; Liu, G.; Yin, Y.; Chen, K.; Yun, Q.; Zhao, D.; Al-Mssallem, I.S.; Yu, J. The complete chloroplast genome sequence of date palm (*Phoenix dactylifera* L.). *PLoS ONE* **2010**, *5*. [CrossRef] [PubMed]

34. Yi, D.K.; Kim, K.J. Complete chloroplast genome sequences of important oilseed crop *Sesamum indicum* L. *PLoS ONE* **2012**, *7*. [CrossRef] [PubMed]

35. Xu, J.; Feng, D.; Song, G.; Wei, X.; Chen, L.; Wu, X.; Li, X.; Zhu, Z. The first intron of rice EPSP synthase enhances expression of foreign gene. *Sci. China Life Sci.* **2003**, *46*, 561–569. [CrossRef] [PubMed]

36. Kim, K.J.; Lee, H.L. Complete chloroplast genome sequences from Korean ginseng (*Panax schinseng* nees) and comparative analysis of sequence evolution among 17 vascular plants. *DNA Res.* **2004**, *11*, 247. [CrossRef] [PubMed]

37. Li, X.; Zhang, T.C.; Qiao, Q.; Ren, Z.; Zhao, J.; Yonezawa, T.; Hasegawa, M.; Crabbe, M.J.; Li, J.; Zhong, Y. Complete chloroplast genome sequence of holoparasite *Cistanche deserticola* (Orobanchaceae) reveals gene loss and horizontal gene transfer from its host *Haloxylon ammodendron* (Chenopodiaceae). *PLoS ONE* **2013**, *8*. [CrossRef] [PubMed]

38. Raubeson, L.A.; Peery, R.; Chumley, T.W.; Dziubek, C.; Fourcade, H.M.; Boore, J.L.; Jansen, R.K. Comparative chloroplast genomics: Analyses including new sequences from the angiosperms *Nuphar advena* and *Ranunculus macranthus*. *BMC Genom.* **2007**, *8*, 174. [CrossRef] [PubMed]

39. Huotari, T.; Korpelainen, H. Complete chloroplast genome sequence of *Elodea canadensis* and comparative analyses with other monocot plastid genomes. *Gene* **2012**, *508*, 96–105. [CrossRef] [PubMed]

40. Zuo, L.H.; Shang, A.Q.; Zhang, S.; Yu, X.Y.; Ren, Y.C.; Yang, M.S.; Wang, J.M. The first complete chloroplast genome sequences of *Ulmus* species by de novo sequencing: Genome comparative and taxonomic position analysis. *PLoS ONE* **2017**, *12*. [CrossRef] [PubMed]

41. Powell, W.; Morgante, M.; McDevitt, R.; Vendramin, G.G.; Rafalski, J.A. Polymorphic simple sequence repeat regions in chloroplast genomes: Applications to the population genetics of pines. *Proc. Natl. Acad. Sci. USA* **1995**, *92*, 7759–7763. [CrossRef] [PubMed]

42. Jiao, Y.; Jia, H.; Li, X.; Chai, M.; Jia, H.; Chen, Z.; Wang, G.; Chai, C.; Weg, E.V.D.; Gao, Z. Development of simple sequence repeat (SSR) markers from a genome survey of Chinese bayberry (*Myrica rubra*). *BMC Genomics* **2012**, *13*, 201. [CrossRef] [PubMed]

43. Xue, J.; Wang, S.; Zhou, S.L. Polymorphic chloroplast microsatellite loci in *Nelumbo* (Nelumbonaceae). *Am. J. Bot.* **2012**, *99*, 240–244. [CrossRef] [PubMed]

44. Yang, A.H.; Zhang, J.J.; Yao, X.H.; Huang, H.W. Chloroplast microsatellite markers in *Liriodendron tulipifera* (Magnoliaceae) and cross-species amplification in *L. chinense*. *Am. J. Bot.* **2011**, *98*, 123–126. [CrossRef] [PubMed]

45. Jansen, R.K.; Cai, Z.; Raubeson, L.A.; Daniell, H.; Depamphilis, C.W.; Leebens-Mack, J.; Müller, K.F.; Guisinger-Bellian, M.; Haberle, R.C.; Hansen, A.K. Analysis of 81 genes from 64 plastid genomes resolves

relationships in angiosperms and identifies genome-scale evolutionary patterns. *Proc. Natl. Acad. Sci. USA* **2007**, *104*, 19369–19374. [CrossRef] [PubMed]

46. Moore, M.J.; Bell, C.D.; Soltis, P.S.; Soltis, D.E. Using plastid genome-scale data to resolve enigmatic relationships among basal angiosperms. *Proc. Natl. Acad. Sci. USA* **2007**, *104*, 19363–19368. [CrossRef] [PubMed]

47. Murata, J.; Ohi, T.; Wu, S.; Darnaedi, D.; Sugawara, T.; Nakanishi, T.; Murata, H. Molecular phylogeny of *Aristolochia* (Aristolochiaceae) inferred from *matK* sequences. *Apg Acta Phyto. Geo.* **2001**, *52*, 75–83.

48. Ohi-Toma, T.; Sugawara, T.; Murata, H.; Wanke, S.; Neinhuis, C.; Jin, M. Molecular phylogeny of *Aristolochia sensu* lato (Aristolochiaceae) based on sequences of *rbcL*, *matK*, and *phyA* genes, with special reference to differentiation of chromosome numbers. *Syst. Bot.* **2006**, *31*, 481–492. [CrossRef]

49. Silva-Brandão, K.L.; Solferini, V.N.; Trigo, J.R. Chemical and phylogenetic relationships among *Aristolochia* L. (Aristolochiaceae) from southeastern Brazil. *Biochem. Syst. Ecol.* **2006**, *34*, 291–302. [CrossRef]

50. Bolger, A.M.; Lohse, M.; Usadel, B. Trimmomatic: A flexible trimmer for Illumina sequence data. *Bioinformatics* **2014**, *30*, 2114–2120. [CrossRef] [PubMed]

51. Luo, R.; Liu, B.; Xie, Y.; Li, Z.; Huang, W.; Yuan, J.; He, G.; Chen, Y.; Pan, Q.; Liu, Y.; et al. SOAPdenovo2: An empirically improved memory-efficient short-read de novo assembler. *GigaScience* **2012**, *1*, 18. [CrossRef] [PubMed]

52. Wyman, S.K.; Jansen, R.K.; Boore, J.L. Automatic annotation of organellar genomes with DOGMA. *Bioinformatics* **2004**, *20*, 3252–3255. [CrossRef] [PubMed]

53. Liu, C.; Shi, L.; Zhu, Y.; Chen, H.; Zhang, J.; Lin, X.; Guan, X. CpGAVAS, an integrated web server for the annotation, visualization, analysis, and GenBank submission of completely sequenced chloroplast genome sequences. *BMC Genom.* **2012**, *13*, 715. [CrossRef] [PubMed]

54. Schattner, P.; Brooks, A.N.; Lowe, T.M. The tRNAscan-SE, snoscan and snoGPS web servers for the detection of tRNAs and snoRNAs. *Nucleic Acids Res.* **2005**, *33*. [CrossRef] [PubMed]

55. Lohse, M.; Drechsel, O.; Bock, R. Organellargenomedraw (OGDRAW): A tool for the easy generation of high-quality custom graphical maps of plastid and mitochondrial genomes. *Curr. Genet.* **2007**, *52*, 267–274. [CrossRef] [PubMed]

56. Sharp, P.M.; Li, W.H. The codon Adaptation Index-a measure of directional synonymous codon usage bias, and its potential applications. *Nucleic Acids Res.* **1987**, *15*, 1281–1295. [CrossRef] [PubMed]

57. Mower, J.P. The PREP suite: Predictive RNA editors for plant mitochondrial genes, chloroplast genes and user-defined alignments. *Nucleic Acids Res.* **2009**, *37*. [CrossRef] [PubMed]

58. Tamura, K.; Stecher, G.; Peterson, D.; Filipski, A.; Kumar, S. MEGA6: Molecular evolutionary genetics analysis version 6.0. *Comput. Mol. Biol. Evol.* **2013**, *30*, 2725–2729. [CrossRef] [PubMed]

59. Kurtz, S.; Choudhuri, J.V.; Ohlebusch, E.; Schleiermacher, C.; Stoye, J.; Giegerich, R. Reputer: The manifold applications of repeat analysis on a genomic scale. *Nucleic Acids Res.* **2001**, *29*, 4633–4642. [CrossRef] [PubMed]

60. Misa-Microsatellite Identification Tool. Available online: http://pgrc.ipk-gatersleben.de/misa/ (accessed on 2 June 2017).

61. Li, X.W.; Gao, H.H.; Wang, Y.T.; Song, J.Y.; Henry, R.; Wu, H.Z.; Hu, Z.G.; Hui, Y.; Luo, H.M.; Luo, K. Complete chloroplast genome sequence of *Magnolia grandiflora* and comparative analysis with related species. *Sci. China Life Sci.* **2013**, *56*, 189–198. [CrossRef] [PubMed]

62. Frazer, K.A.; Pachter, L.; Poliakov, A.; Rubin, E.M.; Dubchak, I. VISTA: Computational tools for comparative genomics. *Nucleic Acids Res.* **2004**, *32*, 273–279. [CrossRef] [PubMed]

63. Thompson, J.D.; Higgins, D.G.; Gibson, T.J. CLUSTAL W: Improving the sensitivity of progressive multiple sequence alignment through sequence weighting, position-specific gap penalties and weight matrix choice. *Nucleic Acids Res.* **1994**, *22*, 4673–4680. [CrossRef] [PubMed]

64. Swofford, D.L. *PAUP**. *Phylogenetic Analysis Using Parsimony (*and Other Methods)*; Version 4.0b10; Sinauer Associates: Sunderland, MA, USA, 2002.

7

Molecular Evolution of Chloroplast Genomes of Orchid Species: Insights into Phylogenetic Relationship and Adaptive Evolution

Wan-Lin Dong †, Ruo-Nan Wang †, Na-Yao Zhang, Wei-Bing Fan, Min-Feng Fang and Zhong-Hu Li *

Key Laboratory of Resource Biology and Biotechnology in Western China, Ministry of Education, College of Life Sciences, Northwest University, Xi'an 710069, China; dongwl@stumail.nwu.edu.cn (W.-L.D.); wangruonan@stumail.nwu.edu.cn (R.-N.W.); 201620812@stumail.nwu.edu.cn (N.-Y.Z.); 201631689@stumail.nwu.edu.cn (W.-B.F.); fangmf@nwu.edu.cn (M.-F.F.)
* Correspondence: lizhonghu@nwu.edu.cn
† These two authors contributed equally to this study.

Abstract: Orchidaceae is the 3rd largest family of angiosperms, an evolved young branch of monocotyledons. This family contains a number of economically-important horticulture and flowering plants. However, the limited availability of genomic information largely hindered the study of molecular evolution and phylogeny of Orchidaceae. In this study, we determined the evolutionary characteristics of whole chloroplast (cp) genomes and the phylogenetic relationships of the family Orchidaceae. We firstly characterized the cp genomes of four orchid species: *Cremastra appendiculata, Calanthe davidii, Epipactis mairei*, and *Platanthera japonica*. The size of the chloroplast genome ranged from 153,629 bp (*C. davidi*) to 160,427 bp (*E. mairei*). The gene order, GC content, and gene compositions are similar to those of other previously-reported angiosperms. We identified that the genes of *ndhC, ndhI*, and *ndhK* were lost in *C. appendiculata*, in that the *ndh I* gene was lost in *P. japonica* and *E. mairei*. In addition, the four types of repeats (forward, palindromic, reverse, and complement repeats) were examined in orchid species. *E. mairei* had the highest number of repeats (81), while *C. davidii* had the lowest number (57). The total number of Simple Sequence Repeats is at least 50 in *C. davidii*, and, at most, 78 in *P. japonica*. Interestingly, we identified 16 genes with positive selection sites (the *psbH, petD, petL, rpl22, rpl32, rpoC1, rpoC2, rps12, rps15, rps16, accD, ccsA, rbcL, ycf1, ycf2*, and *ycf4* genes), which might play an important role in the orchid species' adaptation to diverse environments. Additionally, 11 mutational hotspot regions were determined, including five non-coding regions (*ndhB* intron, *ccsA-ndhD, rpl33-rps18, ndhE-ndhG*, and *ndhF-rpl32*) and six coding regions (*rps16, ndhC, rpl32, ndhI, ndhK*, and *ndhF*). The phylogenetic analysis based on whole cp genomes showed that *C. appendiculata* was closely related to *C. striata* var. *vreelandii*, while *C. davidii* and *C. triplicate* formed a small monophyletic evolutionary clade with a high bootstrap support. In addition, five subfamilies of Orchidaceae, Apostasioideae, Cypripedioideae, Epidendroideae, Orchidoideae, and Vanilloideae, formed a nested evolutionary relationship in the phylogenetic tree. These results provide important insights into the adaptive evolution and phylogeny of Orchidaceae.

Keywords: adaptive variation; chloroplast genome; molecular evolution; Orchidaceae; phylogenetic relationship

1. Introduction

Orchidaceae is the biggest family of monocotyledons and the third largest angiosperm family, containing about five recognized subfamilies (Apostasioideae, Cypripedioideae, Epidendroideae,

Orchidoideae, and Vanilloideae) [1], with over 700 genera and 25,000 species [2–4]. The orchid species are generally distributed in tropical and subtropical regions in the world, while a few species are found in temperate zones. Many orchid species have important ornamental and flowering values, e.g., their flowers are characterized by labella and a column, and they are attractive to humans [5,6]. In recent years, due to the overexploitation and habitat destruction of orchid species, many wild population resources have become rare and endangered [7]. Presently, some scholars have mainly concentrated on the study of Orchidaceae for their morphology and medicinal value, and research on genomes has been relatively scarce [8,9]. Some studies showed that the two subfamilies, Apostasioideae and Cypripedioideae, were clustered into the two respective genetic clades based partial on chloroplast DNA regions and nuclear markers [4,10]. However, the major phylogenetic relationships among the five orchid subfamilies remain unresolved [11].

In recent years, the fast progress of next-generation sequencing technology has provided a good opportunity for the study of genomic evolution and interspecific relationships of organisms based on large-scale genomic dataset resources, such as complete plastid sequences [12,13]. The chloroplast (cp) is made up of multifunctional organelles, playing a critical role in photosynthesis and carbon fixation [5,14–16]. The majority of the cp genomes of angiosperms are circular DNA molecules, ranging from 120 to 160 kb in length, with highly-conserved compositions, in terms of gene content and gene order [17–20].

Generally, the typical cp genome is composed of a large single copy (LSC) region and a small single copy (SSC) region, which are separated by two copies of inverted repeats (IRa/b) [21–23]. Due to its maternal inheritance and conserved structure characteristics [24–27], the cp genomes can provide abundant genetic information for studying species divergence and the interspecific relationships of plants [28–31]. For example, based on complete cp genomes, some studies suggested that *Dactylorhiza viridis* diverged earlier than *Dactylorhiza incarnate* [12]; *Lepanthes* is was distinct from *Pleurothallis* and *Salpistele* [13].

In addition, some researchs based on one nuclear region (*ITS-1*) and five chloroplast DNA fragment variations revealed that *Bolusiella talbotii* and the congeneric *B. iridifolia* were clustered into an earlier diverged lineage [10]. However, up to now, the phylogenetic relationships of some major taxons (e.g., *Cremastra* and *Epipactis*) in the Orchidaceae family remain unclear.

In this study, the complete cp genomes of four orchid species (*Cremastra appendiculata, Calanthe davidii, Epipactis mairei*, and *Platanthera japonica*) were first assembled and annotated. Following this, we analyzed the differences in genome size, content, and structure, and the inverted repeats (IR) contraction and expansion, identifying the sequence divergence, along with variant hotspot regions and adaptive evolution through combination with other available orchid cp genomes. In addition, we also constructed the evolutionary relationships of the Orchidaceae family, based on the large number of cp genome datasets.

2. Results

2.1. The Chloroplast Genome Structures

In this study, the cp genomes of four species displayed a typical quadripartite structure and similar lengths, containing a pair of inverted repeats IR regions (IRa and IRb), one large single-copy (LSC) region, and one small single-copy (SSC) region (Figure 1, Table 2). The cp genome size ranged from 153,629 bp in *C. davidii* to 160,427 bp in *E. mairei*, with *P. japonica* at 154,995 bp and *C. appendiculata* at 155,320 bp. The length of LSC ranged from 85,979 bp (*P. japonica*) to 88,328 bp (*E. mairei*), while the SSC length and IR length ranged from 13,664 bp (*P. japonica*) to 18,513 bp (*E. mairei*), and from 25,956 bp (*C. davidii*) to 27,676 bp (*P. japonica*). In the four species, the GC contents of the LSC and SSC regions (about 34% and 40%) were lower than those of the IR regions (about 43%) (Table 1). There were 37 tRNA genes and eight rRNA genes that were identified in each orchid cp genome, but there were some differences in terms of protein-coding genes. In *C. davidii*, we annotated 86 protein-coding genes. There were no *ndhC*, *ndhI*, and *ndhK* genes in *C. appendiculata*. In *P. japonica* and *E. mairei*, the *ndhI* gene was lost (Tables 1 and 2). Fourteen out of the seventeen genes contained a single intron, while three (*clpP*, *ycf3*, and *rps12*) had two introns (Table 2).

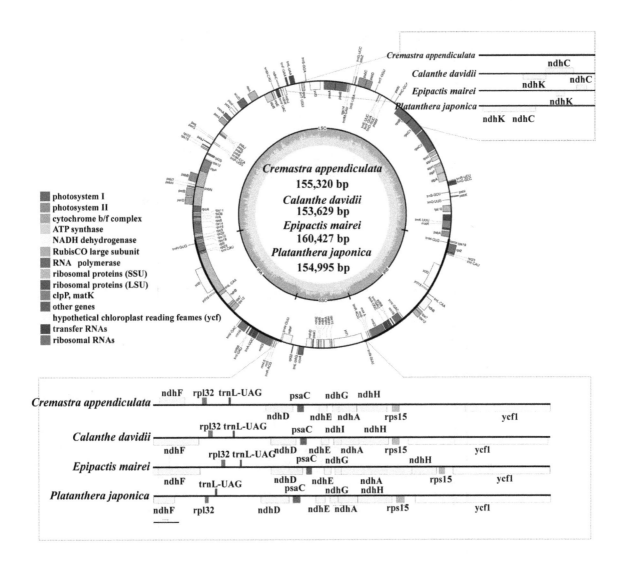

Figure 1. Chloroplast genome maps of the four orchid species. Gene locations outside of the outer rim are transcribed in the counter clockwise direction, whereas genes inside are transcribed in the clockwise direction. The colored bars indicate known different functional groups. The dashed gray area in the inner circle shows the proportional GC content of the corresponding genes. LSC, SSC and IR are large single-copy region, small single-copy region, and inverted repeat region, respectively.

Table 1. Comparison of chloroplast genome features in four orchid species.

Species	Cremastra appendiculata	Calanthe davidii	Epipactis mairei	Platanthera japonica
Accession number	MG925366	MG925365	MG925367	MG925368
Genome size (bp)	155,320	153,629	160,427	154,995
LSC length (bp)	87,098	86,045	88,328	85,979
SSC length (bp)	15,478	15,672	18,513	13,664
IR length (bp)	26,372	25,956	26,790	27,676
Coding (bp)	100,018	104,531	113,915	107,028
Non-coding (bp)	55,302	49,098	46,512	47,967
Number of genes	130 (0)	132 (19)	131 (19)	128 (17)
Number of protein-coding genes	83 (7)	86 (7)	85 (7)	85 (7)
Number of tRNA genes	38 (8)	38 (8)	38 (8)	38 (8)
Number of rRNA genes	8 (4)	8 (4)	8 (4)	8 (4)
GC content (%)	37.2	36.9	37.2	37
GC content in LSC (%)	34.5	34.5	34.9	34.2
GC content in SSC (%)	30.4	30.2	31.0	29
GC content in IR (%)	43.5	43.1	43.1	43.2
Mapped read number	551,680	324,741	230,968	322,259
Chloroplast coverage	544.9	217.4	216	313.6

The numbers in parenthesis indicate the genes duplicated in the IR regions.

Molecular Evolution of Chloroplast Genomes of Orchid Species: Insights into Phylogenetic Relationship... 99

Table 2. List of genes present in four orchid chloroplast genomes.

Category of Genes	Group of Gene	Name of Gene	Name of Gene	Name of Gene	Name of Gene	Name of Gene
Self-replication	Ribosomal RNA genes	rrn16 (×2)	rrn2 (×2)	rrn4.5 (×2)	rrn5 (×2)	
	Transfer RNA genes	trnA-UGC *(×2) trnfM-CAU trnL-GAU *(×2) trnM-CAU trnR-UCU trnT-UGU	trnC-GCA trnG-GCC * trnK-UUU * trnN-GUU (×2) trnS-GCU trnV-GAC (×2)	trnD-GUC trnG-UCC trnL-CAA (×2) trnP-UGG trnS-GGA trnV-UAC (×2)	trnE-UUC trnH-GUG (×2) trnL-UAA * trnQ-UUG trnS-UGA trnW-CCA	trnF-GAA trnI-CAU (×2) trnL-UAG trnR-ACG (×2) trnT-GGU trnY-GUA
	Small subunit of ribosome	rps2 rps11 rps18	rps3 rps12 **(×2) rps19 (×2)	rps4 rps14	rps7 (×2) rps15	rps8 rps16 *
	Large subunit of ribosome	rpl2 *(×2) rpl23 (×2)	rpl14 rpl32	rpl16 * rpl33	rpl20 rpl36	rpl22
	DNA-dependent RNA polymerase	rpoA	rpoB	rpoC1 *	rpoC2	
	Translational initiation factor	infA				
	Subunits of NADH-dehydrogenase	ndhA * ndhF ndhK a	ndhB *(×2) ndhG	ndhC a ndhH	ndhD ndhI a,c,d	ndhE ndhJ
Genes for photosynthesis	Subunits of photosystem I	psaA ycf3 **	psaB ycf4	psaC	psaI	psaJ
	Subunits of photosystem II	psbA psbF psbL	psbB psbH psbM	psbC psbI psbN	psbD psbJ psbT	psbE psbK psZ
	Subunits of cytochrome b/f complex	petA petN	petB*	petD *	petG	petL
	Subunits of ATP synthase	atpA atpI	atpB	atpE	atpF *	atpH
	Subunits of rubisco	rbcL				
Other genes	Maturase	matK				
	Protease	clpP **				
	Envelope membrane protein	cemA				
	Subunit of acetyl-CoA carboxylase	accD				
	C-type cytochrome synthesis gene	ccsA				
Genes of unknown function	Conserved open reading frames	ycf1	ycf2 (×2)			

a gene is no in *Cremastra appendiculata*; c gene is not in *Epipactis mairei*; d gene is not in *Platanthera japonica*; * Gene contains one intron; ** gene contains two introns; (×2) indicates that the number of the repeat unit is 2.

2.2. Repeat Structure and Simple Sequence Repeats

Repeats in cp genomes were analyzed using REPuter (Figure 2a and Table S2). *E. mairei* had the greatest number, including 46 forward, 31 palindromic, three reverse repeats, and 1 complement repeat. This was followed by *C. appendiculata* with 43 forward, 33 palindromic, and 2 reverse repeats. *P. japonica* had 42, 21, 1, and 1 forward, palindromic, reverse, and complement repeats. *C. davidii* had the least number, with only 30 forward and 27 palindromic repeats. The comparison analyses revealed that most of the repeats were 30–90 bp, and that the longest repeats, with a length of 309 bp, were detected in the *E. mairei* cp genome (Figure 2b). Most of the repeats were distributed in non-coding regions. There were 9% repeats in coding sequence and intergenic spacer parts (CDS-IGS) in *E. mairei*, but none in *C. appendiculata* (Figure 2c). The highest number of tandem repeats was 53 in *E. mairei*, and the lowest was 29 in *C. davidii* (Table S3). The total number of SSRs was 51 in *C. appendiculata*, 50 in *C. davidii*, 58 in *E. mairei*, and 78 in *P. japonica* (Table S4). Only one six compound, SSR, was found in *C. appendiculata* (Figure 3a). A large proportion of SSRs were found in the LSC region, and we did not identify C/G mononucleotide repeats, while the majority of the dinucleotide repeat sequences were comprised of AT/TA repeats (Figure 3b).

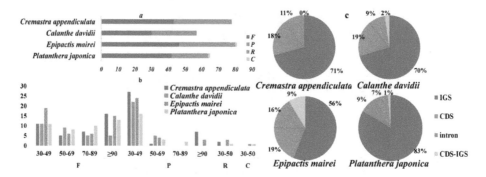

Figure 2. Maps of repeat sequence analyses. Repeat sequences in *C. appendiculata*, *C. davidii*, *E. mairei*, and *P. japonica* chloroplast genomes. (**a**) Number of the four repeat types, F, P, R, and C, indicate the repeat type (F: forward, P: palindrome, R: reverse, and C: complement, respectively). (**b**) Frequency of the four repeat types by length. (**c**) Repeat distribution among four different regions: IGS: intergenic spacer, CDS: coding sequence, intron and CDS-IGS part in CDS and part in IGS.

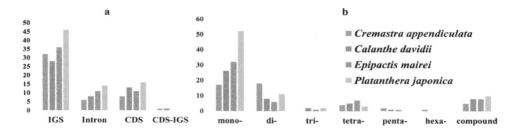

Figure 3. The distribution maps of simple sequence repeats (SSR) in *C. appendiculata*, *C. davidii*, *E. mairei*, and *P. japonica* chloroplast genomes. (**a**) Classification of SSRs in four orchid species. IGS, intergenic spacer; CDS, coding sequence, CDS-IGS, part in CDS and part in IGS. (**b**) Classification of SSRs by repeat type. mono-, mononucleotides; di-, dinucleotides; tri-, trinucleotides; tetra-, tetranucleotides; penta-, pentanucleotides; and hexa-, hexanucleotides.

2.3. IR Contraction and Expansion

We examined the differences between inverted repeat and single-copy (IR/SC) boundary regions among 20 orchid genera, which were classified into several different types (Figure 4). First, the *rps19* gene crossed the large single-copy and inverted repeat b (LSC/IRb) regions within the two parts for eighteen Orchidaceae genera. In *C. crispate* and *C. appendiculata*, the *rps19* gene existed only in the IRb region. Second, in 12 genera, the *ndhF* gene and the *ycf1* pseudogene overlapped in the IRb/SSC region. In *C. appendiculata* and *Dendrobium strongylanthum*, the *ndhF* gene was complete in the SSC region, 8–35 bp away from the IRb region. In *C. crispate, E. pusilla*, and *Phalaenopsis equestris*, the *rpl32* gene was in the SSC region instead of the *ndhF* gene, 280–464 bp away from the IRb region. For the 17 genera mentioned above, the *ycf1* gene crossed the SSC/IRa region. In *C. edavidii* and *Bletilla ochracea*, the *ndhF* gene crossed the IRb/SSC region, and the *ycf1* gene was complete in the SSC region, 101 and 4 bp away from the IRa region. The *trnH-GUG* genes were all located in the LSC region, which was 231 to 1390 bp away from the LSC/IRa boundary. Most specifically, in *Vanilla planifolia*, the *ccsA* gene crossed the IRb/SSC region, as we did not find the *ndhF* and *ycf1* genes where they should be. The SSC/IRa borders were located between the *rpl32* and *ycf1* genes. Thirdly, all 20 genera had the same IRa/LSC borders: the *rps19* gene in the IRa region and the *psbA* gene in the LSC region.

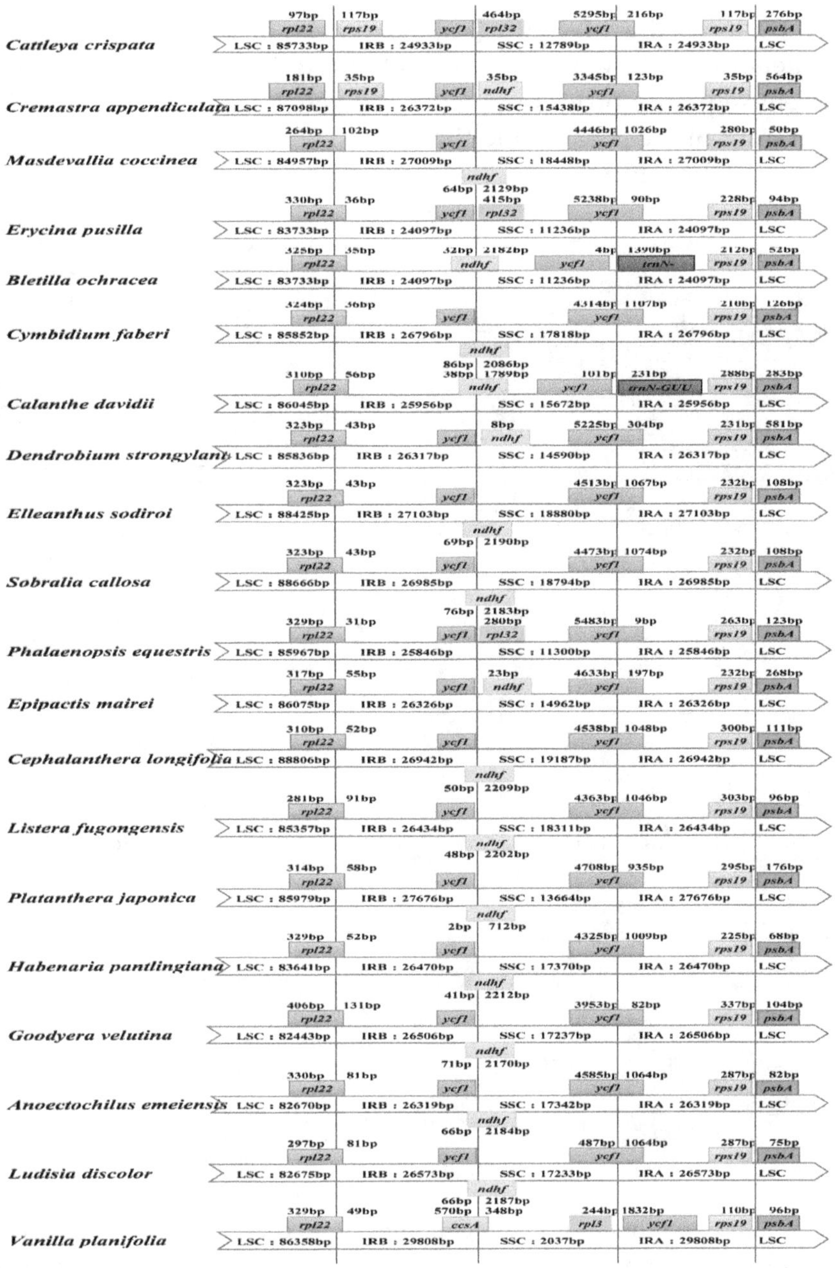

Figure 4. Comparison of the borders of LSC, SSC, and IR regions in 20 orchid complete chloroplast genomes.

2.4. Sequence Divergence and Mutational Hotspot

The whole chloroplast genome sequences of *C. appendiculata*, *C. davidii*, *E. mairei*, and *P. japonica* were compared to 16 other species, using mVISTA [32] (Figures 5 and 6, and Table S5). The comparison analyses showed a high sequence similarity across the cp genomes, with a sequence identity of 82.0%. Interestingly, the proportions of variability in the non-coding regions (introns and intergenic spacers) ranged from 6.77% to 100% with a mean value of 45.97%, i.e., values that are twice as high as in the coding regions (where the range was from 5.80% to 61.76% with a mean value of 24.68%). Five regions within the non-coding regions (*ndhB* intron, *ccsA-ndhD*, *rpl33-rps18*, *ndhE-ndhG*, and *ndhF-rpl32*) and six regions within the coding parts (*rps16*, *ndhC*, *rpl32*, *ndhI*, *ndhK*, and *ndhF*) showed greater levels of variations (percentage of variability >80% and 50%, respectively). In particular, the *ndhB* intron and *ccsA-ndhD* showed a variable percentage of 100%.

In addition, we performed a MAUVE [33] alignment of the 20 orchid chloroplast genomes. The *C. appendiculata* genome is shown at the top as the reference genome (Figure 7). These species maintained a consistent sequence order in most of the genes. However, in *B. ochracea* and *C. faberi*, the *psbM* gene was in front of the *petN*, while the others were upside-down. *Bletilla* and *Cymbidium* actually had the nearest relationship.

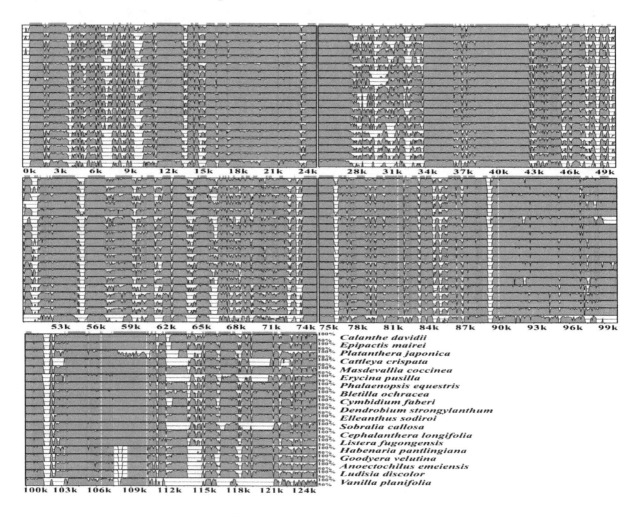

Figure 5. Sequence alignment of chloroplast genomes of 20 orchid species. Sequence identity plot comparing the chloroplast genomes with *C. appendiculata* as a reference using mVISTA. The red color-coded as intergenic spacer regions. The blue color-coded as gene regions. A cut-off of 70% identity was used for the plots, and the Y-scale represents the percent identity between 50% and 100%.

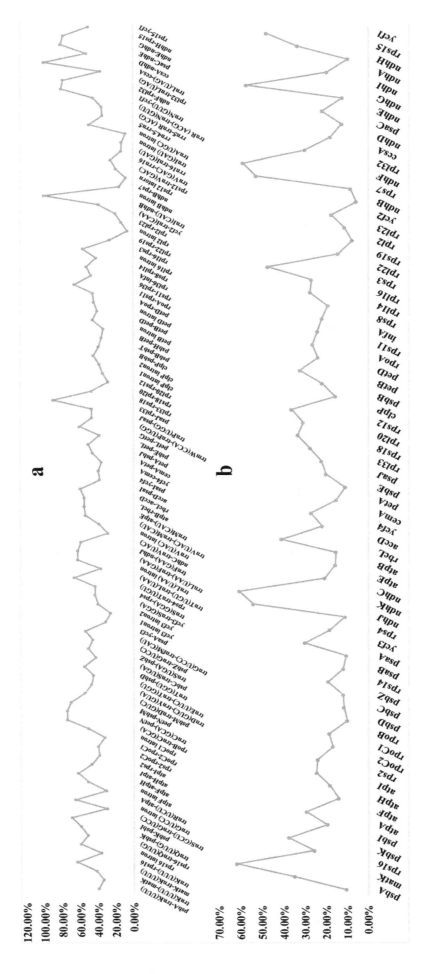

Figure 6. Percentages of variable sites in homologous regions across the 20 orchids with complete chloroplast genomes. (**a**) The introns and spacers (IGS); and (**b**) protein coding sequences (CDS).

Figure 7. MAUVE genome alignments of the 20 orchid chloroplast genomes, with *C. appendiculata* set as a reference genome. The corresponding colored boxes indicate locally-collinear blocks, which present homologous gene clusters. The red vertical line is the location of *atpH* gene. The yellow vertical line is the location of *petN* gene. The green vertical line is the location of *psbM* gene. The blue vertical line is the location of *ycf2* gene.

2.5. Gene Selective Analysis

We compared the rate of nonsynonymous (dN) and synonymous (dS) substitutions for 68 common protein-coding genes between *C. appendiculata*, *C. davidii*, *E. mairei*, and *P. japonica* with 16 other Orchidaceae species (Table S6). Sixteen genes with positive selection sites were identified (Table S7). These genes included one subunit of the photosystem II gene (*psbH*), two genes for cytochrome b/f complex subunit proteins (*petD* and *petL*), two genes for ribosome large subunit proteins (*rpl22* and *rpl32*), two DNA-dependent RNA polymerase genes (*rpoC1* and *rpoC2*), three genes for ribosome small subunit proteins (*rps12*, *rps15*, and *rps16*), and *accD*, *ccsA*, *rbcL*, *ycf1*, *ycf2*, and *ycf4* genes. Interestingly, the *ycf1* gene possesses 13 and 15 positive selective sites, followed by *accD* (8, 10), *rbcL* (4, 7), *ycf2* (2, 3), *rpoC1* (2, 4), *rpoC2* (1, 2), *rpl22* (1, 2), *rps16* (1, 2), *rpl32* (1, 1), *rps12* (1, 1), *ccsA* (0, 2), *petD* (0, 1), *petL* (0, 1), *psbH* (0, 1), and *ycf4* (0, 1). What is more, the likelihood ratio tests (LRTs) of variables under different models were compared in the site-specific models, M0 vs. M3, M1 vs. M2 and M7 vs. M8, in order to support the sites under positive selection ($p < 0.01$) (Table S7).

2.6. Phylogenetic Relationship

In this study, the maximum likelihood (ML) analysis suggested that *C. appendiculata* and the congeneric *C. davidii* clustered into the Epidendroideae subfamily clade with high bootstrap support, and that *E. mairei* and *P. japonica* clustered into Orchidoideae subfamily (Figure 8). Interestingly, five subfamilies of Orchidaceae, Apostasioideae, Cypripedioideae, Epidendroideae, Orchidoideae, and Vanilloideae have a nested evolutionary relationship in the ML tree. Meanwhile, *C. appendiculata* was closely-related to *C. striata* var. *vreelandii*, *C. davidii*, and *C. triplicate*, which formed a small evolutionary clade with a high bootstrap. *P. japonica* and *Habenaria pantlingiana* had a relatively-closer affinity in the Orchidoideae subfamily.

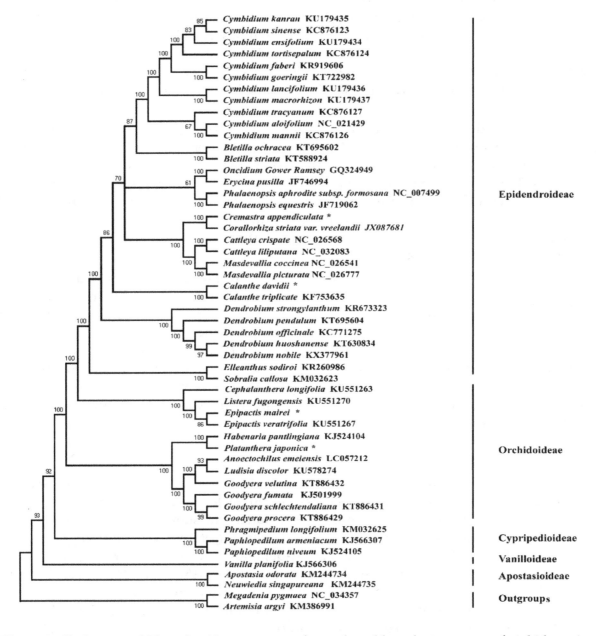

Figure 8. Cladogram of 54 nucleotide sequences of complete chloroplast genomes of orchid species based on the GTRGAMMA model with maximum likelihood (ML) analysis. * The newly generated chloroplast genomes of orchid species.

3. Discussion

3.1. Sequence Variation

In this study, we first determined the whole chloroplast genomes of four orchid species. The cp genome size of *C. davidii* was shorter than that of others, which might be the result of the expansion and contraction of the border positions between the IR and SC regions [21–23]. In addition, the GC contents of the LSC and SSC regions in all the orchid species were much lower than those of the IR regions, which possibly resulted from four rRNA genes (*rrn16*, *rrn23*, *rrn4.5*, and *rrn5*) sequences in the IR regions. In addition, we identified some obvious differences in the protein-coding genes for the orchid chloroplast genomes, despite that the cp genomes of land plants are generally considered to be highly conserved [34]. Interestingly, there were no *ndhC*, *ndhI*, and *ndhK* genes in *C. appendiculata*. In *P. japonica* and *E. mairei*, the *ndhI* gene was lost. Previous studies also found that some orchid species had lost the *ndh* gene, which encodes the subunits of the nicotinamide-adenine dinucleotid (NADH) dehydrogenase-like complex proteins [35–37]. The loss of this gene might have hindered cyclic electron flow around photosystem I and affected the plant photosynthesizing [35–38]. In addition, some studies suggested that the different Orchidaceae species harbored a variable loss or retention of *ndh* genes [35]. For example, *Cymbidium* has the *ndhE*, *ndhJ*, and *ndhC* genes [39], and *Oncidium* has the *ndhB* gene [40]. Nevertheless, the mechanisms that underlie the variable loss or retention of *ndh* genes in orchids remain unclear [11,41].

In addition, we identified 233 SSRs in four orchid species (*C. appendiculata*, *C. davidii*, *E. mairei*, and *P. japonica*); 77.68% of SSRs were distributed in the IGS and intron regions. Generally, microsatellites consist of 1–6 nucleotide repeat units, which are widely distributed across the entire genome and have a great influence on genome recombination and rearrangement [42,43]. The large amount of SSRs also have been identified in *Forsythia suspense* [44], *Dendrobium nobile*, *Dendrobium officinale*, and so on [45]. The majority of these SSRs consisted of mono- and di-nucleotide repeats. Tri-, tetra-, and penta-nucleotide repeat sequences were detected at a much lower frequency in these orchid species and in other organisms [46,47].

Meanwhile, our analyses revealed that the mutational hotspots among orchid genera were highly variable. A diversity of IR contraction and expansion, along with the high level of mutational hotspots, revealed that Orchidaceae had experienced a complex evolution process. Interestingly, in the orchid species, the two IR regions were less divergent than the LSC and SSC regions. Five regions within the non-coding regions (*ndhB* intron, *ccsA-ndhD*, *rpl33-rps18*, *ndhE-ndhG*, and *ndhF-rpl32*) and six regions within the coding regions (*rps16*, *ndhC*, *rpl32*, *ndhI*, *ndhK*, and *ndhF*) showed greater levels of variations (percentage of variability >80% and 50%, respectively). These regions can be used as potential DNA barcodes for the further study of phylogenetic relationships, species identification, and population genetics.

3.2. Adaptive Evolution

We used the site-specific model (seqtype = 1, model = 0, NSsites = 0, 1, 2, 3, 7, 8), one of the codon substitution models, to estimate the selection pressure [48]. Sixteen genes with positive selection sites were identified in these orchid species. These genes included one subunit of the photosystem II gene (*psbH*), two genes for cytochrome b/f complex subunit proteins (*petD* and *petL*), two genes for ribosome large subunit proteins (*rpl22* and *rpl32*), two DNA-dependent RNA polymerase genes (*rpoC1* and *rpoC2*), three genes for ribosome small subunit proteins (*rps12*, *rps15*, and *rps16*), and *accD*, *ccsA*, *rbcL*, *ycf1*, *ycf2*, and *ycf4* genes. We found that the genes with positive selection sites can be divided into four categories: Subunits of photosystem (*psbH* and *ycf4*), subunits of cytochrome (*petD*, *petL*, and *ccsA*), subunits of ribosome (*rpl22*, *rpl32*, *rps12*, *rps15*, and *rps16*) and others (*rpoC1*, *rpoC2*, *accD*, *rbcL*, *ycf1*, and *ycf2*). The plastid *accD* gene, which encodes the β-carboxyl transferase subunit of acetyl-CoA carboxylase, is an essential and required component for plant leaf development [49–53]. In this study, 10 positively-selected sites were identified in *accD* genes for orchid species, suggesting that the *accD* gene played a possible pivotal role in the adaptive evolution of orchids. What is more, the *ycf1* gene is also essential for almost all plant lineages [5,54], except for Gramineae, which lost the *ycf1* gene in its cp genomes [55]. Additionally, *ycf1* is one of the largest cp genes, encoding a component of the chloroplast's inner envelope membrane protein translocon [56]. This gene, which is also highly variable in terms of phylogenetic information at the level of species, has also been shown to be subject to positive selection with 15 sites, as has also been identified in many plant lineages [57–59]. In addition, we found that the *rbcL* gene possessed seven sites under positive selection in orchid species. Generally, *rbcL* is the gene for the Rubisco large subunit protein, and as the result of enzymatic activity of Rubisco, which is an important component as a modulator of photosynthetic electron transport [60,61]. Current research has revealed that positive selection of the *rbcL* gene in land plants may be a common phenomenon [62]. Additionally, the *rbcL* gene is also widely used in the phylogenetic analysis of land plants [63]. In conclusion, these results showed that multiple factors, several of them interconnected (positive selection, heterogeneity environments), have possibly contributed to orchid diversification and adaptation. For example, some positively-selected sites that were identified (e.g., *rbcL*, *ycf1*, and *accD*) were associated in a significant manner with environment adaptation, including factors such as temperature, light, humidity, and atmosphere [49]. Additionally, epiphytism in orchid species is a key innovation which should help generate and maintain high levels of plant diversity. On the other hand, the tropical distributions of orchid species might have increased the rates of speciation relative to those outside of the tropics as a result of more stable climates (e.g., the lack of glaciation and suitable temperatures), the greater habitat area, and together, this possibly provided a greater opportunity for the co-evolution of plants and their mutualists, and for greater adaptation [49,58,59].

3.3. Phylogenetic Relationship

In this study, the maximum likelihood (ML) tree obtained high bootstrap support values, which had 33 nodes with 100% bootstrap support, with 36 of the 46 nodes having values ≥95%. The phylogenetic analyses based on complete cp genomes, suggested that five subfamilies of Orchidaceae (Apostasioideae, Cypripedioideae, Epidendroideae, Orchidoideae, and Vanilloideae) have a nested evolutionary relationship (Figure 8). Apostasioideae is the earliest diverging subfamily of orchids. Some recent molecular studies have shown that the five subfamilies had formed their respective five monophylies [11,41,49]. The generic relationships of the five subfamilies found in our analyses are basically congruent with those of recent studies. However, our finding that Orchidoideae is a nested subfamily is different from the studies of Kim et al. [41] and Givnish et al. [49]. They reconstructed ML trees using the concatenated coding sequences of plastid genes, resulting in large amounts of missing data for these orchid taxa. In this study, we sampled these newly orchid species (*C. longifolia*, *L. fugongensis*, *E. mairei*, and *E. veratrifolia*) to construct a more widespread Orchidaceae phylogenetic tree, through which we obtained the different species relationships. However, some molecular phylogenetic studies, to date, have failed to identify the placement of Cypripedioideae and Vanilloideae [8,25,64,65]. Recently, Givnish et al. [49] and Niu et al. [11] reconstructed ML trees from 39 and 53 orchids species, respectively, using the sequence variations in 75 genes and 67 genes from the plastid genomes. Their results showed that five orchid subfamilies clustered into the five monophyletic clades: Epidendroideae–Orchidoideae–Cypripedioideae–Vanilloideae–Apostasioideae. However, the current study found that *C. appendiculata* and the congeneric *C. davidii* clustered into the Epidendroideae subfamily clade, and that *E. mairei* and *P. japonica* clustered into the Orchidoideae subfamily. These results were largely consistent with traditional morphological evidence [66–68]. However, the inconsistent phylogenetic relationships for the five subfamilies may be due to the differences in the collected samples used in different studies [11,49,64,65], which need to be further explored by sampling a much higher number of orchid species.

4. Materials and Methods

4.1. Plant Material, DNA Extraction, Library Construction, and Sequencing

Fresh leaf tissues were collected from *Cremastra appendiculata*, *Calanthe davidii*, *Epipactis mairei*, and *Platanthera japonica* in the Qinling Mountains, Shaanxi Province, China. The leaves were cleaned and preserved in a −80 °C refrigerator at Northwest University. The voucher specimens of the four species materials were deposited into the Northwest University Herbarium (NUH). The total genomic DNA was isolated using the modified Cetyltrimethyl Ammonium Bromide (CTAB) method [69], which added the EDTA buffer (Amresco, Washington, DC, USA) (1.0 mol/L Tris-HCl (Amresco, Washington, DC, USA) (pH 8.0), 0.5 mol/L EDTA-Na$_2$ (Amresco, Washington, DC, USA), 5.0 mol/L NaCl) solution before isolating the high-quality DNA with the CTAB solution (1.0 mol/L Tris-HCl (pH 8.0), 0.5 mol/L EDTA-Na$_2$, 2% CTAB). Following this, we constructed a pair-end (PE) library with 350 bp insert size fragments using TruSeq DNA sample preparation kits (Sangon, Shanghai, China). Subsequently, we sequenced at least 4.5 GB of clean data for each orchid species. The detailed next-generation sequencing was conducted on the Illumina Hiseq 2500 platform by Sangon Biotech (Shanghai, China).

4.2. Chloroplast Genome Assembly and Annotation

First, we used the software, NGSQCToolkit v2.3.3 [70], to trim the low-quality reads. After removing the low-quality sequences, the clean reads were assembled using MIRA v4.0.2 [70] and MITObim v1.8 [71] with the cp genome of a closely-related species, *Dendrobium nobile* (KX377961), as reference. The programs, DOGMA (http://dogma.ccbb.utexas.edu/) [72] and Geneious v8.0.2 [73] were used to annotate the chloroplast genome. Finally, we obtained four high-quality, complete chloroplast genome sequences. The Circle maps of the four species were drawn using OGDRAW v1.1 [74].

4.3. Repeat Sequence Analyses

The REPuter program (Available online: https://bibiserv.cebitec.uni-bielefeld.de/reputer/manual.html) was used to identify repeats, including forward, reverse, palindrome, and complement sequences. The maximum computed repeats and the minimal repeat size were limited to 50 and 30, respectively, with a Hamming distance equal to 3 [75]. The tandem repeats finder welcome page (http://tandem.bu.edu/trf/trf.html) was used to identify tandem repeats sequences [76]. The alignment parameters match, mismatch, and indels, were 2, 7, and 7, respectively. The minimum alignment score to report repeat, maximum period size and maximum TR array Size (bp, millions) are limited to 80, 500, and 2, respectively. A Perl script MISA (MIcroSAtellite identification tool, http://pgrc.ipk-gatersleben.de/misa/) was used to search for simple sequence repeat (SSR or microsatellite) loci in the chloroplast genomes [77]. Tandem repeats of 1–6 nucleotides were viewed as microsatellites. The minimum number of repeats were set to 10, 5, 4, 3, 3, and 3, for mono-, di-, tri-, tetra-, penta-, and hexanucleotides, respectively.

4.4. Genome Structure and Mutational Hotspot

In order to compare the genome structures and divergence hotspots in a broad manner, we used 16 cp genomes (available in Genbank https://www.ncbi.nlm.nih.gov/) representing each orchid genus, and added the four newly-sequenced ones (Table 3). The boundaries between the IR and SC regions of *C. appendiculata*, *C. davidii*, *E. mairei*, and *P. japonica* and other 16 sequences were compared and analyzed. Meanwhile, the whole-genome alignment of the chloroplast genomes of the 20 species were performed and plotted using the mVISTA program [32]. Following this, we selected the regions within non-coding and coding regions that had a greater level of variation (percentage of variability >80% and 50%, respectively) as mutational hotspots. The formula was as follows: percentage of variable = (number of nucleotide substitutions + the number of indels)/(the length of aligned sites−the length of indels + the number of indels) × 100%.

Table 3. List of taxa sampled in the study and species accessions numbers (GenBank).

Subfamily	Species	Accession Number
Orchidaceae subfamily. Epidendroideae	Cattleya crispata	KP168671
	Cremastra appendiculata	MG925366
	Masdevallia coccinea	KP205432
	Erycina pusilla	JF746994
	Phalaenopsis equestris	JF719062
	Bletilla ochracea	KT695602
	Cymbidium faberi	KR919606
	Calanthe davidii	MG925365
	Dendrobium strongylanthum	KR673323
	Elleanthus sodiroi	KR260986
	Sobralia callosa	KM032623
Orchidaceae subfamily. Orchidoideae	Epipactis mairei	MG925367
	Cephalanthera longifolia	KU551263
	Listera fugongensis	KU551270
	Platanthera japonica	MG925368
	Habenaria pantlingiana	KJ524104
	Goodyera velutina	KT886432
	Anoectochilus emeiensis	LC057212
	Ludisia discolor	KU578274
Orchidaceae subfamily. Vanilloideae	Vanilla planifolia	KJ566306

4.5. Gene Selective Pressure Analysis

The codon substitution models in the Codeml program, PAML3.15 [46] were used for calculating the non-synonymous (dN) and synonymous (dS) substitution rates, along with their ratios (ω = dN/dS). We analyzed all CDS gene regions, except *ndh*, due to there being too many losses there. These unique CDS gene sequences were separately extracted and aligned using Geneious v8.0.2 [73]. A maximum likelihood phylogenetic tree was built based on the complete cp genomes of the 20 species using RAxML [78]. We used the site-specific model (seqtype = 1, model = 0, NSsites = 0, 1, 2, 3, 7, 8) to estimate the selection pressure [79]. This model allowed the ω ratio to vary among sites, with a fixed ω ratio in all the branches. Comparing the site-specific model, M1 (nearly neutral) vs. M2 (positive selection), M7 (β) vs. M8 (β and ω) and M0 (one-ratio) vs. M3 (discrete) were calculated in order to detect positive selection [79].

4.6. Phylogenetic Analysis

In order to deeply detect the evolutionary relationship of the Orchidaceae family, 50 available complete chloroplast genomes were downloaded from the NCBI Organelle Genome Resources database (Table S1). In addition, *Artemisia argyi* and *Megadenia pygmaea* were used as outgroups. In total, 54 nucleotide sequences of complete chloroplast genomes were aligned using MAFFT [73]; the detailed parameters were as follows: 200 PAM/K = 2 and 1.53 gap open penalty [73]. The choice of the best nucleotide sequence substitution model (GTRGAMMA model) was determined using the Modeltest v3.7 [80]. We constructed a maximum likelihood phylogenetic tree based on these complete plastomes using MAGA7 [34] with 1000 bootstrap replicates under the GTRGAMMA model [80].

Acknowledgments: This research was co-supported by the National Natural Science Foundation of China (31470400) and Shaanxi Provincial Key Laboratory Project of Department of Education (grant no. 17JS135).

Author Contributions: Zhong-Hu Li conceived the work. Wan-Lin Dong and Ruo-Nan Wang performed the experiments. Zhong-Hu Li, Wan-Lin Dong, Ruo-Nan Wang, Na-Yao Zhang, Wei-Bing Fan and Min-Feng Fang contributed materials/analysis tools. Wan-Lin Dong and Zhong-Hu Li wrote the paper. Zhong-Hu Li and Wan-Lin Dong revised the paper. All authors approved the final paper.

References

1. Chase, M.W.; Cameron, K.M.; Barrett, R.L.; Freudenstein, J.V. DNA data and Orchidaceae systematics: A new phylogenetic classification. In *Orchid Conservation*; Dixon, K.W., Kell, S.P., Barrett, R.L., Cribb, P.J., Eds.; Natural History Publications: Kota Kinabalu, Malaysia, 2003; pp. 69–89.

2. Dressler, R.L. *The Orchids: Natural History and Classification*; Harvard University Press: Cambridge, MA, USA, 1990.

3. Chase, M.W. Classification of Orchidaceae in the age of DNA data. *Curtis's Bot. Mag.* **2005**, *22*, 2–7. [CrossRef]

4. Luo, J.; Hou, B.W.; Niu, Z.T.; Liu, W.; Xue, Q.Y.; Ding, X.Y. Comparative chloroplast genomes of photosynthetic orchids: Insights into evolution of the Orchidaceae and development of molecular markers for phylogenetic applications. *PLoS ONE* **2014**, *9*, e99016. [CrossRef] [PubMed]

5. Raubeson, L.A.; Jansen, R.K. Chloroplast genomes of plants. In *Plant Diversity and Evolution: Genotypic and Phenotypic Variation in Higher Plants*; Henry, R.J., Ed.; CAB International: Wallingford, UK, 2005; pp. 45–68.

6. Van den Berg, C.; Goldman, D.H.; Freudenstein, J.V.; Pridgeon, A.M.; Cameron, K.M.; Chase, M.W. An overview of the phylogenetic relationships within Epidendroideae inferred from multiple DNA regions and recircumscription of Epidendreae and Arethuseae (Orchidaceae). *Am. J. Bot.* **2005**, *92*, 13–24. [CrossRef] [PubMed]

7. Mendonca, M.P.; Lins, L.V. *Revisao das Listas das Especies da Flora eda Fauna Ameaçadas de Extincao do Estado de Minas Gerais*; Fundacao Biodiversitas: BeloHorizonte, Brazil, 2007.

8. Cameron, K.M.; Chase, M.W.; Whitten, W.M.; Kores, P.J.; Jarrell, D.C.; Albert, V.A.; Yukawa, T.; Hills, H.G.; Goldman, D.H. A phylogenetic analysis of the Orchidaceae: Evidence from *rbcL* nucleotide. *Am. J. Bot.* **1999**, *86*, 8–24. [CrossRef]

9. Van den Berg, C.; Higgins, W.E.; Dressler, R.L.; Whitten, W.M.; Soto-Arenas, M.; Chase, M.W. A phylogenetic study of *Laeliinae* (Orchidaceae) based on combined nuclear and plastid DNA sequences. *Ann. Bot.* **2009**, *104*, 17–30. [CrossRef] [PubMed]

10. Verlynde, S.; D'Haese, C.A.; Plunkett, G.M.; Simo-Droissart, M.; Edwards, M.; Droissart, V.; Stévart, T. Molecular phylogeny of the genus *Bolusiella* (Orchidaceae, Angraecinae). *Plant Syst. Evol.* **2017**, *304*, 269–279. [CrossRef]

11. Niu, Z.T.; Xue, Q.Y.; Zhu, S.Y.; Sun, J.; Liu, W.; Ding, X.Y. The complete plastome sequences of four orchid species: Insights into the evolution of the Orchidaceae and the utility of plastomic mutational hotspots. *Front. Plant. Sci.* **2017**, *8*, 1–11. [CrossRef] [PubMed]

12. Bateman, R.M.; Rudall, P.J. Clarified relationship between *Dactylorhiza viridis* and *Dactylorhiza iberica* renders obsolete the former genus *Coeloglossum* (Orchidaceae: Orchidinae). *Kew Bull.* **2018**, *73*, 1–17. [CrossRef]

13. Wilson, M.; Frank, G.S.; Lou, J.; Pridgeon, A.M.; Vieira-Uribe, S.; Karremans, A.P. Phylogenetic analysis of *Andinia* (Pleurothallidinae; Orchidaceae) and a systematic re-circumscription of the genus. *Phytotaxa* **2017**, *295*, 101–131. [CrossRef]

14. Neuhaus, H.E.; Emes, M.J. Nonphotosynthetic metabolism in plastids. *Annu. Rev. Plant Biol.* **2000**, *51*, 111–140. [CrossRef] [PubMed]

15. Rodríguezezpeleta, N.; Brinkmann, H.; Burey, S.C.; Roure, B.; Burger, G.; Löffelhardt, W.; Bohnert, H.J.; Philippe, H.; Lang, B.F. Monophyly of primary photosynthetic eukaryotes: Green plants, red algae, and glaucophytes. *Curr. Biol.* **2005**, *15*, 1325–1330. [CrossRef] [PubMed]

16. Yap, J.Y.; Rohner, T.; Greenfield, A.; Van Der Merwe, M.; McPherson, H.; Glenn, W.; Kornfeld, G.; Marendy, E.; Pan, A.Y.; Wilton, A.; et al. Complete chloroplast genome of the Wollemi pine (*Wollemia nobilis*): Structure and evolution. *PLoS ONE* **2015**, *106*, 126–128. [CrossRef] [PubMed]

17. Wicke, S.; Schneeweiss, G.M.; Müller, K.F.; Quandt, D. The evolution of the plastid chromosome in land plants: Gene content, gene order, gene function. *Plant Mol. Biol.* **2011**, *76*, 273–297. [CrossRef] [PubMed]

18. Dong, W.; Liu, H.; Xu, C.; Zuo, Y.J.; Chen, Z.J.; Zhou, S.L. A chloroplast genomic strategy for designing taxon specific DNA mini-barcodes: A case study on ginsengs. *BMC Genet.* **2014**, *15*, 138–145. [CrossRef] [PubMed]

19. Zhang, Y.; Li, L.; Yan, T.L.; Liu, Q. Complete chloroplast genome sequences of Praxelis (*Eupatorium catarium* Veldkamp), an important invasive species. *Gene* **2014**, *549*, 58–69. [CrossRef] [PubMed]

20. Xu, C.; Dong, W.P.; Li, W.Q.; Lu, Y.Z.; Xie, X.M.; Jin, X.B.; Shi, J.; He, K.; Suo, Z. Comparative analysis of six *Lagerstroemia* complete chloroplast genomes. *Front. Plant Sci.* **2017**, *8*, 15–26. [CrossRef] [PubMed]

21. Jer, J.D. Plastid chromosomes: Structure and evolution. In *Cell Culture and Somatic Cell Genetics in Plants, the Molecular Biology of Plastids 7A*; Vasil, I.K., Bogorad, L., Eds.; Academic Press: San Diego, CA, USA, 1991; pp. 5–53.

22. Bendich, A.J. Circular chloroplast chromosomes: The grand illusion. *Plant Cell* **2004**, *16*, 1661–1666. [CrossRef] [PubMed]

23. Jansen, R.K.; Raubeson, L.A.; Boore, J.L.; Pamphilis, C.W.; Chumley, T.W.; Haberle, R.C.; Wyman, S.K.; Alverson, A.J.; Peery, R.; Herman, S.J.; et al. Methods for obtaining and analyzing whole chloroplast genome sequences. *Methods Enzymol.* **2015**, *395*, 348–384.

24. Burke, S.V.; Grennan, C.P.; Duvall, M.R. Plastome sequences of two new world bamboos-*Arundinaria gigantea* and *Cryptochloa strictiflora* (Poaceae)-extend phylogenomic understanding of Bambusoideae. *Am. J. Bot.* **2012**, *99*, 1951–1961. [CrossRef] [PubMed]

25. Civan, P.; Foster, P.G.; Embley, M.T.; Séneca, A.; Cox, C.J. Analyses of charophyte chloroplast genomes help characterize the ancestral chloroplast genome of land plants. *Genome Biol. Evol.* **2014**, *6*, 897–911. [CrossRef] [PubMed]

26. Guo, W.; Grewe, F.; Cobo-Clark, A.; Fan, W.; Duan, Z.; Adams, R.P.; Schwarzbach, A.E.; Mower, J.P. Predominant and substoichiometric isomers of the plastid genome coexist within *Juniperus* plants and have shifted multiple times during cupressophyte evolution. *Genome Biol. Evol.* **2014**, *6*, 580–590. [CrossRef] [PubMed]

27. Ruhfel, B.R.; Gitzendanner, M.A.; Soltis, P.S.; Soltis, D.E.; Burleigh, J.G. From algae to angiosperms-inferring the phylogeny of green plants (Viridiplantae) from 360 plastid genomes. *BMC Evol. Biol.* **2014**, *14*, 385–399. [CrossRef] [PubMed]

28. Moore, M.J.; Bell, C.D.; Soltis, P.S.; Soltis, D.E. Using plastid genome-scale data to resolve enigmatic relation-ships among basal angiosperms. *Proc. Natl. Acad. Sci. USA* **2007**, *104*, 19363–19368. [CrossRef] [PubMed]

29. Huang, H.; Shi, C.; Liu, Y.; Mao, S.Y.; Gao, L.Z. Thirteen Camellia chloroplast genome sequences determined by high-throughput sequencing: Genome structure and phylogenetic relationships. *BMC Evol. Biol.* **2014**, *14*, 4302–4315. [CrossRef] [PubMed]

30. Walker, J.F.; Zanis, M.J.; Emery, N.C. Comparative analysis of complete chloroplast genome sequence and inversion variation in *Lasthenia burkei* (Madieae, Asteraceae). *Am. J. Bot.* **2014**, *101*, 722–729. [CrossRef] [PubMed]

31. Oldenburg, D.J.; Bendich, A.J. The linear plastid chromosomes of maize: Terminal sequences, structures, and implications for DNA replication. *Curr. Genet.* **2015**, *62*, 1–12. [CrossRef] [PubMed]

32. Frazer, K.A.; Pachter, L.; Poliakov, A.; Rubin, E.M.; Dubchak, I. VISTA: Computational tools for comparative genomics. *Nucleic Acids Res.* **2004**, *32*, 273–279. [CrossRef] [PubMed]

33. Doose, D.; Grand, C.; Lesire, C. MAUVE Runtime: A Component-Based Middleware to Reconfigure Software Architectures in Real-Time. In Proceedings of the IEEE International Conference on Robotic Computing (IRC), Taichung, Taiwan, 10–12 April 2017; pp. 208–211.

34. Kumar, S.; Stecher, G.; Tamura, K. Mega7: Molecular evolutionary genetics analysis version 7.0 for bigger datasets. *Mol. Biol. Evol.* **2016**, *33*, 1870–1874. [CrossRef] [PubMed]

35. Clegg, M.T.; Gaut, B.S.; Learn, G.H., Jr.; Morton, B.R. Rates and patterns of chloroplast DNA evolution. *Proc. Natl. Acad. Sci. USA* **1994**, *91*, 6795–6801. [CrossRef] [PubMed]

36. Delannoy, E.; Fujii, S.; Colas des Francs-Small, C.; Brundrett, M.S. Rampant gene loss in the underground orchid *Rhizanthella gardneri* highlights evolutionary constraints on plastid genomes. *Mol. Biol. Evol.* **2011**, *28*, 2077–2086. [CrossRef] [PubMed]

37. Logacheva, M.D.; Schelkunov, M.I.; Penin, A.A. Sequencing and analysis of plastid genome in mycoheterotrophic orchid Neottia nidus-avis. *Genome Biol. Evol.* **2011**, *3*, 1296–1303. [CrossRef] [PubMed]

38. Barrett, C.F.; Davis, J.I. The plastid genome of the mycoheterotrophic *Corallorhiza striata* (Orchidaceae) is in the relatively early stages of degradation. *Am. J. Bot.* **2012**, *99*, 1513–1523. [CrossRef] [PubMed]

39. Yang, J.B.; Tang, M.; Li, H.T.; Zhang, Z.R.; Li, D.Z. Complete chloroplast genome of the genus *Cymbidium*: Lights into the species identification, phylogenetic implications and population genetic analyses. *BMC Evol. Biol.* **2013**, *13*, 84. [CrossRef] [PubMed]

40. Wu, F.H.; Chan, M.T.; Liao, D.C.; Hsu, C.T.; Lee, Y.W.; Daniell, H.; Duvall, M.R.; Lin, C.S. Complete chloroplast genome of *Oncidium Gower Ramsey* and evaluation of molecular markers for identification and breeding in Oncidiinae. *BMC Plant Biol.* **2010**, *10*, 68. [CrossRef] [PubMed]

41. Kim, H.T.; Kim, J.S.; Moore, M.J.; Neubig, K.M.; Williams, N.H.; Whitten, W.M.; Kim, J.H. Seven new complete plastome sequences reveal rampant independent loss of the *ndh* gene family across orchids and associatedinstability of the inverted repeat/small single-copy region boundaries. *PLoS ONE* **2015**, *10*, e0142215.

42. Ni, L.; Zhao, Z.; Xu, H.; Chen, S.; Dorje, G. The complete chloroplast genome of *Gentiana straminea* (Gentianaceae), an endemic species to the Sino-Himalayan subregion. *Gene* **2016**, *577*, 281–288. [CrossRef] [PubMed]

43. Ni, L.; Zhao, Z.; Xu, H.; Chen, S.; Dorje, G. Chloroplast genome structures in *Gentiana* (Gentianaceae), based on three medicinal alpine plants used in Tibetan herbal medicine. *Curr. Genet.* **2017**, *63*, 241–252. [CrossRef] [PubMed]

44. Wang, W.B.; Yu, H.; Wang, J.H.; Lei, W.J.; Gao, J.H.; Qiu, X.P.; Wang, J.S. The complete chloroplast genome sequences of the medicinal plant *Forsythia suspensa* (Oleaceae). *Int. J. Mol. Sci.* **2017**, *18*, 2288. [CrossRef] [PubMed]

45. Kanga, J.Y.; Lua, J.J.; Qiua, S.; Chen, Z.; Liu, J.J.; Wang, H.Z. *Dendrobium* SSR markers play a good role in genetic diversity and phylogenetic analysis of Orchidaceae species. *Sci. Hortic.* **2015**, *183*, 160–166. [CrossRef]

46. Song, Y.; Wang, S.; Ding, Y.; Xu, J.; Li, M.F.; Zhu, S.; Chen, N. Chloroplast genomic resource of Paris for species discrimination. *Sci. Rep.* **2017**, *7*, 3427–3434. [CrossRef] [PubMed]

47. Yu, X.Q.; Drew, B.T.; Yang, J.B.; Gao, L.M.; Li, D.Z. Comparative chloroplast genomes of eleven *Schima* (Theaceae) species: Insights into DNA barcoding and phylogeny. *PLoS ONE* **2017**, *12*, e0178026. [CrossRef] [PubMed]

48. Wang, B.; Jiang, B.; Zhou, Y.; Su, Y.; Wang, T. Higher substitution rates and lower dN/dS for the plastid genes in Gnetales than other gymnosperms. *Biochem. Syst. Ecol.* **2015**, *59*, 278–287. [CrossRef]

49. Givnish, T.J.; Spalink, D.; Ames, M.; Lyon, S.P.; Hunter, S.J.; Zuluaga, A.; Iles, W.J.; Clements, M.A.; Arroyo, M.T.; Leebens-Mack, J.; et al. Orchid phylogenomics and multiple drivers of their extraordinary diversification. *Proc. Biol. Sci. B* **2015**, *282*, 2108–2111. [CrossRef] [PubMed]

50. Sasaki, Y.; Hakamada, K.; Suama, Y.; Nagano, Y.; Furusawa, I.; Matsuno, R. Chloroplast encoded protein as a subunit of acetyl-COA carboxylase in pea plant. *J. Biol. Chem.* **1993**, *268*, 25118–25123. [PubMed]

51. Konishi, T.; Shinohara, K.; Yamada, K.; Sasaki, Y. Acetyl-CoA carboxylase in higher plants: Most plants other than Gramineae have both the prokaryotic and the eukaryotic forms of this enzyme. *Plant Cell Physiol.* **1996**, *37*, 117–122. [CrossRef] [PubMed]

52. Kode, V.; Mudd, E.A.; Iamtham, S.; Day, A. The tobacco plastid *accD* gene is essential and is required for leaf development. *Plant J.* **2005**, *44*, 237–244. [CrossRef] [PubMed]

53. Nakkaew, A.; Chotigeat, W.; Eksomtramage, T.; Phongdara, A. Cloning and expression of a plastid-encoded subunit, beta-carboxyltransferase gene (*accD*) and a nuclear-encoded subunit, biotin carboxylase of acetyl-CoA carboxylase from oil palm (*Elaeis guineensis* Jacq.). *Plant Sci.* **2008**, *175*, 497–504. [CrossRef]

54. Drescher, A.; Ruf, S.; Calsa, T.J.; Carrer, H.; Bock, R. The two largest chloroplast genome-encoded open reading frames of higher plants are essential genes. *Plant J.* **2000**, *22*, 97–104. [CrossRef] [PubMed]

55. Asano, T.; Tsudzuki, T.; Takahashi, S.; Shimada, H.; Kadowaki, K. Complete nucleotide sequence of the sugarcane (*Saccharum officinarum*) chloroplast genome: A comparative analysis of four monocot chloroplast genomes. *DNA Res.* **2004**, *11*, 93–99. [CrossRef] [PubMed]

56. Kikuchi, S.; Bédard, J.; Hirano, M.; Hirabayashi, Y.; Oishi, M.; Imai, M.; Takase, M.; Ide, T.; Nakai, M. Uncovering the protein translocon at the chloroplast inner envelope membrane. *Science* **2013**, *339*, 571–574. [CrossRef] [PubMed]

57. Greiner, S.; Wang, X.; Herrmann, R.G.; Rauwolf, U.; Mayer, K.; Haberer, G.; Meurer, J. The complete nucleotide sequences of the 5 genetically distinct plastid genomes of *Oenothera*, subsection *Oenothera*: II. A microevolutionary view using bioinformatics and formal genetic data. *Mol. Biol. Evol.* **2008**, *25*, 2019–2030. [CrossRef] [PubMed]

58. Carbonell-Caballero, J.; Alonso, R.; Ibañez, V.; Terol, J.; Talon, M.; Dopazo, J. A phylogenetic analysis of 34 chloroplast genomes elucidates the relationships between wild and domestic species within the genus *Citrus*. *Mol. Biol. Evol.* **2015**, *32*, 2015–2035. [CrossRef] [PubMed]

59. Hu, S.; Sablok, G.; Wang, B.; Qu, D.; Barbaro, E.; Viola, R.; Li, M.; Varotto, C. Plastome organization and evolution of chloroplast genes in *Cardamine* species adapted to contrasting habitats. *BMC Genom.* **2015**, *16*, 1. [CrossRef] [PubMed]

60. Allahverdiyeva, Y.; Mamedov, F.; Mäenpää, P.; Vass, I.; Aro, E.M. Modulation of photosynthetic electron transport in the absence of terminal electron acceptors: Characterization of the *rbcL* deletion mutant of tobacco. *Biochim. Biophys. Acta Bioenerg.* **2005**, *1709*, 69–83. [CrossRef] [PubMed]

61. Piot, A.; Hackel, J.; Christin, P.A.; Besnard, G. One-third of the plastid genes evolved under positive selection in PACMAD grasses. *Planta* **2018**, *247*, 255–266. [CrossRef] [PubMed]

62. Kapralov, M.V.; Filatov, D.A. Widespread positive selection in the photosynthetic Rubisco enzyme. *BMC Evol. Biol.* **2007**, *7*, 73–82. [CrossRef] [PubMed]

63. Ivanova, Z.; Sablok, G.; Daskalova, E.; Zahmanova, G.; Apostolova, E.; Yahubyan, G.; Baev, V. Chloroplast genome analysis of resurrection tertiary relict *Haberlea rhodopensis* highlights genes important for desiccation stress response. *Front. Plant Sci.* **2017**, *8*, 1–15. [CrossRef] [PubMed]

64. Shi, Y.; Yang, L.F.; Yang, Z.Y.; Ji, Y.H. The complete chloroplast genome of *Pleione bulbocodioides* (Orchidaceae). *Conserv. Genet. Resour.* **2017**, 1–5. [CrossRef]

65. Górniak, M.; Paun, O.; Chase, M.W. Phylogenetic relationships with Orchidaceae based on a low-copy nuclear-coding gene, Xdh: Congruence with organellar and nuclear ribosomal DNA results. *Mol. Phylogenet. Evol.* **2010**, *56*, 784–795. [CrossRef] [PubMed]

66. Lin, C.S.; Chen, J.J.; Huang, Y.T.; Chan, M.T.; Daniell, H.; Chang, W.J.; Hsu, C.T.; Liao, D.C.; Wu, F.H.; Lin, S.Y.; et al. The location and translocation of *ndh* genes of chloroplast origin in the Orchidaceae family. *Sci. Rep.* **2015**, *5*, 9040. [CrossRef] [PubMed]

67. Rasmussen, F.N. The families of the monocotyledones—Structure, evolution and taxonomy. In *Orchids*; Dahlgren, R., Cliford, H.T., Yeo, P.F., Eds.; Springer: Berlin/Heidelberg, Germany, 1985; pp. 249–274.

68. Szlachetko, D.L. Systema orchidalium. *Fragm. Florist. Geobot. Pol.* **1995**, *3*, 1–152.

69. Doyle, J.J. A rapid DNA isolation procedure from small quantities of fresh leaf tissues. *Phytochem. Bull.* **1987**, *19*, 11–15.

70. Chevreux, B.; Pfisterer, T.; Drescher, B.; Driesel, A.J.; Müller, W.E.; Wetter, T.; Suhai, S. Using the mira EST assembler for reliable and automated mRNA transcript assembly and SNP detection in sequenced ESTs. *Genome Res.* **2004**, *14*, 1147–1159. [CrossRef] [PubMed]

71. Hahn, C.; Bachmann, L.; Chevreux, B. Reconstructing mitochondrial genomes directly from genomic next-generation sequencing reads-abaiting and iterative mapping approach. *Nucleic Acids Res.* **2013**, *41*, e129. [CrossRef] [PubMed]

72. Wyman, S.K.; Jansen, R.K.; Boore, J.L. Automatic annotation of organellar genomes with DOGMA. *Bioinformatics* **2004**, *20*, 3252–3255. [CrossRef] [PubMed]

73. Kearse, M.; Moir, R.; Wilson, A.; Steven, S.H.; Matthew, C.; Shane, S.; Simon, B.; Alex, C.; Markowitz, S.; Duran, C.; et al. Geneious Basic: An integrated and extendable desktop software platform for the organization and analysis of sequence data. *Bioinformatics* **2012**, *12*, 1647–1649. [CrossRef] [PubMed]

74. Lohse, M.; Drechsel, O.; Kahlau, S.; Bock, R. Organellar Genome DRAW-a suite of tools for generating physical maps of plastid and mitochondrial genomes and visualizing expression data sets. *Nucleic Acids Res.* **2013**, *41*, 575–581. [CrossRef] [PubMed]

75. Kurtz, S.; Schleiermacher, C. REPuter: Fast computation of maximal repeats incomplete genomes. *Bioinformatics* **1999**, *15*, 426–427. [CrossRef] [PubMed]

76. Benson, G. Tandem repeats finder: A program to analyze DNA sequences. *Nucleic Acids Res.* **1999**, *27*, 573. [CrossRef] [PubMed]

77. Thiel, T.; Michalek, W.; Varshney, R.K.; Graner, A. Exploiting EST databases for the development and characterization of gene-derived SSR-markers in barley (*Hordeum vulgare* L.). *Theor. Appl. Genet.* **2003**, *106*, 411–422. [CrossRef] [PubMed]

78. Stamatakis, A. RAxML-VI-HPC: Maximum likelihood-based phylogenetic analyses with thousands of taxa and mixed models. *Bioinformatics* **2006**, *22*, 2688–2690. [CrossRef] [PubMed]

79. Yang, Z.; Nielsen, R. Codon-substitution models for detecting molecular adaptation at individual sites along specific lineages. *Mol. Biol. Evol.* **2002**, *19*, 908–917. [CrossRef] [PubMed]

80. Posada, D.; Crandall, K.A. Modeltest: Testing the model of DNA substitution. *Bioinformatics* **1998**, *14*, 817–818. [CrossRef] [PubMed]

8

Stable Membrane-Association of mRNAs in Etiolated, Greening and Mature Plastids

Julia Legen and Christian Schmitz-Linneweber *

Institut of Biology, Department of the Life Sciences, Humboldt-Universität Berlin, Philippstraße 11–13, Grüne Amöbe, 10115 Berlin, Germany; legenjul@staff.hu-berlin.de
* Correspondence: smitzlic@rz.hu-berlin.de

Abstract: Chloroplast genes are transcribed as polycistronic precursor RNAs that give rise to a multitude of processing products down to monocistronic forms. Translation of these mRNAs is realized by bacterial type 70S ribosomes. A larger fraction of these ribosomes is attached to chloroplast membranes. This study analyzed transcriptome-wide distribution of plastid mRNAs between soluble and membrane fractions of purified plastids using microarray analyses and validating RNA gel blot hybridizations. To determine the impact of light on mRNA localization, we used etioplasts, greening plastids and mature chloroplasts from *Zea mays* as a source for membrane and soluble extracts. The results show that the three plastid types display an almost identical distribution of RNAs between the two organellar fractions, which is confirmed by quantitative RNA gel blot analyses. Furthermore, they reveal that different RNAs processed from polycistronic precursors show transcript-autonomous distribution between stroma and membrane fractions. Disruption of ribosomes leads to release of mRNAs from membranes, demonstrating that attachment is likely a direct consequence of translation. We conclude that plastid mRNA distribution is a stable feature of different plastid types, setting up rapid chloroplast translation in any plastid type.

Keywords: chloroplast; etioplast; membrane; organelle; ribosome; RNA processing; translation; *Zea mays*

1. Introduction

Subcellular RNA localization is an important means of gene regulation in eukaryotic organisms [1], but also in bacteria [2]. In plants, insights into intracellular RNA localization are limited and have been mostly studied for the localization of mRNAs for endosperm storage proteins [3]. Plant cells do not only contain RNAs in the nucleo-cytosolic compartments, but also within the two DNA-containing organelles, the mitochondria and chloroplasts. Chloroplast genomes of angiosperms code for about 80 proteins. In addition, there are genes for tRNAs and rRNAs; rRNAs are 70S ribosomes that translate all chloroplast mRNAs.

The sub-organellar localization of chloroplast RNAs has rarely been addressed, but there are a few notable exceptions. In the green algae *Chlamydomonas reinhardtii*, several chloroplast mRNAs (e.g., *psbC*, *psbD*, *rbcL*) have been demonstrated to be enriched around the chloroplast pyrenoid [4,5] by fluorescence in situ hybridization (FISH). The pyrenoid seems to be a dominant site of mRNA translation and thylakoid membrane complex assembly in this algae [6]. A large fraction of all

Chlamydomonas chloroplast 70S ribosomes is attached to thylakoid membranes, as evidenced both by electron microscopy as well as cell fractionation [7–9]. This suggested early on that chloroplast mRNAs are translated in close association with chloroplast membranes. Since many chloroplast mRNAs encode integral membrane proteins, it was assumed that translating ribosomes get trapped on membranes because the nascent protein chain is translated directly into the thylakoid membrane. This is supported by the finding that membrane attachment of RNAs depends on active translation [10] and is proportional to overall protein synthesis within thylakoid membranes [11]. These findings imply that the membrane-bound ribosomes are associated with mRNAs. This is supported by in vitro experiments, where washed thylakoids are capable of synthesizing proteins when supplemented with soluble factors [12,13]. Recently, ribosome-protected mRNA fragments in the soluble fraction of chloroplasts were compared with ribosome-protected mRNA fragments in the membrane fraction. This demonstrated transcriptome-wide association of mRNAs coding for membraneous proteins to chloroplast membranes [14]. Noteworthy, mRNA association via nascent-chain bound ribosomes with membranes is astonishingly stable and occurs even in the absence of chlorophyll production [15].

Chloroplast mRNAs are produced as polycistronic units and are subsequently processed [16]. RNA splicing, endonucleolytic cleavage and exonucleolytic trimming and decay gives rise to complex transcript patterns. For most transcript units, the biological importance of this complexity is unclear. Some processing events have been shown to be required for producing translatable messages (e.g., [17]). In maize chloroplast development, a correlation of the accumulation of monocistronic transcript isoforms with their translational efficiency has been noted [18]. In general, however, both unprocessed polycistronic as well as processed monocistronic forms are supposed to be translated [14,19]. How prevalent differential translation of longer or shorter transcript isoforms is in the chloroplast transcriptome is largely unknown.

We set out to determine which mRNAs are associated with chloroplast membranes versus free mRNAs in the chloroplast stroma fraction. Our results suggests that individual transcripts from a single operon show preference for either stromal or membrane association. In rare cases such associations are subject to developmental change, but in general, membrane-enrichment of plastid RNAs is constant during chloroplast development.

2. Results and Discussion

2.1. Preparation of Sub-Organellar Fractions Highly Enriched in Membrane and Stroma Marker Proteins

We prepared membranes and stroma extracts from isolated plastids from 9-day-old *Zea mays* seedlings that were either grown in the dark (i.e., are etiolated), from greening seedlings (grown in the dark and then irradiated for 16 h, then kept in the dark for an 8 h night and harvested next morning after an additional 75 min of light exposure), or from plants grown under standard long-day conditions. The three tissue types represent etioplasts, greening plastids, and mature chloroplasts, respectively. The fractions were analyzed immunologically for the presence of the marker proteins PetD and RbcL (Figure 1). PetD is a membrane protein of the cytochrome b_6f complex; RbcL is a stromal protein. There is a slight signal for PetD in stroma from etiolated plants, while the protein is beyond the detection limit in the two other stroma preparations. A small fraction of RbcL can be found within each membrane fraction, indicating a slight contamination of our membrane preparation with stromal proteins. Nevertheless, this immunological analysis shows successful enrichment for both, membrane and stroma proteins, in the two fractions.

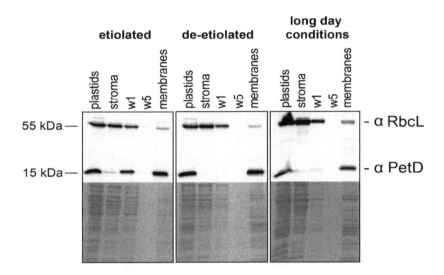

Figure 1. Enrichment of marker proteins in chloroplast membrane and stroma fractions. Western blot analysis was performed from each type of tissue using RbcL and PetD antisera as markers for stroma and membrane, respectively. RNAs from these fractions were used for microarray analysis. A seven hundredth of each sample was analyzed by sodium dodecyl sulfate polyacrylamide gel electrophoresis (SDS-PAGE). The membrane fraction was washed five times prior to RNA extraction. Aliquots from the supernatants of the first and the last wash were analyzed here as well (W1 and W5).

2.2. Global Association of Chloroplast mRNAs for Membrane Proteins with Chloroplast and Etioplast Membranes

To investigate the association of RNA with membranes on a global level, we labelled RNA from stroma and membrane fractions with fluorescent dyes and competitively hybridized them to a whole-genome tiling array of the maize chloroplast genome [20]. We opted for stroma preparations versus total chloroplast RNA as a control sample, since at least some RNA degradation is expected to occur during membrane/stroma preparation, but is largely absent from total chloroplast RNA preparations. The ratio of membrane-signal over stroma signal (the membrane enrichment value (MEV)) was calculated for all probes. MEVs from four biological replicates were normalized, combined and plotted against the genome position of the probes (Figure 2a). A gene ontology (GO) analysis was carried out for the top 10% probes with the highest MEVs in comparison to all probes on the array (Figure 2b; full data set can be found in Table S1). This revealed that probes representing components of photosystem I, II and the cytochrome b_6f complex are overrepresented in the top 10% MEV probes. Several operons containing genes for these complexes are represented by multiple probes of similar MEVs, for example the *psbB–petB* and the *psbD/C* operons, which is further support for the validity of the assay. Membrane-based translation of mRNAs coding for photosynthetic proteins has been demonstrated before by ribosome profiling [14]. By contrast, probes for ribosomal proteins, the NADH dehydrogenase (NDH) complex and the ATP synthase are underrepresented in the top 10% MEV probes. Probes for *rbcL*, tRNAs, the plastid RNA polymerase and uncharacterized open reading frames (ORFs) are not found at all among the top 10% MEV probes. This distribution of MEVs was found for each of the three tissues analyzed, i.e., noteworthy also for etiolated tissue (Figure 2 and Table S2).

While most RNAs within the top 10% membrane-enriched fraction are encoding proteins with photosynthetic functions, there are also interesting exceptions, namely *rps14* and *cemA*. Both are represented by multiple probes within the top 10% MEV probes. *rps14* is the only mRNA coding for a ribosomal protein that shows such high enrichment in membranes (Table S2). It is suggested that the *cemA* gene product is required for the import of inorganic carbon across the envelope membrane, but it was not found to be attached to membranes in recent ribosome profiling analysis [14]. Both *rps14* and *cemA* are part of main peaks in this analysis, represented by multiple probes corresponding to the *psaA/B* and the *cemA–petA* operon (Figure 2a). *rps14* is accumulating as part of a tricistronic transcript together with *psaA* and *psaB* and *cemA* is located within an operon adjacent to *petA*. *petA* and *psaA/B* encode core membrane proteins of photosystem I and the *cytochrome b$_6$f* complex. Thus, *rps14* and *cemA* are likely drawn to the membrane as part of a larger transcript tethered to the thylakoid membrane via translated *psaA/B* and *petA*, respectively.

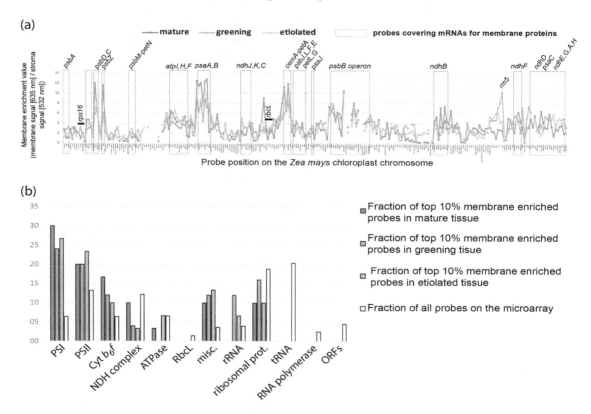

Figure 2. Microarray analysis of RNAs enriched at chloroplast membranes. (**a**) Membrane-enrichment of RNA along the chloroplast chromosome. Isolated chloroplast from maize seedlings grown at three different light regimes (leading to mature, greening, and etiolated plastids, respectively) were processed into stroma and membrane fractions. Two µg of RNA isolated from each fraction were differentially labeled and hybridized to a microarray representing the maize chloroplast genome in a tiling fashion. The ratio of membrane versus stroma signal was plotted against the genomic position on the *Zea mays* chloroplast genome (acc. NC_001666). The graphs shown represent four biological replicates. Normalization between conditions is based on the sum of membrane enrichment values (MEVs) of all probes for each condition. Probes corresponding to mRNAs coding for known membrane proteins are highlighted by dashed boxes. Selected soluble RNAs are labelled as well. The data underlying this chart are deposited in Table S1; (**b**) Gene ontology (GO) term analysis of membrane-enriched RNAs in the chloroplast. The probes showing the top 10 percent MEVs binned into 12 functional categories (PSI = photosystem I; PSII = photosystem II; Prot. = proteins; ORFs = open reading frames of unknown function; misc. = miscellaneous). Probes covering more than one category were counted in each relevant bin. The numbers of probes within each bin are expressed as the fraction of the total number of probes for each condition (in %). This is compared to the distribution of all probes on the array across the bins (open bars). Probes for photosystems I and II are over-represented in the top 10% MEV probes of all three tissue types.

2.3. Membrane Association of Chloroplast RNAs via Ribosomes

The suggested mechanism of membrane association of chloroplast mRNAs is via actively translating ribosomes [7,9,14]. To test directly for the importance of ribosomes for mRNA-membrane association, we treated membranes with ethylenediaminetetraacetic acid (EDTA) to separate the small and large subunit of the ribosome and thus disrupt membrane and mRNA association. RNA retrieved from supernatants and membranes after EDTA treatment was analyzed by RNA gel blot hybridization using a probe specific for the *psbD* gene (Figure 3a). We observed that in mock-treated samples, the signal for *psbD* mRNAs is almost completely retained in the membrane fractions. Only shorter transcripts of 1.0 to 1.5 kb are found in the stroma of etiolated and greening tissue. These transcripts likely represent monocistronic *psbD* (the *psbD* coding region is 1.0 kb in length) and are in excess of longer precursors. Possibly, such abundant shorter transcripts are not all engaged with ribosomes and can thus partially remain in the stroma. Larger transcripts encompassing *psbD*, *psbC* and further genes are only found in membrane fractions prior to EDTA treatment, which could be explained by their more efficient translation relative to the smaller monocistronic RNAs. After EDTA treatment, no *psbD* remains detectable in membranes and the RNA is instead found in the supernatant. This is further evidence for ribosomes being required for membrane attachment of chloroplast mRNAs. Still, it cannot be ruled out that larger ribonucleoprotein (RNP) granules that would co-enrich with membranes despite a physical interaction, dissolve due to EDTA treatment and are thus released in the supernatant. Such granules have been described in *Chlamydomonas reinhardtii* [21].

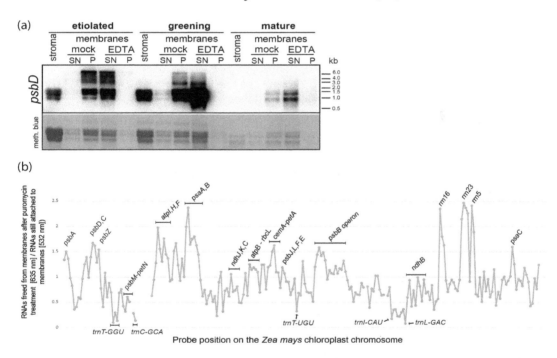

Figure 3. RNAs are associated to chloroplast membranes via translating ribosomes. (**a**) Disruption of ribosomes by ethylenediaminetetraacetic acid (EDTA) releases *psbD* mRNAs into the soluble fraction. Membrane fractions from the same number of plastids from three different tissue types were either treated with EDTA in order to separate the large and small subunit of the ribosome or were treated with buffer (mock). Membranes were spun down after treatment (P) and the supernatant was collected as well (S). RNA was extracted from all fractions and fractionated on 1.2% agarose gel and analyzed by hybridization to a radiolabeled *psbD* probe (see Table S4 for primer sequences). As a quality control, the rRNAs were stained with methylene blue; (**b**) Release of mRNAs from chloroplast membranes after puromycin treatment. Membranes were isolated from mature chloroplasts that were purified from four-week-old plants grown under long-day conditions. Membranes were treated with puromycin to release nascent chains and ribosomes from the mRNAs and thus secondarily free mRNAs from the membrane. The membrane was washed multiple times. RNA from the supernatant after puromycin treatment and from the washed membranes was analyzed by a whole genome tiling array of the maize chloroplast genome. The ratio of freed RNA over membrane-bound RNA was calculated for each probe and plotted against genome position. Selected probes are labelled by the gene-names they represent. The isoacceptor tRNAs are named by their respective codon (GGU, GCA, UGU, CAU and GAC, respectively).

Therefore, to validate the hypothesis that nascent chains emerging from ribosomes tether polysomes to chloroplast membranes, we treated mature membranes with puromycin. Puromycin leads to premature peptide chain termination and thus releases ribosomes from mRNAs. We analyzed RNAs released into the supernatant versus membrane-bound RNAs on our microarray after puromycin treatment. As expected, puromycin led to a massive release of ribosomal RNAs from membranes, demonstrating that the antibiotic treatment was successful (Figure 3b). This rRNA release was accompanied by a loss of mRNAs from membranes. mRNAs that showed high MEV values in our previous analysis were particularly prone to detachment from membranes after puromycin treatment. This was for example seen for the *psbD/C* and *psaA/B* operons (Figures 2a and 3b). A similar set of RNAs was found when we treated membranes with EDTA and analyzed the RNAs on microarrays (Figure S1). By contrast, tRNAs were released to a far lesser extent (Figure 3b). tRNAs and also the known stromal mRNAs *rbcL* were found in membranes even after extensive washing. This could be caused by the protection of RNAs within vesicles that are forming from membranes during preparation. In sum, association of mRNAs with membranes was mediated globally by translating ribosomes tethered to membranes via nascent peptide chains.

2.4. tRNAs, mRNAs for Stromal Proteins and Short mRNAs for Membrane Proteins Are Enriched in the Soluble Fraction of Plastids

As expected, many tRNAs show low MEVs, in fact the lowest of the entire analysis (Table S2). 79% of all genes represented in the 10% least enriched RNAs across all three tissue types are tRNA genes. An additional 16% of the low MEVs correspond to ORFs of unknown function (Table S2). In order to visualize, which mRNAs show low MEVs, we removed probes for tRNAs and rRNAs from the analysis (Table S3). Expectedly, mRNAs like *rbcL*, *clpP*, *infA* and *matK* that code for soluble proteins do show low MEVs. Similarly, many mRNAs for ribosomal proteins display low MEVs. In addition, few probes for photosynthetic mRNAs coding for membrane proteins have low MEVs as well. Of the latter, almost all represent short reading frames, including *petN* (29 codons), *petL* (31), *petG* (37), *psaI* (36), *psaJ* (42), *psbM* (34) and *psbJ* (40). As has recently been discussed for ribosome profiling, such short mRNAs are fully translated before a sufficiently long peptide chain emerges from the large subunit's exit channel that would enable contact with the membrane [14]. An exception within this group is *psbA*, which encodes the 40 kDa membrane protein D1 and is 354 codons long. Several probes for *psbA* appear at the bottom of the MEV list for all three tissue types. mRNA for *psbA* is accumulating in large amounts in chloroplasts, mostly untranslated [22,23], and is believed to be required as a reservoir during stress-induced demand for D1 production [24]. Thus, the untranslated excess of *psbA* mRNA likely leads to the observed low membrane enrichment values. In sum, our analysis demonstrates on a transcriptome-wide scale that mRNAs coding for soluble proteins localize to the stroma. In addition, short mRNAs for membrane proteins as well as the *psbA* mRNA are predominantly stroma-localized as well. Paralleling what we observed for membrane-associated RNAs, the list of soluble RNAs is similar between all three tissue types examined.

2.5. Etioplasts Display Membrane Enrichment of RNAs Similar to Chloroplasts

Etioplasts differ dramatically in their internal structure from structures known in developing and mature chloroplasts, showing a para-crystalline prolammelar body (PLB) instead of thylakoid membranes. Thus, it came as a surprise that the membrane enrichment curves of plastid RNAs are similar between the three plastid types analyzed (Figure 2a). To uncover potential differences, we did a pairwise comparison of the MEV datasets (Figure S2). This corroborated the large similarities found in the operon (Figure 2a) and GO term analysis (Figure 2b). Intriguingly, greatest similarities are not found between the two green plastid types, but between etioplasts and plastids from greening tissue (Pearson correlation $r_{E/DE} = 0.83$; Figure S2). By contrast, transcript enrichment at membranes in mature chloroplasts are less similar to the two other plastid preparations ($r_{DE/G} = 0.66$; $r_{E/G} = 0.72$, Figure S2). Possibly, the short time span of irradiation has not sufficed to erase the etioplast transcript pattern in greening plastids, thus making these two plastid types more similar on the RNA level and thus also membrane enrichment level than either is to mature chloroplasts. By contrast, the translational activity will have much increased in the greening plastids, but has not led to stronger membrane attachment in our dataset nor in a previous analysis of selected mRNAs [25].

Overall, the most pronounced difference was found for the aforementioned ORFs. The region encompassing ORF46 to ORF137 shows a higher membrane enrichment in etiolated and greening versus mature tissue (Figure S2). The ORFs in question are remnants of a long reading frame of unknown function called *ycf2* [26]. Their low spot count, indicative of a low expression level, suggests that further, more sensitive assays will be necessary to ascertain the differential membrane association seen here.

Why should there be mRNA association with membranes in etioplasts at all? 95% of all prolammelar body (PLB) proteins are protochlorophyllid oxidoreductases (PORA and PORB), which are encoded in the nuclear genome [27–31]. In addition, a number of proteins involved in the light reactions of photosynthesis has been identified. This includes the soluble chloroplast-encoded proteins AtpA, AtpB, RbcL, as well as membranous, chloroplast-encoded proteins like the subunits of the cytochrome b6f complex, and the PSII subunit PsbE [28,30,32–34]. Thus, etioplast do have an active translation system. We conclude, that the observed association of many mRNAs coding for membrane proteins with etioplast membranes is reflecting this translational activity (Figures 2 and 3; [14,35]). It remains however unclear, to which membrane type these RNAs are bound, since we did not differentiate between etioplast envelope membranes, the PLB and lamellar prothylakoid membranes present in various cell types of etiolated maize leafs [36]. Also, we cannot exclude that the green light we utilize during etioplast preparation affects translational activity and thus mRNA localization.

In contrast to the examples listed above, many other of the chloroplast-encoded proteins of the thylakoid membrane complexes have not been detected in etioplasts. These appear to be only accumulating upon illumination. For example, the photosynthetic membrane proteins D1, PsbB, PsbC or the PSI reaction core proteins are not or only barely detectable in etioplasts, but have been shown to exhibit a massive increase in expression during photomorphogenesis [25,33,37]. We show here on a genome-wide scale that this increased expression is not reflected by a comparable increase in association of the corresponding mRNAs to chloroplast membranes in greening versus etiolated tissue. Rather, membrane association of plastid mRNAs appears stable across the three plastid types analyzed here. Even *psbA* and *psbB*, which show induction of mRNA levels in greening tissue, exhibit only a mild change in the ratio of membrane-bound versus stroma-bound RNAs. This is in line with previous analyses that show only a slight increase in *psbA* membrane attachment during de-etiolation in barley [38].

2.6. Quantitative Analysis of Membrane-Bound and Stromal mRNAs on a "Per-Chloroplast" Base Uncovers Transcript-Autonomous Localization Patterns

Our array analysis indicates association of RNAs with chloroplast membranes, but the resolution of this approach is limited by the number and length of probes on the microarray. Thus, the microarray analysis misses the great variety and complexity of transcripts representing individual genes within each operon. To understand which RNA species are bound to membranes, we performed quantitative RNA gel blot hybridization assays. We again extracted chloroplasts from the three maize cultures and followed the same procedure for RNA preparation from membrane and stroma fractions as for the microarray approach, with the exception that we counted chloroplasts prior to lysis and cell fractionation. This way, we could normalize the RNA amount from stroma and membrane preparations to chloroplast numbers and thus represent the quantitative differences in RNA abundance between the two fractions more accurately than when total RNA amounts are used for normalization. We utilized probes for eight different chloroplast genes that can be divided into two groups: a group of genes coding for membrane-proteins; and a second group including mRNAs coding for soluble proteins and RNAs that are not translated. *atpH, petA, psbB, psbA* represent the former group, while *psaC, rps16, rbcL*, *rrn5* and *rrn16* represent the latter. The RNAs we recover from the chloroplast fractions are intact: there are no bands found that would not also be seen in total RNA preparations (Figure 4).

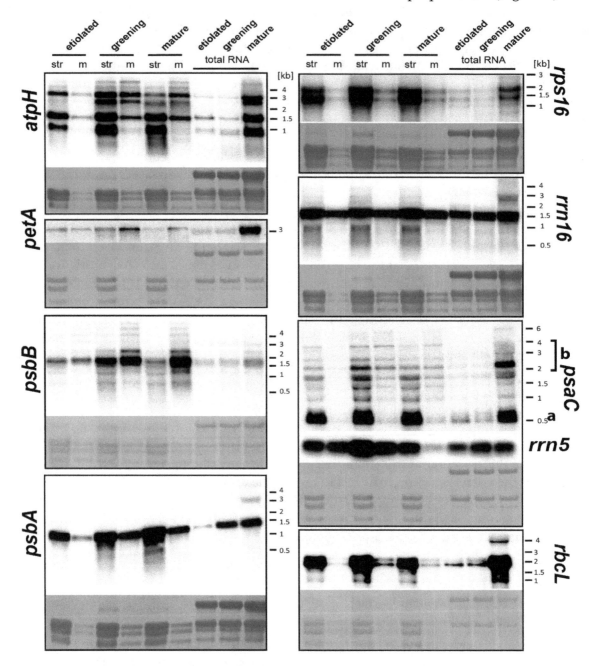

Figure 4. Analysis of transcript accumulation in membrane and stroma preparations on a per-chloroplast base. Equal aliquots of RNA from membrane and stroma fractions of purified plastids from three different tissues were extracted from the same number of chloroplasts. In addition, five micrograms of total leaf RNA was analyzed as well. The RNAs were fractionated on 1.2% agarose gel and analyzed by hybridization to radiolabeled probes for the plastid RNAs indicated (see Table S4 for primer sequences). As a quality control, the rRNAs were stained with methylene blue. Str = RNA from stroma fractions; m = RNA from membrane fractions.

2.6.1. The Analysis of Equal Amounts of Total Cellular RNA from Different Tissues Does Not Accurately Reflect per Organelle RNA Levels for Most Plastid Genes

All RNAs analyzed are found to be expressed in etiolated, greening and mature plastids. In etiolated tissue, there is overall less RNA per plastid than in the two photosynthetic tissues examined

(Figure 4). This is most pronounced for RNAs related to the light reactions of photosynthesis (*psaC*, *petA*, *psbB*, *atpH*, *psbA*) and to a lesser extent also for *rbcL* and genes of the genetic apparatus (*rps16*, *rrn5*). 16S rRNA accumulation remains constant across all three conditions. For a number of RNAs, peak accumulation levels are found for greening tissue rather than for mature tissue (e.g., *rrn5*, *rbcL*, *atpH*). Hence, in line with previous analyses, there is a global dependency of chloroplast RNA accumulation on light signals and chloroplast stage. Whether the observed smaller differences in the extent and timing of induction are biologically relevant, is unclear at present.

When considering total RNA preparations (lanes loaded with equal total cellular RNA amounts), there is dramatically less chloroplast RNAs in etiolated and greening tissue than in mature tissue for most genes analyzed. Apparently, the amount of RNA per chloroplast remains constant during initial chloroplast development while, later, the ratio of plastid RNA to cytoplasmic ribosomal RNAs increases towards mature chloroplasts. Again, only the 16S ribosomal RNAs displays an approximately equal accumulation in total RNA across the three conditions. In addition, *psbA* and *rrn5* display induction of RNA levels already in total RNA preparations of greening tissue. Apparently, RNA accumulation is reacting faster to irradiation for these two genes than for the other genes assessed here. A rapid response of *psbA* on all levels of gene expression has indeed been described in various plant species [33,39]. In general, our analysis demonstrates that changes observed for chloroplast-encoded transcripts in different tissues or conditions are pronouncedly dependent on whether normalization is made according to total cellular or total chloroplast RNAs.

2.6.2. mRNAs Are Enriched at Chloroplast Membranes in a Transcript-Autonomous Fashion

The RNA gel blot hybridizations corroborate microarray findings. For example, RNAs for *psbB* and *petA* that display the most pronounced membrane-bias in the northern analysis do also display high MEVs in microarrays. RNAs found enriched in the stroma in RNA gel blots like *rbcL* and *rps16* have comparatively low MEVs (compare these four mRNAs in Figures 2a and 4). In line with this, mRNAs for *atpH*, *psbA*, and the extrinsic membrane proteins *psaC*, display low membrane enrichment values in the microarray analysis and are also found predominantly in stroma fractions in RNA gel blots. This demonstrates a qualitative congruence of the two assays.

For all probes, we noted that the qualitative transcript patterns are similar between stroma and membrane fractions: All transcripts found in the stroma can also be detected in the membrane fraction for any of the probes used (Figure 4). Intriguingly, different transcripts detected with the same probe display distinct distributions between membrane and stroma. A striking example for this is the *psaC* probing, which shows several transcripts highly enriched in the stroma fraction, and in addition larger transcripts that are approximately equally distributed. Most prominent is a band of less than 0.5 kb (labelled "a" in Figure 4), which corresponds to the monocistronic form of *psaC* [40] and is found almost exclusively in the stroma fraction. This RNA species has been described as the major translatable RNA species in tobacco [40,41]. Given that PsaC is translated as a soluble protein and only later assembled into PSI [42,43], it is not surprising that its translated mRNA is found in the stroma. Longer transcripts are however found in membrane fractions (summarized as "b" in Figure 4). These are co-transcripts including cistrons encoding membraneous subunits of the NDH complex (e.g., NdhD). These likely draw the *psaC*-cistron to membranes in the process of their translation. Similarly, monocistronic *atpH* transcripts are found almost exclusively in the stroma, while the long, polycistronic precursors are co-fractioning also considerably with membranes. An additional cistrons in the longer transcripts is represented by *atpF*, which is translated on membranes [14]. Thus, the membrane-attachment of polycistronic *atpH–atpF* transcripts can be explained by the affinity of the *atpF* mRNA for membranes. As a general trend, different transcripts from complex operons can show differential association with membranes in chloroplasts.

2.6.3. Differential Membrane-Association of rRNA Species

If a number of chloroplast mRNAs display membrane-association and if this association is mediated by ribosomes, then we should find rRNA attached to chloroplast membranes as well. Indeed, we do observe that 16S rRNA and 5S rRNA co-fractionate with membranes (Figure 4). The unexpected finding is, that different rRNA species show different MEVs. While the ratios of membrane versus stroma signal of 16S and 23S rRNAs remain constant across the three conditions analyzed, 5S rRNA displays a strong decline of membrane association in tissue with mature chloroplasts, which is also reflected by our microarray analysis (see *rrn5* in Figure 2a). 5S rRNA and 23S rRNA are part of the large ribosomal subunit. Thus, the finding that a subpopulation of 5S rRNA localizes to the stroma in mature chloroplasts suggests that it does so independently of the ribosome. Alternatively, a recently discovered antisense RNA to 5S may cause this discrepancy [44]. We cannot distinguish between sense and antisense transcripts with our microarray nor with the double-stranded probe used to detect 5S rRNA on RNA gel blots. Whether the tissue-dependent localization of 5S rRNA is of functional significance remains to be determined.

2.6.4. Constant Association of Plastid RNAs with Chloroplast Membranes Suggests Altered Translational Rates during Chloroplast Greening rather than Increased Accumulation or Improved Recruitment of mRNAs to Membranes

Consistent with the microarray analyses, the RNA gel blot hybridizations also shows a striking similarity in RNA distribution between membranes and stroma in the three tissue types (Figure 4). mRNAs for the soluble proteins *rbcL*, *rps16* and *psaC* display almost identical preferential localization to the stroma fractions with only minor amounts within the membrane preparation in all three tissues. A noticeable, albeit small shift of RNAs towards membrane pools in mature chloroplasts was observed for *psbB* and *petA*, and a minor shift also for *psbA*. *psbA* translation is massively induced after irradiation [33,39], far exceeding the minor increase in membrane attachment seen here. Given that in general, translational induction in the light is known to be massive for many mRNAs [33], we have to assume that RNA tethering to membranes does not directly mirror translational activity. This is reflecting previous analyses in *Chlamydomonas* that found EF-Tu and *psaA/B* mRNAs are strongly and consistently attached to chloroplast membranes throughout dark-light cycles, while translation occurs only in the light [45].

Similarly, the barley *psaA/B* mRNA associates with chloroplast membranes in etioplasts without noticeable *psaA/B* translation [25,38]. It was suggested that there is active repression of *psaA/B* translation in the dark since translation off of isolated etioplast membranes could be initiated in vitro after adding a translation-competent extract [25]. This is supported by studies in pea that report that membrane-bound, non-polysomal RNAs are moving within 8 min into polysomes after illumination, suggesting that the initial association with membranes is either mediated by one or few ribosomes or by other means [10]. Increasing translation from low levels to high levels would not necessarily lead to an increased detection of mRNAs in membranes, since at least in theory, it would not matter for our approach if few or many ribosomes tether an mRNA to thylakoid membranes. Quantitative ribosome profiling experiments in etioplasts versus chloroplasts could solve this problem.

In sum, our analyses demonstrate that etioplasts are poised for a rapid translational answer to light signals on the level of global mRNA attachment to chloroplast membranes. This likely goes hand in hand with parallel processes that stage the necessary production of mRNAs and the corresponding expression factors in etioplasts [30,32]. It remains to be shown, how important constant attachment of mRNAs to membranes is for rapid photomorphogenesis versus other processes, like the provision of RNA binding proteins for translational initiation.

3. Materials and Methods

3.1. Plant Material

Wild type maize seedlings were grown on soil for 8 days with 16 h light/8 h dark cycle at 25 °C. Etiolated tissue was grown without light for 8 days and harvested directly. Greening tissue was generated by growing plants for 6 days without light and the light-exposure on the seventh day for 16 h and subsequently return into darkness for 8 h. On the 8th day the tissue was exposed for a further hour and 15 min prior to harvest, which was carried out in safety green light. Tissue with mature chloroplasts was generated by growing plants for 8 days under a standard light regime (16 h light/8 h dark). On the 8th day, mature seedlings were exposed for one additional hour to light after which the tissue was harvested. During plastid isolation from etiolated and greening tissue, safety green light was used, so additional exposure to photomorphogenic light was eliminated. All plants were grown at 26 °C.

3.2. Extraction of Stroma and Membrane Fractions

Intact plastids from three different tissue types were isolated from 8-day-old seedlings [46] with the following modifications. After Percoll gradient purification, washed plastids were resuspended in resuspension buffer (50 mM HEPES/KOH pH 8, 330 mM sorbitol) by gentle agitation at 4 °C. Dilution of extracted plastids was used for estimation of the number of plastids per microliter microscopically. Plastids from etiolated, deetiolated and green tissue was adjusted to the same number of plastids per microliter. The plastid pellet was resuspended in hypotonic polysomal buffer [19] without detergents, containing heparin, chloramphenicol, cycloheximide, β-mercatoethanol and protease inhibitor cocktail without EDTA. Chloramphenicol stabilizes ribosomes on mRNAs. Lysis of plastids was performed by passing the extract for 40 to 50 times through a 0.5 mm × 25 mm syringe in a hypotonic buffer designed to keep polysomes intact [19,47]. Membranes and stroma were separated by centrifugation at $40,000 \times g$ at 4 °C for 30 min. The membrane pellets were washed five to six times in polysomal buffer, and finally resuspended in the same volume as the stroma volume obtained. From equal volume proportions, RNA was isolated. For RNA gel blot analyses, RNAs were loaded on volume basis.

3.3. EDTA and Puromycin Treatments of Purified Plastid Membranes

Plastid membrane fractions from all three tissue types were obtained from 8-day-old seedlings. Aliquots (300 µL) were incubated at room temperature with 100 mM EDTA (final concentration) for 15 min with gentle shaking. Control reactions were performed without addition of EDTA. After incubation, RNAs released from membrane particles, were separated by centrifugation for 10 min at $21,000 \times g$ at 4 °C. Membrane pellets were resuspended in lysis buffer in same volume as the one of the corresponding supernatant. From each of the fractions RNA was extracted by phenol/chloroform/isoamylalcohol (24:24:1).

For puromycin treatment, membrane fractions were incubated in 500 ng/µL Puromycin in polysome buffer [19] for 15 min at room temperature. Afterwards, membranes were pelleted by centrifugation at $21,000 \times g$ for 10 min at 4 °C. The membrane pellet was washed five times in polysome buffer. RNA was extracted from supernatant (freed RNA) and pellet (membrane-bound RNA) fractions. Two µg of RNA from each fraction were labelled and subjected to microarray analysis see below.

3.4. RNA Gel Blot Analyses

Total RNA was extracted from etiolated, deetiolated and green tissue with TRIzol reagent (Invitrogen, Carlsbad, CA, USA) according to the manufacturer's instructions. For each individual sample a pool of collected leaves was used. Five micrograms of total RNAs from tissue or RNAs extracted from stroma and membrane fractions were separated on an agarose gel containing 1.2%

formaldehyde and transferred to uncharged nylon membranes (Hybond N, GE Healthcare, Little Chalfont, UK). After transfer, the blots were UV cross-linked (150 mJ/cm^2) and then stained with methylene blue to check RNA integrity and loading.

Polymerase chain reaction (PCR) based probes were used for in vitro transcription via T7 polymerase (Thermo Fisher, Waltham, MA, USA) in the presence of 32P-UTP according to the manufacturer's instructions. Primers used are listed in Table S4. For all northern blots, except *rrn16* and *rrn5*, single stranded in vitro transcripts were used for hybridization. The *rrn16* and *rrn5* RNA gel blots were hybridized with a double-stranded PCR probe labelled in presence of 32P-CTP via a Klenow exo- fragment (Thermo Fisher, Waltham, MA, USA) according to the manufacturer's instructions. After pre-incubation in Church buffer (0.5 M sodium phosphate buffer, pH 7.2 and 7% sodium dodecyl sulfate (SDS) for 1 h at 68 °C, hybridization of radiolabelled probe was performed overnight at 68 °C in the same buffer, followed by at least three 15-min washes in $1\times$ SSC; 0.1% SDS, $0.5\times$ SSC/0.1% SDS and $0.2\times$ SSC/0.1% SDS, respectively. Ribosomal RNA related probes, *psbA*, and *rbcL* probes were washed additionally for 15 min at 68 °C with $0.1\times$ SSC/0.1% SDS and $0.05\times$ SSC/0.1% SDS. Signals were detected by autoradiography with the Personal Molecular Imager system (Bio-Rad, Hercules, CA, USA).

3.5. Immunoblot Analyses

Proteins from stroma and membrane fractions were loaded on a volume-per-volume basis. Proteins were separated by sodium dodecyl sulfate polyacrylamide gel electrophoresis (SDS-PAGE) and transferred to Hybond-C Extra Nitrocellulose membranes (GE healthcare). Integrity and loading of protein samples was detected by Ponceau S stain of the membrane prior incubation with antibodies. Antibody hybridization was performed for 1 h in 2% skim milk powder in TBST buffer for primary antibody and secondary horseradish peroxidase coupled antibody in TBST.

3.6. Tiling Microarray Design

Overlapping PCR products covering the whole maize chloroplast genome were generated by using self-made Taq-polymerase and were purified via the QIAquick PCR purification kit (Qiagen, Hilden, Germany). A total of 500 ng of each PCR product was transferred into a 384-well plate, dried and resuspended in 5 µL of 1M betaine in $3\times$ SSC buffer. DNA fragments were transferred on to silanated glass slides (Vantage Silanated Amine Slides; CEL Associates, Los Angeles, CA, USA) using an OmniGrid Accent microarrayer (GeneMachines, San Francisco, CA, USA). Array design as described [20].

3.7. Microarray Hybridisation

Microarrays were cross-linked at 250 mJ/cm^2 in a UV cross linker (GS Gene Linker, Bio-Rad), and blocked with BSA buffer (1% BSA, 0.1% SDS and $5\times$ SSC) for 1 h at 42 °C. The labelled RNAs from stroma and membrane fractions (approximately 15 µL) were mixed, loaded on to array and covered with a cover slip. Hybridisation was performed overnight in a 42 °C water bath in Corning microarray hybridization chambers. Unspecifically bound RNA was washed off the slide by incubation in $1\times$ SSC, $0.2\times$ SSC, and $0.05\times$ SSC for 8 min each on a horizontal shaker at 180 rpm. Slides were dried by centrifugation at 1500 rpm in plate centrifuge and scanned using the ScanArrayGx Plus microarray scanner (Perkin Elmer, Shelton, CT, USA).

3.8. Microarray Analysis

Data from four replicate experiments (this totals 48 spots per probe since we were using 12 replicate spots per probe per array) were filtered against elements with low signal-to-noise ratios, and local background was calculated according to default parameters in Genepix Pro 7.0 software (Molecular Devices, Sunnyvale, CA, USA). Only spots with a signal-to-background ratio >4 and for which 60% of pixels have a F532 fluorescent signal >2 SD above background were chosen for further analysis. Fragments for which fewer than half the number of all spots (i.e., less than 24 spots) per array passed these cutoffs were not used for subsequent analyses and appear as gaps when enrichment ratios are plotted according to chromosomal position. Background-subtracted data were used to calculate the median (median of ratios (membrane RNA = 635 nm: stroma RNA = 532 nm)) as described in [21]. This value is called the membrane enrichment value (MEV). Normalization for Figure 2a and Table S1 was done according to the median (median of ratios (F635/F532)) value for all probes with signal above background on each array.

Acknowledgments: Generous funding by the DFG to CSL (as part of the collaborative research centre TRR175 project A02) is gratefully acknowledged. The antisera against PetD were kindly provided by Alice Barkan (University of Oregon). We thank Alice Barkan for critical reading and comments on the manuscript. The authors have no conflict of interest to declare.

Author Contributions: Julia Legen and Christian Schmitz-Linneweber conceived and designed the experiments; Julia Legen performed the experiments; Julia Legen and Christian Schmitz-Linneweber analyzed the data; Christian Schmitz-Linneweber wrote the paper.

Abbreviations

MEV Membrane Enrichment Value
NDH NADH Dehydrogenase

References

1. Jung, H.; Gkogkas, C.G.; Sonenberg, N.; Holt, C.E. Remote control of gene function by local translation. *Cell* **2014**, *157*, 26–40. [CrossRef] [PubMed]
2. Keiler, K.C. RNA localization in bacteria. *Curr. Opin. Microbiol.* **2011**, *14*, 155–159. [CrossRef] [PubMed]
3. Doroshenk, K.A.; Crofts, A.J.; Morris, R.T.; Wyrick, J.J.; Okita, T.W. Ricerbp: A resource for experimentally identified RNA binding proteins in *Oryza sativa*. *Front. Plant Sci.* **2012**, *3*, 90. [CrossRef] [PubMed]
4. Uniacke, J.; Zerges, W. Chloroplast protein targeting involves localized translation in *Chlamydomonas*. *Proc. Natl. Acad. Sci. USA* **2009**, *106*, 1439–1444. [CrossRef] [PubMed]
5. Uniacke, J.; Zerges, W. Photosystem II assembly and repair are differentially localized in *Chlamydomonas*. *Plant Cell* **2007**, *19*, 3640–3654. [CrossRef] [PubMed]
6. Schottkowski, M.; Peters, M.; Zhan, Y.; Rifai, O.; Zhang, Y.; Zerges, W. Biogenic membranes of the chloroplast in *Chlamydomonas reinhardtii*. *Proc. Natl. Acad. Sci. USA* **2012**, *109*, 19286–19291. [CrossRef] [PubMed]
7. Chua, N.-H.; Blobel, G.; Siekevitz, P.; Palade, G.E. Attachment of chloroplast polysomes to thylakoid membranes in *Chlamydomonas reinhardtii*. *Proc. Natl. Acad. Sci. USA* **1973**, *70*, 1554–1558. [CrossRef] [PubMed]
8. Chua, N.H.; Blobel, G.; Siekevitz, P.; Palade, G.E. Periodic variations in the ratio of free to thylakoid-bound chloroplast ribosomes during the cell cycle of *Chlamydomonas reinhardtii*. *J. Cell Biol.* **1976**, *71*, 497–514. [CrossRef] [PubMed]
9. Margulies, M.M.; Michaels, A. Ribosomes bound to chloroplast membranes in *Chlamydomonas reinhardtii*. *J. Cell Biol.* **1974**, *60*, 65–77. [CrossRef] [PubMed]
10. Fish, L.E.; Jagendorf, A.T. Light-induced increase in the number and activity of ribosomes bound to pea chloroplast thylakoids in vivo. *Plant Physiol.* **1982**, *69*, 814–824. [CrossRef] [PubMed]
11. Hurewitz, J.; Jagendorf, A.T. Further characterization of ribosome binding to thylakoid membranes. *Plant Physiol.* **1987**, *84*, 31–34. [CrossRef] [PubMed]
12. Michaels, A.; Margulies, M.M. Amino acid incorporation into protein by ribosomes bound to chloroplast thylakoid membranes: Formation of discrete products. *Biochim. Biophys. Acta* **1975**, *390*, 352–362. [CrossRef]
13. Alscher-Herman, R.; Jagendorf, A.T.; Grumet, R. Ribosome-thylakoid association in peas: Influence of anoxia. *Plant Physiol.* **1979**, *64*, 232–235. [CrossRef] [PubMed]
14. Zoschke, R.; Barkan, A. Genome-wide analysis of thylakoid-bound ribosomes in maize reveals principles of cotranslational targeting to the thylakoid membrane. *Proc. Natl. Acad. Sci. USA* **2015**, *112*, E1678–E1687. [CrossRef] [PubMed]
15. Zoschke, R.; Chotewutmontri, P.; Barkan, A. Translation and co-translational membrane engagement of plastid-encoded chlorophyll-binding proteins are not influenced by chlorophyll availability in maize. *Front. Plant Sci.* **2017**, *8*, 385. [CrossRef] [PubMed]

16.	Barkan, A. Expression of plastid genes: Organelle-specific elaborations on a prokaryotic scaffold. *Plant Physiol.* **2011**, *155*, 1520–1532. [CrossRef] [PubMed]

17.	Adachi, Y.; Kuroda, H.; Yukawa, Y.; Sugiura, M. Translation of partially overlapping *psbD–psbC* mRNAs in chloroplasts: The role of 5′-processing and translational coupling. *Nucleic Acids Res.* **2012**, *40*, 3152–3158. [CrossRef] [PubMed]

18.	Chotewutmontri, P.; Barkan, A. Dynamics of chloroplast translation during chloroplast differentiation in maize. *PLoS Genet.* **2016**, *12*, e1006106. [CrossRef] [PubMed]

19.	Barkan, A. Proteins encoded by a complex chloroplast transcription unit are each translated from both monocistronic and polycistronic mRNAs. *EMBO J.* **1988**, *7*, 2637–2644. [PubMed]

20.	Schmitz-Linneweber, C.; Williams-Carrier, R.; Barkan, A. RNA immunoprecipitation and microarray analysis show a chloroplast pentatricopeptide repeat protein to be associated with the 5′ region of mRNAs whose translation it activates. *Plant Cell* **2005**, *17*, 2791–2804. [CrossRef] [PubMed]

21.	Zerges, W.; Rochaix, J.D. Low density membranes are associated with RNA-binding proteins and thylakoids in the chloroplast of *Chlamydomonas reinhardtii*. *J. Cell Biol.* **1998**, *140*, 101–110. [CrossRef] [PubMed]

22.	Klaff, P.; Gruissem, W. A 43 kd light-regulated chloroplast RNA-binding protein interacts with the *psbA* 5′ non-translated leader RNA. *Photosyn. Res.* **1995**, *46*, 235–248. [CrossRef] [PubMed]

23.	Hotto, A.M.; Huston, Z.E.; Stern, D.B. Overexpression of a natural chloroplast-encoded antisense RNA in tobacco destabilizes 5S rRNA and retards plant growth. *BMC Plant Biol.* **2010**, *10*, 213. [CrossRef] [PubMed]

24.	Mulo, P.; Sakurai, I.; Aro, E.M. Strategies for *psbA* gene expression in cyanobacteria, green algae and higher plants: From transcription to psii repair. *Biochim. Biophys. Acta* **2012**, *1817*, 247–257. [CrossRef] [PubMed]

25.	Klein, R.R.; Mason, H.S.; Mullet, J.E. Light-regulated translation of chloroplast proteins. I. Transcripts of *psaA–psaB*, *psbA*, and *rbcL* are associated with polysomes in dark-grown and illuminated barley seedlings. *J. Cell Biol.* **1988**, *106*, 289–301. [CrossRef] [PubMed]

26.	Maier, R.M.; Neckermann, K.; Igloi, G.L.; Kossel, H. Complete sequence of the maize chloroplast genome: Gene content, hotspots of divergence and fine tuning of genetic information by transcript editing. *J. Mol. Biol.* **1995**, *251*, 614–628. [CrossRef] [PubMed]

27.	Blomqvist, L.A.; Ryberg, M.; Sundqvist, C. Proteomic analysis of highly purified prolamellar bodies reveals their significance in chloroplast development. *Photosynth. Res.* **2008**, *96*, 37–50. [CrossRef] [PubMed]

28.	Blomqvist, L.A.; Ryberg, M.; Sundqvist, C. Proteomic analysis of the etioplast inner membranes of wheat (*Triticum aestivum*) by two-dimensional electrophoresis and mass spectrometry. *Physiol. Plant.* **2006**, *128*, 368–381. [CrossRef]

29.	Ikeuchi, M.; Murakami, S. Behaviour of the 36,000-dalton protein in the integral membranes of squash etioplasts during greening. *Plant Cell Physiol.* **1982**, *23*, 575–583.

30.	Von Zychlinski, A.; Kleffmann, T.; Krishnamurthy, N.; Sjolander, K.; Baginsky, S.; Gruissem, W. Proteome analysis of the rice etioplast: Metabolic and regulatory networks and novel protein functions. *Mol. Cell. Proteom.* **2005**, *4*, 1072–1084. [CrossRef] [PubMed]

31.	Selstam, E.; Sandelius, A.S. A comparison between prolamellar bodies and prothylakoid membranes of etioplasts of dark-grown wheat concerning lipid and polypeptide composition. *Plant Physiol.* **1984**, *76*, 1036–1040. [CrossRef] [PubMed]

32.	Kleffmann, T.; von Zychlinski, A.; Russenberger, D.; Hirsch-Hoffmann, M.; Gehrig, P.; Gruissem, W.; Baginsky, S. Proteome dynamics during plastid differentiation in rice. *Plant Physiol.* **2007**, *143*, 912–923. [CrossRef] [PubMed]

33.	Kanervo, E.; Singh, M.; Suorsa, M.; Paakkarinen, V.; Aro, E.; Battchikova, N.; Aro, E.M. Expression of protein complexes and individual proteins upon transition of etioplasts to chloroplasts in pea (*Pisum sativum*). *Plant Cell Physiol.* **2008**, *49*, 396–410. [CrossRef] [PubMed]

34.	Lonosky, P.M.; Zhang, X.; Honavar, V.G.; Dobbs, D.L.; Fu, A.; Rodermel, S.R. A proteomic analysis of maize chloroplast biogenesis. *Plant Physiol.* **2004**, *134*, 560–574. [CrossRef] [PubMed]

35.	Margulies, M.M.; Tiffany, H.L.; Hattori, T. Photosystem I reaction center polypeptides of spinach are synthesized on thylakoid-bound ribosomes. *Arch. Biochem. Biophys.* **1987**, *254*, 454–461. [CrossRef]

36.	Mackender, R.O. Etioplast development in dark-grown leaves of *Zea mays* L. *Plant Physiol.* **1978**, *62*, 499–505. [CrossRef] [PubMed]

37.	Klein, R.R.; Mullet, J.E. Control of gene expression during higher plant chloroplast biogenesis. Protein synthesis and transcript levels of *psbA*, *psaA–psaB*, and *rbcL* in dark-grown and illuminated barley seedlings. *J. Biol. Chem.* **1987**, *262*, 4341–4348. [PubMed]

38. Laing, W.; Kreuz, K.; Apel, K. Light-dependent, but phytochrome-independent, translational control of the accumulation of the p700 chlorophyll-A protein of photosystem I in barley (*Hordeum vulgare* L.). *Planta* **1988**, *176*, 269–276. [CrossRef] [PubMed]

39. Klein, R.R.; Mullet, J.E. Regulation of chloroplast-encoded chlorophyll-binding protein translation during higher plant chloroplast biogenesis. *J. Biol. Chem.* **1986**, *261*, 11138–11145. [PubMed]

40. Ruf, S.; Kossel, H.; Bock, R. Targeted inactivation of a tobacco intron-containing open reading frame reveals a novel chloroplast-encoded photosystem I-related gene. *J. Cell Biol.* **1997**, *139*, 95–102. [CrossRef] [PubMed]

41. Hirose, T.; Sugiura, M. Both RNA editing and RNA cleavage are required for translation of tobacco chloroplast *NdhD* mRNA: A possible regulatory mechanism for the expression of a chloroplast operon consisting of functionally unrelated genes. *EMBO J.* **1997**, *16*, 6804–6811. [CrossRef] [PubMed]

42. Antonkine, M.L.; Jordan, P.; Fromme, P.; Krauss, N.; Golbeck, J.H.; Stehlik, D. Assembly of protein subunits within the stromal ridge of photosystem I. Structural changes between unbound and sequentially PSI-bound polypeptides and correlated changes of the magnetic properties of the terminal iron sulfur clusters. *J. Mol. Biol.* **2003**, *327*, 671–697. [CrossRef]

43. Li, N.; Zhao, J.D.; Warren, P.V.; Warden, J.T.; Bryant, D.A.; Golbeck, J.H. PsaD is required for the stable binding of PsaC to the photosystem I core protein of synechococcus sp. Pcc 6301. *Biochemistry* **1991**, *30*, 7863–7872. [CrossRef] [PubMed]

44. Sharwood, R.E.; Hotto, A.M.; Bollenbach, T.J.; Stern, D.B. Overaccumulation of the chloroplast antisense RNA *as5* is correlated with decreased abundance of 5S rRNA in vivo and inefficient 5S rRNA maturation in vitro. *RNA* **2011**, *17*, 230–243. [CrossRef] [PubMed]

45. Breidenbach, E.; Leu, S.; Michaels, A.; Boschetti, A. Synthesis of EF-Tu and distribution of its mRNA between stroma and thylakoids during the cell cycle of *Chlamydomonas reinhardii*. *Biochim. Biophys. Acta* **1990**, *1048*, 209–216. [CrossRef]

46. Voelker, R.; Barkan, A. Two nuclear mutations disrupt distinct pathways for targeting proteins to the chloroplast thylakoid. *EMBO J.* **1995**, *14*, 3905–3914. [PubMed]

47. Finster, S.; Eggert, E.; Zoschke, R.; Weihe, A.; Schmitz-Linneweber, C. Light-dependent, plastome-wide association of the plastid-encoded RNA polymerase with chloroplast DNA. *Plant J.* **2013**, *76*, 849–860. [CrossRef] [PubMed]

9

Phylogenomic and Comparative Analyses of Complete Plastomes of *Croomia* and *Stemona* (Stemonaceae)

Qixiang Lu [†], Wenqing Ye [†], Ruisen Lu, Wuqin Xu and Yingxiong Qiu *

Key Laboratory of Conservation Biology for Endangered Wildlife of the Ministry of Education, and College of Life Sciences, Zhejiang University, Hangzhou 310058, China; 0016616@zju.edu.cn (Q.L.); yewenqing@zju.edu.cn (W.Y.); reason@zju.edu.cn (R.L.); 21707105@zju.edu.cn (W.X.)
* Correspondence: qyxhero@zju.edu.cn
† These authors have contributed equally to this work.

Abstract: The monocot genus *Croomia* (Stemonaceae) comprises three herbaceous perennial species that exhibit EA (Eastern Asian)–ENA (Eastern North American) disjunct distribution. However, due to the lack of effective genomic resources, its evolutionary history is still weakly resolved. In the present study, we conducted comparative analysis of the complete chloroplast (cp) genomes of three *Croomia* species and two *Stemona* species. These five cp genomes proved highly similar in overall size (154,407–155,261 bp), structure, gene order and content. All five cp genomes contained the same 114 unique genes consisting of 80 protein-coding genes, 30 tRNA genes and 4 rRNA genes. Gene content, gene order, AT content and IR/SC boundary structures were almost the same among the five Stemonaceae cp genomes, except that the *Stemona* cp genome was found to contain an inversion in *cem*A and *pet*A. The lengths of five genomes varied due to contraction/expansion of the IR/SC borders. A/T mononucleotides were the richest Simple Sequence Repeats (SSRs). A total of 46, 48, 47, 61 and 60 repeats were identified in *C. japonica*, *C. heterosepala*, *C. pauciflora*, *S. japonica* and *S. mairei*, respectively. A comparison of pairwise sequence divergence values across all introns and intergenic spacers revealed that the *ndh*F–*rpl*32, *psb*M–*trn*D and *trn*S–*trn*G regions are the fastest-evolving regions. These regions are therefore likely to be the best choices for molecular evolutionary and systematic studies at low taxonomic levels in Stemonaceae. Phylogenetic analyses of the complete cp genomes and 78 protein-coding genes strongly supported the monophyly of *Croomia*. Two Asian species were identified as sisters that likely diverged in the Early Pleistocene (1.62 Mya, 95% HPD: 1.125–2.251 Mya), whereas the divergence of *C. pauciflora* dated back to the Late Miocene (4.77 Mya, 95% HPD: 3.626–6.162 Mya). The availability of these cp genomes will provide valuable genetic resources for further population genetics and phylogeographic studies on *Croomia*.

Keywords: *Croomia*; *Stemona*; chloroplast genome; comparative genomics; phylogeny; biogeography

1. Introduction

Croomia Torr. ex Torr. et Gray belongs to the monocot family Stemonaceae Engl (Pandanales, Liliidae) and comprises three herbaceous perennial species: *C. pauciflora* (Nutt.) Torr., *C. japonica* Miq. and *C. heterosepala* (Bak.) Oku. Of these three species, *C. japonica* and *C. heterosepala* are endemic to warm-temperate deciduous forests in East Asia, while *C. pauciflora* grows in temperate-deciduous forests in North America [1–3]. There is a considerable difference in morphological traits among this genus. For example, the four tepals of *C. japonica* are homomorphic with a re-curved edge, while those of *C. heterosepala* have a flat edge, and one outside tepal is much larger than the other three [4,5]. Compared to two Asian species, *C. pauciflora* has a smaller flower, shorter petiole, denser underground stem nodes and a more obvious heart-shape leaf base [1]. As the roots of *Croomia* species contain compounds such as pachysamine, didehydrocroomine and croomine groups, they are used as folk medicine to treat cough and injuries [6,7]. *Croomia* can reproduce sexually through seed formation via cross-pollination and asexually through underground rhizomes [1,8]. Due to their limited distribution and small population sizes, the three extant species of *Croomia* are listed as "threatened" or "endangered" in China, Japan and the Americas [9–11]. The other three genera of Stemonaceae are *Pentastemona*, *Stemona* and *Stichoneuron*. The species of *Stichoneuron* are located in India, Thailand and Peninsular Malaysia, while those of *Pentastemona* are only in Sumatra [3,8]. The genus *Stemona* comprises ca. 25 species with the widest distribution from Northeast Asia to Southeast Asia and Australia. The roots of *Stemona* species contain similar medicine compounds as *Croomia* [7]. Although *Croomia* and *Stemona* species have important pharmacological and ecological value, limited molecular markers were available for the utilization, conservation and breeding of these species in the context of population genetics and phylogenetic studies [12].

Croomia exhibits a well-known classic intercontinental disjunct distribution between Eastern Asia (EA) and Eastern North America (ENA) [1,8,13,14]. This continental disjunction pattern was suggested to have resulted from fragmentation of the mid-Tertiary mesophytic forest flora throughout a large part of the Northern Hemisphere, as global temperature cooled down in the late Tertiary and Quaternary [15,16]. For the two East Asian endemics, *C. japonica* is distributed in East China and southern Japan, while *C. heterosepala* is in northern Japan, and they have adjacent ranges in South Japan [17,18]. Therefore, *Croomia* is well suited for testing biogeographic hypotheses about the evolution of both the eastern Asian–eastern North American and eastern Asian–Japanese Archipelago floristic disjunctions. Based on previous molecular phylogenetic analyses using cpDNA sequence variation of the *trn*L-F region, the two Asian species were identified as sister species that likely diverged in the Mid-to-Late Pleistocene (0.84–0.13 million years ago, Mya), whereas the divergence of *C. pauciflora* dates back to the Late Plio-/Pleistocene (<2.6 Mya) [12]. However, the previous cpDNA analysis based on a few parsimony informative sites yielded low bootstrap values for the majority of clades [12]. Thus, it is necessary to develop more highly variable genetic markers for determining the phylogenetic relationships and divergence times for *Croomia*. Nowadays, many phylogenetic relationships that remained unresolved with few loci have been clarified by using whole cp genome sequences [19–21]. Thus, whole cp genome sequences are increasingly being used in phylogeny reconstruction and providing hypervariable genetic markers for population genetic studies, especially in a group of recently-diverged species [22,23].

Here, we sequenced three *Croomia* and two *Stemona* cp genomes using the next-generation Illumina genome analyzer platform. We compared the cp genomes of two Stemonaceae genera to characterize their structural organization and variations and identify the most variable regions. This information on interspecific variability of each region will help guide further systematic and evolutionary studies of Stemonaceae. In addition, we used the whole cp genomes to resolve the phylogenetic relationships of *Croomia* and infer the historical biogeography of the genus.

2. Results and Discussion

2.1. Genome Assembly and Features

Illumina paired-end sequencing yielded 14,163,520–31,094,272-bp clean reads after trimming, and the de novo assembly generated 50,369–123,479 contigs for five Stemonaceae species. With the cp genome of *C. palmata* as a reference, contigs were combined to generate the draft cp genome for each species. The lengths of determined nucleotide sequences were 154,672, 154,407, 155,261, 154,224 and 154,307 bp for *C. japonica*, *C. heterosepala*, *C. pauciflora*, *S. japonica* and *S. mairei*, respectively. (Figure 1, Table S1). All five cp genomes exhibited the typical quadripartite structure of angiosperms, consisting of a pair of IR regions (27,082–27,243 bp) separated by an LSC region (81,844–82,429 bp) and an SSC region (17,889–18,346 bp). The cp genomes of three *Croomia* species and two *Stemona* species were deposited in GenBank (MH177871, MH191379–MH191382).

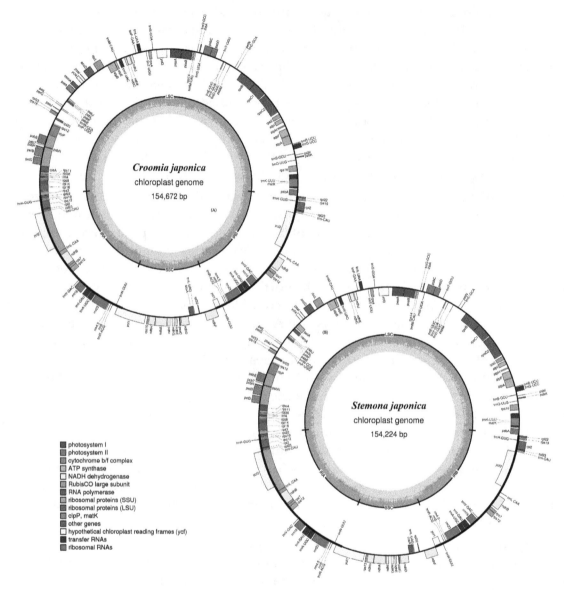

Figure 1. Gene maps of *Croomia* and *Stemona* chloroplast genomes. (**A**) *Croomia japonica*; (**B**) *Stemona japonica*.

These five cp genomes contained 134 genes identically, of which 114 were unique and 20 were duplicated in IR regions (Table S1). Those 134 genes were arranged almost in the same order except *cem*A and *pet*A, which were inverted at the LSC region of two *Stemona* species. Gene inversions at LSC were also reported in other angiosperm, such as *Silene* [24], *Cymbidium* [19] and *Acacia dealbata* [25]. The 114 unique genes included 80 protein-coding genes, 30 tRNA genes and 4 rRNA genes. In *Croomia* species, the overall GC content was 38.3%, and the GC contents of the LSC, SSC and IR regions were 36.6%, 32.3–32.5% and 42.8–42.9%, respectively, while those of *Stemona* were 38.0%, 36.2%, 32.1% and 42.7% (Table S1). In all five genomes, nine of the protein-coding genes and six of the tRNA genes possessed a single intron, while three genes (*rps*12, *clp*P and *ycf*3) contained two introns (Table 1). The *rps*12 gene was trans-spliced; the 5′ end exon was located in the LSC region, and the 3′ end exon and intron were located in the IR regions. Compared to many other species, such as *Salvia miltiorrhiza* [26] and *Cornales* [27], the SSC region of the five studied species was found to have a different (reverse) orientation. The reverse orientation of the SSC region has also been reported in a wide variety of plant species [28–30]. This phenomenon is sometimes interpreted as a major inversion existing within the species [29,31,32]. In fact, the two orientations of the SSC region have been found to occur regularly during the course of chloroplast DNA replication within individual plant cells [33,34]. Thus, the reverse orientation of the SSC region found in the five Stemonaceae cp genomes may represent a form of plastid heteroplasmy [30,35].

Table 1. List of genes in Stemonaceae chloroplast genomes.

Category of Genes	Groups of Genes	Names of Genes
Self-replication	rRNA genes	*rrn16*(×2), *rrn23*(×2), *rrn4.5*(×2), *rrn 5*(×2)
	tRNA genes	*trnA-UGC* *(×2), *trnC-GCA*, *trnD-GUC*, *trnE-UUC*, *trnF-GAA*, *trnfM-CAU*, *trnG-GCC*, *trnG-UCC* *, *trnH-GUG*(×2), *trnI-CAU*(×2), *trnI-GAU* *(×2), *trnK-UUU* *, *trnL-CAA*(×2), *trnL-UAA* *, *trnL-UAG*, *trnM-CAU*, *trnN-GUU*(×2), *trnP-UGG*, *trnQ-UUG*, *trnR-ACG*(×2), *trnR-UCU*, *trnS-GCU*, *trnS-GGA*, *trnS-UGA*, *trnT-GGU*, *trnT-UGU*, *trnV-GAC*(×2), *trnV-UAC* *, *trnW-CCA*, *trnY-FUA*
	Small subunit of ribosome	*rps2*, *rps3*, *rps4*, *rps7*(×2), *rps8*, *rps11*, *rps12* **(×2), *rps14*, *rps15*, *rps16* *, *rps18*, *rps19*(×2)
	Large subunit of ribosome	*rpl2* *(×2), *rpl14*, *rpl16* *, *rpl20*, *rpl22*(×2), *rpl23*(×2), *rpl32*, *rpl33*, *rpl36*
	DNA-dependent RNA polymerase	*rpoA*, *rpoB*, *rpoC1* *, *rpoC2*
Genes for photosynthesis	Subunit of NADH-dehydrogenase	*ndhA* *, *ndhB* *(×2), *ndhC*, *ndhD*, *ndhE*, *ndhF*, *ndhG*, *ndhI*, *ndhH*, *ndhJ*, *hdhK*
	Subunit of Photosystem 1	*psaA*, *psaB*, *psaC*, *psaI*, *psaJ*, *ycf3* **
	Subunit of Photosystem 2	*psbA*, *psbB*, *psbC*, *psbD*, *psbE*, *psbF*, *psbH*, *psbI*, *psbJ*, *psbK*, *psbL*, *psbM*, *psbN*, *psbT*
	Subunits of cytochrome b/f complex	*petA*, *petB* *, *petD* *, *petG*, *petL*, *petN*
	Subunits of ATP synthase	*atpA*, *atpB*, *atpE*, *atpF* *, *atpH*, *atpI*
	Large subunit of rubisco	*rbcL*
Other genes	Maturase	*matK*,
	Protease	*clpP* **
	Envelope membrane protein	*cemA*
	Subunit of Acetyl-CoA-carboxylase	*accD*
	c-type cytochrome synthesis gene	*ccsA*
	Translation initiation factor IF-1	*infA*
Genes of unknown function	Open reading frames (ORF, ycf)	*ycf1*, *ycf2*(×2), *ycf4*, *lhbA*

* Gene with one intron, ** gene with two introns; (×2) indicates genes duplicated in the IR region.

2.2. Contraction and Expansion of Inverted Repeats

Length variation in angiosperm cp genomes is due most typically to the expansion or contraction of the IR into or out of adjacent single-copy regions and/or changes in sequence complexity due to insertions or deletions of novel sequences [36,37]. Compared to reference cp genome *C. palmata*, all five species exhibited IR expansion at the IRb/LSC border, leading to entire *rpl22* duplication. In a previous study, a partial duplication of the *rpl22* gene was reported in some monocot species of Asparagales and Commelinales [38]. Although the gene number and gene order were conserved across these five Stemonaceae species, minor differences were still observed at the boundaries (Figure 2). At the IRa/LSC border, the spacer from *rpl22* to this border of *Stemona* (65 bp) was longer than that of *Croomia* (24–25 bp), except *C. pauciflora*. As for the *ycf1* gene, there were 4580–4662-bp sequences located at SSC in *Croomia* and 4374–4383 bp in *Stemona*, while the pseudogene fragment duplications in IRb were 970 bp and 1206 bp in *Croomia* and *Stemona*, respectively. The *ndhF* gene exhibited variable sequences in SSC of *Croomia* (2215–2226 bp), while invariable in *Stemona* with a 2190-bp length. At the border of

IRb/LSC, the spacer from *psb*A to this border of *Croomia* ranged from 91 bp–99 bp, while it ranged from 94 bp–104 bp in *Stemona*. These differences between the five cp genomes led to the length variation of their whole genome sequences.

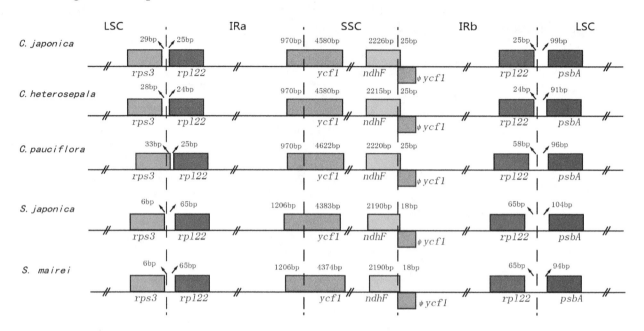

Figure 2. Comparison of LSC, IR and SSC junction positions among five Stemonaceae chloroplast genomes.

2.3. Divergence Hotspot Regions

To elucidate the level of sequence divergence, the three *Croomia* and two *Stemona* cp genome sequences were compared and plotted using the mVISTA program (Figure 3). Like most angiosperms, the sequence divergence of IR regions was lower than that of the LSC and SSC region [39,40], which may involve copy correction of IRs as a mechanism [41]. We identified 140 regions in total with more than a 200-bp length (68 protein-coding regions (CDS), 53 Intergenic Spacers (IGS) and 19 introns). The nucleotide variability (Pi) of these 140 regions ranged from 0.080% (*rrn*16) to 9.565% (IGS *petN–psbM*) among the five cp genomes. The average Pi of the non-coding region was 3.644%, much higher than coding regions (1.587%), as found in most angiosperms [42,43]. For the 68 CDS, the Pi values for each region ranged from 0.231% (*rpl*2 CDS1) to 4.047% (*ycf*1), whereby 10 regions (i.e., *matK*, *rpl*33, *rps*15, *psbH*, *rps*18, *rps*3, *rpl*20, *ccs*A, *ndhF*, *accD*) had remarkably high values (*pi* > 2.5%). For the 53 IGS regions, Pi values ranged from 0.185% (*trnN–ycf*1) to 9.565% (*petN–psbM*). Again, ten of those regions showed considerably high values (*pi* > 5.7%; i.e., *rpl*32–*ndhF*, *trnS–trnG*, *ndhE–psaC*, *ndhD–ccs*A, *atpF–atpH*, *psbM–trnD*, *trnE–trnT*, *petL–petG*, *rps*16–*trnQ*, *accD–psaI*; see Figure 4). A comparison of DNA sequence divergence revealed that three of these ten noncoding regions, *ndhF–rpl*32 (PICs = 96), *psbM–trnD* (PICs = 73) and *trnS–trnG* (PICs = 49), are the most variable regions across Stemonaceae (Figure A1). Thus, these three regions may be good candidates for resolving future low-level phylogeny and phylogeography in Stemonaceae. In a previous study, the availability of plastid noncoding regions was compared across 10 major lineages of angiosperms (such as Nymphaeales, monocots, eurosids) [44]. However, only five families of monocots represented by five species pairs were included, without Stemonaceae. The three variable regions predicted here are among the top 13 regions of monocots in the research by Shaw et al. [44], with *ndhF–rpl*32, *psbM–trnD* and *trnS–trnG* ranked first, third and 11th, respectively. Of these regions, *ndhF–rpl*32 has long been a popular region in phylogenetic studies of angiosperms [44,45]. Meanwhile, *psbM–trnD* and *trnS–trnG* are also noted as highly variable in Liliaceae [46] and occasionally used in low-level phylogenetic analyses (*Scabiosa*: [47]; *Solms-laubachia*: [48]). The resolution of recent divergences in monocots would benefit considerably by the inclusion of any or all of these highly variable regions.

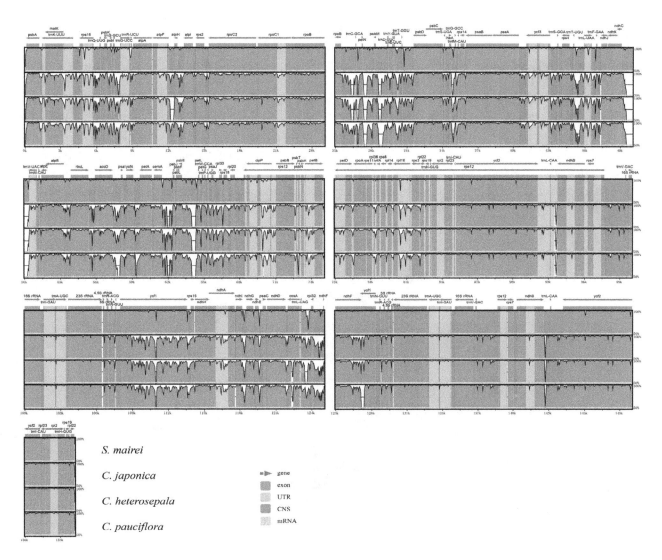

Figure 3. Sequence identity plots among five Stemonaceae chloroplast genomes, with *Stemona japonica* as a reference. CNS: conserved non-coding sequences; UTR: untranslated region.

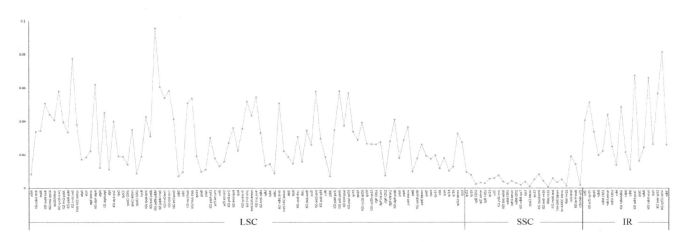

Figure 4. The nucleotide variability (Pi) values were compared among five Stemonaceae species.

2.4. Repetitive Sequences and SSR Polymorphisms

With the criterion of a copy size of 30 bp or longer and a sequence identity >90%, REPUTER [49] identified 47, 49, 48, 61 and 60 repeats (including forward, palindromic, complement and reverse repeats) in five cp genome sequences of *C. japonica*, *C. heterosepala*, *C. pauciflora*, *S. japonica* and *S. mairei*, respectively (Figure 5A). *C. japonica* contained 21 forward repeats, 24 palindromic repeats, 1 complement repeat and 1 reverse repeat, and *S. japonica* contained 27, 25, 1 and 8 repeats, correspondingly. The other two *Croomia* species and *S. mairei* contained no complement repeats. The numbers of forward repeats, palindromic and reverse repeats were, respectively, 21, 27 and 1 in *C. japonica*, 25, 21 and 2 in *C. heterosepala* and 27, 25 and 8 in *S. mairei* (Figure 5A). The lengths of majority repeats were 30, 31 and 43 bp in size (Figure 5B). For *Croomia*, the repeats were mainly located in *ycf*2 (46.8–58.3%) and non-coding regions (27.1–38.3%). As for *Stemona*, the repeats were mostly located in non-coding regions (58.3–59.0%) and *ycf*2 (33.3–34.4%). Only one repeat was across IGS (*psbC–trnS*) and CDS (*trnS*^{UGA}). The remaining repeats were found located in genes such as *ccs*A, *ycf*1, *trn*G^{UGA}, *trn*S^{GGA}, *trn*S^{GCU} and *psa*B.

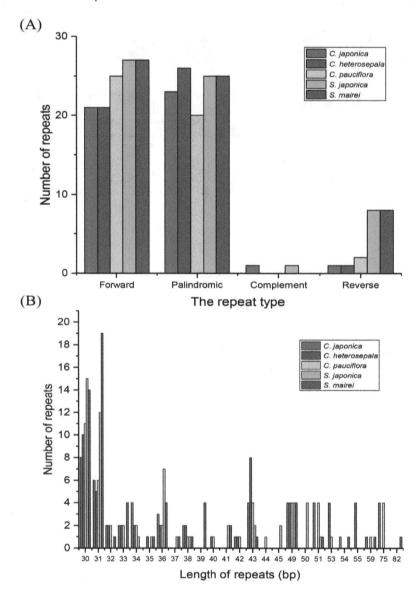

Figure 5. Analysis of repeated sequences in five Stemonaceae chloroplast genomes. (**A**) Frequency of repeats by length; (**B**) frequency of repeat types.

SSRs in the cp genome present high diversity in copy numbers, and they are important molecular markers for plant population genomics and evolutionary history [50,51]. SSRs ($\geq$10 bp) were detected in these five Stemonaceae cp genomes by MIcroSAtellite (MISA) analysis [52], ranging from 90–116 in total. Among these SSRs, the mononucleotide repeat unit (A/T) occupied the highest proportion, with

71.2% in *C. japonica*, 70.9% in *C. heterosepala*, 63.3% in *C. pauciflora*, 64.0% in *S. japonica* and 62.4% in *S. mairei* (Figure 6A). SSR loci were mainly located in IGS (71.4%) (Figure 6B) and were also detected in introns (16.5%) and CDS (12.1%), such as *mat*K, *atp*A, *rpo*C2, *rpo*B, *cem*A, *psb*F, *ycf*2, *ycf*1 and *ndh*D. In general, the SSRs of these five cp genomes showed great variation, which can be used in population genetic studies of *Croomia* and *Stemona* species.

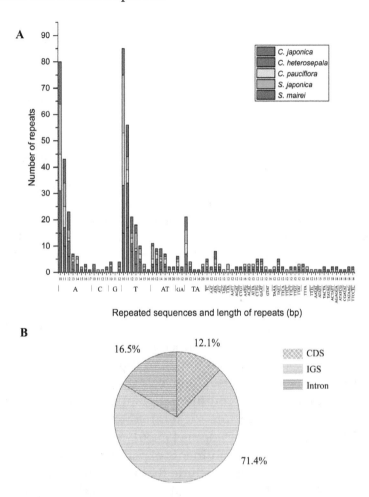

Figure 6. Simple Sequence Repeats (SSRs) in five Stemonaceae chloroplast genomes. (**A**) Numbers of SSRs by length; (**B**) distribution of SSR loci. IGS: intergenic spacer region; CDS: protein-coding regions.

2.5. Phylogenetic Analysis, Divergence Time and Ancestral Area Reconstruction

CP genome sequences have been successfully used in angiosperm phylogenetic studies [22,53]. The Maximum Likelihood (ML) and Bayesian Inference (BI) analyses of both whole sequences and protein-coding region of three *Croomia* and two *Stemona* cp genomes yielded nearly identical tree topologies, with 100% bootstrap and 1.0 Bayesian posterior probabilities at each node (Figure 7). This phylogenetic tree supports the monophyly of *Croomia*. Two Asian species *C. japonica* and *C. heterosepala* formed a clade, being strongly recovered as sisters of the North American species *C. pauciflora*. This tree topology is largely congruent with that inferred from *trn*L–F [12], but obtained much higher bootstrap support values. Using average substitution rates of whole cp genomes, the divergence time between the two Asian species, *C. japonica* and *C. heterosepala*, was estimated as ca. 1.621 Mya (1.125–2.251 Mya) and, thus, compatible with the early-Pleistocene event. By contrast, the divergence time between North American *C. pauciflora* and Asian species was estimated as ca. 4.774 Mya (3.626–6.162 Mya) (i.e., the Late Miocene). The divergence times estimated in this paper are much older than that estimated by the strict molecular clock method (*C. pauciflora*/the East Asian lineage: ca. 2.61–0.41 Mya; *C. japonica*/*C. heterosepala*: ca. 0.84–0.13 Mya) [12].

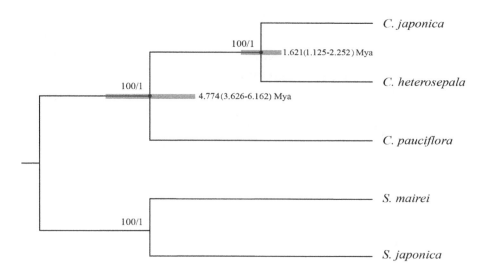

Figure 7. Phylogenetic relationships of three *Croomia* species inferred from Maximum Likelihood (ML) and Bayesian Inference (BI) and divergence time of three *Croomia* species estimated using Bayesian Evolutionary Analysis Sampling Trees (BEAST) analysis. Numbers above the lines represent ML bootstrap values and BI posterior probability. Blue bars indicate the 95% highest posterior density (HPD) credibility intervals for node ages (Mya). Numbers at the node represent divergence time (Mya) and 95% highest posterior density intervals. The phylogenetic tree based on 74 protein-coding genes is completely consistent with this topology.

The divergence between *C. pauciflora* and two Asian species coincides with the first sundering of the Bering Land Bridge (BLB) between the late Miocene and early Pliocene, most approximately at 5.4–5.5 Mya (Milne and Abbott, 2002) [54]. The Bayesian Binary MCMC (BBM) analysis of ancestral area reconstruction identified Asia as the most likely ancestral range (Node III, marginal probability: 0.93; Figure A2), indicating a possible intercontinental plant migration from Asia to North America. Indeed, the BLB served as an important route for temperate floristic exchanges between Asia and North America from the Eocene to the early Pliocene [55,56]. Subsequently, as a member of the Tertiary relict flora [15], *Croomia* species on the two continents experienced disjunct distribution and evolved separately after the Late Miocene. Thus, we conclude that the current distribution and differentiation of *Croomia* species in eastern Asia and eastern North America likely resulted from a combination of ancient migration and vicariant events. The divergence time between *C. japonica* and *C. heterosepala* fell into the Early Pleistocene. Habitat fragmentation resulting from the climatic vicissitudes of the (Late) Quaternary likely led to the speciation of *C. japonica* and *C. heterosepala* [12]. The above inferences seem to be consistent with the palaeovegetational and climatic history of eastern Asia and eastern America. However, considering that the cp genome is a haploid, uniparentaly-inherited and single locus [57], a nuclear (biparental) marker is also needed to elucidate the diversification process and demography history of *Croomia* species.

3. Materials and Methods

3.1. Sample Preparation, Sequencing, Assembly and Validation

Fresh leaves of *C. japonica* from China, *C. heterosepala* from Japan, *C. pauciflora* from North America and two outgroup species *Stemona japonica* (Bl.) Miq. and *S. mairei* (Levl.) Krause from China were sampled and dried with silica gel. The voucher specimens were deposited in the Herbarium of Zhejiang University (HZU). Total genomic DNA was extracted from ~3 mg materials using DNA Plantzol Reagent (Invitrogen, Carlsbad, CA, USA) following the manufacturer's protocol. The quality and concentration of the DNA were detected using agarose gel electrophoresis. Purified DNA was sheared into ~500-bp fragments, and the fragmentation quality was checked on a Bioanalyzer 2100

(Agilent Technologies, Santa Clara, CA, USA). Paired-end sequencing libraries were constructed according to the Illumina standard protocol (Illumina, San Diego, CA, USA). Genomic DNAs of five species were sequenced using an Illumina HiSeq™ 2000 (Illumina, San Diego, CA, USA) at Beijing Genomics Institute (BGI; Shenzhen, China). Plastome sequences were assembled using a combination of de novo and reference-guided assembly [58]. Firstly, to obtain clean reads, the CLC-quality trim tool was used to remove low-quality bases ($Q < 20$, 0.01 probability error). Secondly, we assembled the clean reads into contigs on the CLC de novo assembler. Thirdly, all the contigs were aligned with the reference cp genome of *Carludovica palmate* Ruiz. & Pav. (NC_026786.1) using local BLAST (http://blast.ncbi.nlm.nih.gov/) (27 December 2016), and aligned contigs were ordered according to the reference cp genome with ≥90% similarity and query coverage. Then, to construct the draft cp genome of each species, the ordered contigs usually representing the whole reconstructed genome were imported into GENEIOUS v9.0.5 software (http://www.geneious.com) (18 March 2017), where the clean reads were remapped onto the contigs.

3.2. Genome Annotation and Whole Genome Comparison

The annotation of five species was performed using the Dual Organellar GenoMe Annotator (DOGMA) [59]. The start and stop codons and intron/exon boundaries were manually corrected by comparison to homologous genes from the reference genome of *C. palmate*. We also verified the transfer RNAs (tRNAs) using tRNAscan-SE v1.21 with default settings [60]. The circular genome maps were drawn using the OrganellarGenome DRAW tool (OGDRAW) [61], followed by manual modification.

Genome comparison among the five Stemonaceae cp genomes was analyzed using mVISTA [62] with *C. palmate* as a reference. Six genome sequences were aligned in Shuffle-LAGAN mode with default parameters, and the conservation region was visualized in an mVISTA plot. To identify the divergence hotspot regions in the five Stemonaceae cp genomes, the nucleotide variability of protein coding genes, introns and intergenic spacer sequences of five species were evaluated using DNASP v5.10 [63]. The above regions were extracted following two criteria: (a) total number of mutation (Eta) >0; and (b) the aligned length >200 bp. The inverted regions in *cem*A, *cem*A–*pet*A and *pet*A were excluded. The top ten most variable noncoding regions with a high Pi value were counted by Potentially Informative Characters (PICs) across species pair of *C. japonica* and *S. japonica* following Shaw et al. [64]. Any large structural event of the cp genome, such as gene order rearrangements or IR expansion/contractions, were recorded.

3.3. Characterization of Repeat Sequence and SSRs

REPUTER [49] was used to find the location and length of repeat sequences, including forward, palindrome, complement and reverse repeats in the five cp genomes. The minimum repeat size was set to 30 bp, and the sequence identity of repeats was no less than 90% or greater sequence identity with the Hamming distance equal to 3. The MISA perl script was used to detect simple sequence repeats (SSRs) [52] with thresholds of 10 bp in length for mono-, di-, tri, tetra-, penta- and hexa-nucleotide SSRs.

3.4. Phylogenetic Analysis, Divergence Time and Ancestral Area Reconstruction

The five cp genome sequences were aligned using MAFFT v7 [65]. Two *Stemona* species were used as outgroups. ML and BI analysis were used to reconstruct the phylogenetic trees. In order to examine the phylogenetic utility of different regions, two datasets were used: (1) the complete cp genome sequences; (2) 78 protein-coding genes shared by the five cp genomes (two inverted genes of *cem*A and *pet*A in *Stemona* species were excluded). Gaps (indels) were treated as missing data. The Akaike Information Criterion (AIC) in JMODELTEST v2.1.4 [66] was used to determine the best-fitting model of nucleotide substitutions. The GTR + I + G model was used for two datasets. The ML tree was constructed using RAXML-HPC v8.2.10 with 1000 replicates on the Cyberinfrastructure for Phylogenetic Research (CIPRES) Science Gateway website (http://www.phylo.org/) (10 May 2017) [67]. BI analysis was conducted in MRBAYES v3.2 [68]. The Markov chain Monte Carlo (MCMC) was set to run 1,000,000 generations and sampled every 1000 generations. The first 25% of generations was discarded as burn-in.

Due to the lack of fossil records, we used the average substitution rate 0.51952×10^{-9} per site per year (s/s/y) of the whole cp genome in Brassicaceae [69,70] to estimate interspecific divergence time of *Croomia*. The Bayesian analysis was implemented in BEAST v1.8.4 [71] using the GTR + I + G substitution model. MCMC analysis of 20,000,000 generations was implemented, in which every 1000 generations were sampled, under an uncorrelated lognormal relaxed clock approach using the Yule speciation tree prior with the substitution rate. TRACER v1.6 [72] was used to check the effective population size (ESS) >200. TREEANNOTATOR v.1.8.4 [73] was used to produce maximum clade credibility trees from the trees after burning-in of 25%. The final tree was visualized in FIGTREE v1.4.3 (http://tree.bio.ed.ac.uk/software/figtree/) (13 May 2017).

To reconstruct the historical biogeography of *Croomia*, we performed Bayesian Binary MCMC (BBM) analysis as implemented in RASP v3.1 [74] using trees retained from the BI analysis (see above). According to the distribution of *Croomia*, we defined the following two areas: A, Asia (East Asia/South Asia); and B, North America. Accounting for phylogenetic uncertainty, we used 500 trees randomly chosen across all post-burn-in trees generated from BEAST analysis and ran the BBM analysis. A fixed JC + G (Jukes–Cantor + Gamma) model was chosen with a null root distribution. The MCMC chains were run for 500,000 generations, and every 100 generations were sampled. The ancestral ranges obtained were projected onto the MCC tree.

4. Conclusions

Here, we sequenced the first five complete cp genomes in Stemonaceae. Each genome possesses the typical structure shared with other angiosperm species. Several highly variable noncoding cpDNA regions were identified, which should be the best choices for future phylogenetic, phylogeographic and population-level genetic studies in Stemonaceae. The phylogenomic and biogeographic analyses of *Croomia* reveal that ancient migration and vicariance-driven allopatric speciation resulting from historical climate oscillations most likely played roles in the formation of the disjunct distributions and divergence of these three *Croomia* species.

Author Contributions: Y.Q. conceived of the idea. R.L. contributed to the sampling. Q.L. performed the experiment. Q.L., W.Y., R.L. and W.X. analyzed the data. The manuscript was written by Q.L., W.Y. and Y.Q.

Acknowledgments: The authors thank Shota Sakaguchi from Kyoto University for collecting plant materials in Japan.

Abbreviations

cp	Complete chloroplast
IR	Inverted Repeat
lSC	Large Single Copy
SSC	Small Single Copy
Pi	Nucleotide variability
SSR	Simple Sequence Repeat
PIC	Potentially Informative Character
ML	Maximum likelihood
BI	Bayesian Inference
MCMC	Markov chain Monte Carlo

Appendix A

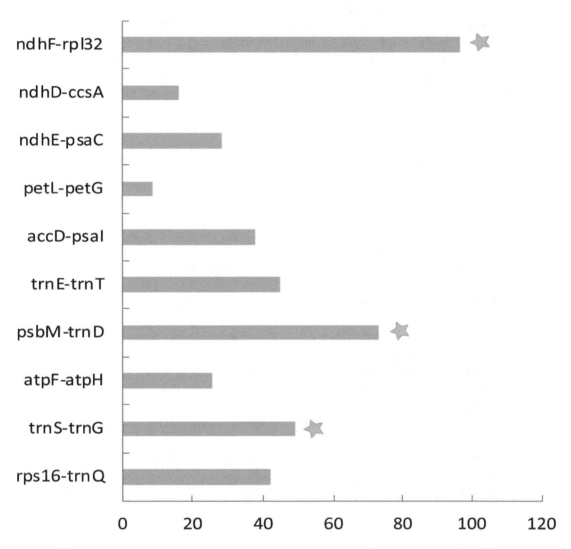

Figure A1. PIC values of the top ten most variable noncoding regions in Stemonaceae.

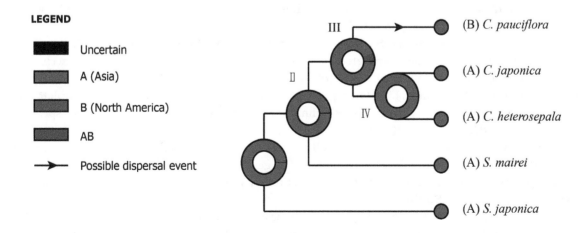

References

1.   Rogers, G.K. The Stemonaceae in the southeastern United States. *J. Arnold Arboretum* **1982**, *63*, 327–336.

2.   Whetstone, R.D. Notes on *Croomia* pauciflora (Stemonaceae). *Rhodora* **1984**, *25*, 131–137.

3.   Li, E.X. Studies on Phylogeography of *Croomia* and Phylogeny of *Croomia* and Its Allies. Ph.D. Thesis, Zhejiang University, Hangzhou, China, 2006.

4.   Okuyama, S. On the Japanese species of *Croomia*. *J. Jpn. Bot.* **1944**, *20*, 31–32.

5.   Ohwi, J. *Croomia*. In *Flora of Japan*; Smithsonian Institution: Washington, DC, USA, 1965; p. 279.

6.   Lin, W.; Cai, M.; Ying, B.; Feng, R. Studies on the chemical constituents of *Croomia japonica* Miq. *Yao Xue Xue Bao* **1993**, *28*, 202–206. [PubMed]

7.   Pilli, R.A.; Ferreira, M.C. Recent progress in the chemistry of the *Stemona* alkaloids. *Nat. Prod. Rep.* **2000**, *17*, 117–127. [CrossRef] [PubMed]

8.   Ji, Z.H.; Duyfjes, B.E. "Stemonaceae", in Flora of China, Flagellariaceae through Marantaceae. In *Flora of China*; Wu, Z.Y., Raven, P.H., Eds.; Missouri Botanical Garden Press: Beijing, China, 2000; Volume 24, pp. 70–72.

9.   Patrick, T.S.; Allison, J.; Krakow, G. Protected plants of Georgia. Georgia Department of Natural Resources, Natural Heritage Program. *Soc. Circ.* **1995**, *25*, 36–38.

10. Estill, J.C.; Cruzan, M.B. Phytogeography of rare plant species endemic to the southeastern United States. *Castanea* **2001**, *36*, 3–23.

11. Sung, W.; Yan, X. *China Species Red List*; Higher Education Press: Beijing, China, 2004.

12. Li, E.; Yi, S.; Qiu, Y.; Guo, J.; Comes, H.P.; Fu, C. Phylogeography of two East Asian species in *Croomia* (Stemonaceae) inferred from chloroplast DNA and ISSR fingerprinting variation. *Mol. Phylogenet. Evol.* **2008**, *49*, 702–714. [CrossRef] [PubMed]

13. Xiang, Q.Y.; Soltis, D.E.; Soltis, P.S. The eastern Asian and eastern and western North American floristic disjunction: Congruent phylogenetic patterns in seven diverse genera. *Mol. Phylogenet. Evol.* **1998**, *10*, 178–190. [CrossRef] [PubMed]

14. Wen, J. Evolution of eastern Asian and eastern North American disjunct distributions in flowering plants. *Annu. Rev. Ecol. Syst.* **1999**, *30*, 421–455. [CrossRef]

15. Wolfe, J.A. Some aspects of plant geography in the Northern Hemisphere during the late Cretaceous and Tertiary. *Ann. Mo. Bot. Gard.* **1975**, *62*, 264–279. [CrossRef]

16. Tiffney, B.H.; Manchester, S.R. The use of geological and paleontological evidence in evaluating plant phylogeographic hypotheses in the Northern Hemisphere Tertiary. *Int. J. Plant Sci.* **2001**, *162*, 48–52. [CrossRef]

17. Li, H.L. Floristic relationships between eastern Asia and eastern North America. *Trans. Am. Philos. Soc.* **1952**, *42*, 371–429. [CrossRef]

18. Fukuoka, N.; Kurosaki, N. Phytogeographical notes on some species of west Honshu, Japan 5. *Shoei Jr. Col. Ann. Rep. Stud.* **1985**, *17*, 61–71.

19. Yang, J.B.; Tang, M.; Li, H.T.; Zhang, Z.R.; Li, D.Z. Complete chloroplast genome of the genus *Cymbidium*: Lights into the species identification, phylogenetic implications and population genetic analyses. *BMC Evol. Boil.* **2013**, *13*, 84–98. [CrossRef] [PubMed]

20. Ruhsam, M.; Rai, H.S.; Mathews, S.; Ross, T.G.; Graham, S.W.; Raubeson, L.A.; Mei, W.; Thomas, P.I.; Gardner, M.F.; Ennos, R.A. Does complete plastid genome sequencing improve species discrimination and phylogenetic resolution in *Araucaria*? *Mol. Ecol. Res.* **2015**, *15*, 1067–1078. [CrossRef] [PubMed]

21. Firetti, F.; Zuntini, A.R.; Gaiarsa, J.W.; Oliveira, R.S.; Lohmann, L.G.; Van Sluys, M.A. Complete chloroplast genome sequences contribute to plant species delimitation: A case study of the *Anemopaegma* species complex. *Am. J. Bot.* **2017**, *104*, 1493–1509. [CrossRef] [PubMed]

22. Jansen, R.K.; Cai, Z.; Raubeson, L.A.; Daniell, H.; Claude, W.; Leebensmack, J.; Guisingerbellian, M.; Haberle, R.C.; Hansen, A.; Chumley, T.W. Analysis of 81 genes from 64 plastid genomes resolves relationships in angiosperms and identifies genome-scale evolutionary patterns. *Proc. Natl. Acad. Sci. USA* **2007**, *104*, 19369–19374. [CrossRef] [PubMed]

23. Cai, J.; Ma, P.; Li, H.; Li, D. Complete plastid genome sequencing of four *Tilia* species (Malvaceae): A comparative analysis and phylogenetic implications. *PLoS ONE* **2015**, *10*, e0142705. [CrossRef] [PubMed]

24. Sloan, D.B.; Alverson, A.J.; Wu, M.; Palmer, J.D.; Taylor, D.R. Recent acceleration of plastid sequence and

structural evolution coincides with extreme mitochondrial divergence in the angiosperm genus *Silene*. *Genome Biol. Evol.* **2012**, *4*, 294–306. [CrossRef] [PubMed]

25. Wang, Y.; Qu, X.; Chen, S.; Li, D.; Yi, T. Plastomes of Mimosoideae: Structural and size variation, sequence divergence, and phylogenetic implication. *Tree Genet. Genom.* **2017**, *13*, 41–56. [CrossRef]

26. Qian, J.; Song, J.; Gao, H.; Zhu, Y.; Xu, J.; Pang, X.; Yao, H.; Sun, C.; Li, X.E.; Li, C. The complete chloroplast genome sequence of the medicinal plant *Salvia miltiorrhiza*. *PLoS ONE* **2013**, *3*, e57607. [CrossRef] [PubMed]

27. Fu, C.; Li, H.; Milne, R.I.; Zhang, T.; Ma, P.; Yang, J.; Li, D.; Gao, L. Comparative analyses of plastid genomes from fourteen Cornales species: Inferences for phylogenetic relationships and genome evolution. *BMC Genom.* **2017**, *18*, 956–963. [CrossRef] [PubMed]

28. Hansen, D.R.; Dastidar, S.G.; Cai, Z.; Penaflor, C.; Kuehl, J.V.; Boore, J.L.; Jansen, R.K. Phylogenetic and evolutionary implications of complete chloroplast genome sequences of four early-diverging angiosperms: *Buxus* (Buxaceae), *Chloranthus* (Chloranthaceae), *Dioscorea* (Dioscoreaceae), and *Illicium* (Schisandraceae). *Mol. Phylogenet. Evol.* **2007**, *45*, 547–563. [CrossRef] [PubMed]

29. Liu, Y.; Huo, N.; Dong, L.; Wang, Y.; Zhang, S.; Young, H.A.; Feng, X.; Gu, Y.Q. Complete chloroplast genome sequences of mongolia medicine *Artemisia frigida* and phylogenetic relationships with other plants. *PLoS ONE* **2013**, *8*, e57533. [CrossRef] [PubMed]

30. Walker, J.F.; Zanis, M.J.; Emery, N.C. Comparative analysis of complete chloroplast genome sequence and inversion variation in *Lasthenia burkei* (Madieae, Asteraceae). *Am. J. Bot.* **2014**, *101*, 722–729. [CrossRef] [PubMed]

31. Yang, M.; Zhang, X.; Liu, G.; Yin, Y.; Chen, K.; Yun, Q.; Zhao, D.G.; Almssallem, I.S.; Yu, J. The complete chloroplast genome sequence of date palm (*Phoenix dactylifera* L.). *PLoS ONE* **2010**, *5*, e12762. [CrossRef] [PubMed]

32. Wang, M.; Cui, L.; Feng, K.; Deng, P.; Du, X.; Wan, F.; Weining, S.; Nie, X. Comparative analysis of asteraceae chloroplast genomes: Structural organization, RNA editing and evolution. *Plant Mol. Biol. Rep.* **2015**, *33*, 1526–1538. [CrossRef]

33. Palmer, J.D. Chloroplast DNA exists in two orientations. *Nature* **1983**, *301*, 92–93. [CrossRef]

34. Wolfe, A.D.; Randle, C.P. Recombination, heteroplasmy, haplotype polymorphism, and paralogy in plastid genes: Implications for plant molecular systematics. *Syst Bot.* **2004**, *29*, 1011–1020. [CrossRef]

35. Walker, J.F.; Jansen, R.K.; Zanis, M.J.; Emery, N.C. Sources of inversion variation in the small single copy (SSC) region of chloroplast genomes. *Am. J. Bot.* **2015**, *102*, 1751–1752. [CrossRef] [PubMed]

36. Kim, K.J.; Lee, H.L. Complete chloroplast genome sequences from Korean ginseng (*Panax schinseng Nees*) and comparative analysis of sequence evolution among 17 vascular plants. *DNA Res.* **2004**, *11*, 247–261. [CrossRef] [PubMed]

37. Downie, S.R.; Jansen, R.K. A comparative analysis of whole plastid genomes from the apiales: Expansion and contraction of the inverted repeat, mitochondrial to plastid transfer of DNA, and identification of highly divergent noncoding regions. *Sys. Bot.* **2015**, *40*, 336–351. [CrossRef]

38. Wang, R.J.; Cheng, C.L.; Chang, C.C.; Wu, C.L.; Su, T.M.; Chaw, S. Dynamics and evolution of the inverted repeat-large single copy junctions in the chloroplast genomes of monocots. *BMC Evol. Biol.* **2008**, *8*, 6–8. [CrossRef] [PubMed]

39. Nazareno, A.G.; Carlsen, M.; Lohmann, L.G. Complete chloroplast genome of tanaecium tetragonolobum: The first *Bignoniaceae plastome*. *PLoS ONE* **2015**, *10*, e0129930. [CrossRef] [PubMed]

40. Yao, X.; Tang, P.; Li, Z.; Li, D.; Liu, Y.; Huang, H. The first complete chloroplast genome sequences in Actinidiaceae: Genome structure and comparative analysis. *PLoS ONE* **2015**, *10*, e0129347. [CrossRef] [PubMed]

41. Khakhlova, O.; Bock, R. Elimination of deleterious mutations in plastid genomes by gene conversion. *Plant J.* **2006**, *46*, 85–94. [CrossRef] [PubMed]

42. Zhang, Y.; Ma, P.; Li, D. High-throughput sequencing of six bamboo chloroplast genomes: Phylogenetic implications for temperate woody bamboos (Poaceae: Bambusoideae). *PLoS ONE* **2011**, *6*, e20596. [CrossRef] [PubMed]

43. Choi, K.S.; Chung, M.G.; Park, S. The Complete Chloroplast Genome Sequences of Three Veroniceae Species (Plantaginaceae): Comparative Analysis and Highly Divergent Regions. *Front. Plant Sci.* **2016**, *7*, 355–394. [CrossRef] [PubMed]

44. Shaw, J.; Shafer, H.L.; Leonard, O.R.; Kovach, M.J.; Schorr, M.S.; Morris, A.B. Chloroplast DNA sequence utility for the lowest phylogenetic and phylogeographic inferences in angiosperms: The tortoise and the hare IV. *Am. J. Bot.* **2014**, *101*, 1987–2004. [CrossRef] [PubMed]

45. Shaw, J.; Lickey, E.B.; Schilling, E.E.; Small, R.L. Comparison of whole chloroplast genome sequences to choose noncoding regions for phylogenetic studies in angiosperms: The tortoise and the hare III. *Am. J. Bot.* **2007**, *94*, 275–288. [CrossRef] [PubMed]

46. Li, P.; Lu, R.; Xu, W.; Ohitoma, T.; Cai, M.; Qiu, Y.; Cameron, K.M.; Fu, C. Comparative genomics and phylogenomics of east Asian tulips (amana, Liliaceae). *Front. Plant Sci.* **2017**, *8*, 35–39. [CrossRef] [PubMed]

47. Carlson, S.E.; Linder, H.P.; Donoghue, M.J. The historical biogeography of *Scabiosa* (Dipsacaceae): Implications for Old World plant disjunctions. *J. Biol.* **2012**, *39*, 1086–1100. [CrossRef]

48. Yue, J.; Sun, H.; Baum, D.A.; Jianhua, L.I.; Alshehbaz, I.A.; Ree, R.H. Molecular phylogeny of *Solms-laubachia* (Brassicaceae) s.l., based on multiple nuclear and plastid DNA sequences, and its biogeographic implications. *J. Syst. Evol.* **2009**, *47*, 402–415. [CrossRef]

49. Kurtz, S.; Schleiermacher, C. REPuter: Fast computation of maximal repeats in complete genomes. *Bioinformatics* **1999**, *15*, 426–427. [CrossRef] [PubMed]

50. Huang, H.; Shi, C.; Liu, Y.; Mao, S.Y.; Gao, L.Z. Thirteen *Camellia* chloroplast genome sequences determined by high-throughput sequencing: Genome structure and phylogenetic relationships. *BMC Evol. Biol.* **2014**, *14*, 151–162. [CrossRef] [PubMed]

51. Zhao, Y.; Yin, J.; Guo, H.; Zhang, Y.; Xiao, W.; Sun, C.; Wu, J.; Qu, X.; Yu, J.; Wang, X. The complete chloroplast genome provides insight into the evolution and polymorphism of *Panax ginseng*. *Front. Plant Sci.* **2015**, *5*, 696–698. [CrossRef] [PubMed]

52. Thiel, T.; Michalek, W.; Varshney, R.; Graner, A. Exploiting EST databases for the development and characterization of gene-derived SSR-markers in barley (*Hordeum vulgare* L.). *Theor. Appl. Genet.* **2003**, *106*, 411–422. [CrossRef] [PubMed]

53. Kim, K.; Lee, S.C.; Lee, J.; Yu, Y.; Yang, K.; Choi, B.S.; Koh, H.J.; Waminal, N.E.; Choi, H.I.; Kim, N.H. Complete chloroplast and ribosomal sequences for 30 accessions elucidate evolution of Oryza AA genome species. *Sci. Rep.* **2015**, *5*, 15–16. [CrossRef] [PubMed]

54. Milne, R.I.; Abbott, R.J. The origin and evolution of tertiary relict floras. *Adv. Bot. Res* **2002**, *38*, 281–314.

55. Tiffney, B.H. Perspectives on the origin of the floristic similarity between eastern Asia and eastern North America. *J. Arn. Arb.* **1985**, *66*, 73–94. [CrossRef]

56. Xiang, Q.; Soltis, D.E.; Soltis, P.S.; Manchester, S.R.; Crawford, D.J. Timing the eastern Asian-eastern North American floristic disjunction: Molecular clock corroborates paleontological estimates. *Mol. Phylogenet. Evol.* **2000**, *15*, 462–472. [CrossRef] [PubMed]

57. Birky, C.W. Uniparental inheritance of mitochondrial and chloroplast genes: Mechanisms and evolution. *Proc. Natl. Acad. Sci. USA* **1995**, *92*, 11331–11338. [CrossRef] [PubMed]

58. Cronn, R.; Liston, A.; Parks, M.; Gernandt, D.S.; Shen, R.; Mockler, T. Multiplex sequencing of plant chloroplast genomes using Solexa sequencing-by-synthesis technology. *Nucleic Acids Res.* **2008**, *36*, 122–125. [CrossRef] [PubMed]

59. Wyman, S.K.; Jansen, R.K.; Boore, J.L. Automatic annotation of organellar genomes with DOGMA. *Bioinformatics* **2004**, *20*, 3252–3255. [CrossRef] [PubMed]

60. Schattner, P.; Brooks, A.N.; Lowe, T.M. The tRNAscan-SE, snoscan and snoGPS web servers for the detection of tRNAs and snoRNAs. *Nucleic Acids Res.* **2005**, *33*, 686–689. [CrossRef] [PubMed]

61. Lohse, M.; Drechsel, O.; Bock, R. OrganellarGenomeDRAW (OGDRAW): A tool for the easy generation of high-quality custom graphical maps of plastid and mitochondrial genomes. *Curr. Genet.* **2007**, *52*, 267–274. [CrossRef] [PubMed]

62. Frazer, K.A.; Pachter, L.; Poliakov, A.; Rubin, E.M.; Dubchak, I. VISTA: Computational tools for comparative genomics. *Nucleic Acids Res.* **2004**, *32*, 273–279. [CrossRef] [PubMed]

63. Librado, P.; Rozas, J. DnaSP v5. *Bioinformatics* **2009**, *25*, 1451–1452. [CrossRef] [PubMed]

64. Shaw, J.; Lickey, E.B.; Beck, J.T.; Farmer, S.B.; Liu, W.; Miller, J.; Siripun, K.C.; Winder, C.T.; Schilling, E.E.; Small, R.L. The tortoise and the hare II: Relative utility of 21 noncoding chloroplast DNA sequences for phylogenetic analysis. *Am. J. Bot.* **2005**, *92*, 142–166. [CrossRef] [PubMed]

65. Katoh, K.; Standley, D.M. MAFFT multiple sequence alignment software version 7: Improvements in performance and usability. *Mol. Bio. Evol.* **2013**, *30*, 772–780. [CrossRef] [PubMed]

66. Posada, D. jModelTest: Phylogenetic model averaging. *Mol. Biol. Evol.* **2008**, *25*, 1253–1256. [CrossRef] [PubMed]
67. Miller, M.A.; Pfeiffer, W.; Schwartz, T. Creating the CIPRES Science Gateway for inference of large phylogenetic trees. *Gatew. Comput. Environ. Workshop* **2010**, *3*, 1–8.
68. Ronquist, F.; Huelsenbeck, J.P. MrBayes 3: Bayesian phylogenetic inference under mixed models. *Bioinformatics* **2003**, *19*, 1572–1574. [CrossRef] [PubMed]
69. Hohmann, N.; Wolf, E.M.; Lysak, M.A.; Koch, M.A. A time-calibrated road map of Brassicaceae species radiation and evolutionary history. *Plant Cell* **2015**, *27*, 2770–2784. [CrossRef] [PubMed]
70. Hu, H.; Hu, Q.; Alshehbaz, I.A.; Luo, X.; Zeng, T.; Guo, X.; Liu, J. Species delimitation and interspecific relationships of the genus *Orychophragmus* (Brassicaceae) inferred from whole chloroplast genomes. *Front. Plant Sci.* **2016**, *7*, 1826–2884. [CrossRef] [PubMed]
71. Drummond, A.J.; Suchard, M.A.; Xie, D.; Rambaut, A. Bayesian phylogenetics with BEAUti and the BEAST 1.7. *Mol. Biol. Evol.* **2012**, *29*, 1969–1973. [CrossRef] [PubMed]
72. Rambaut, A.; Drummond, A.J. Tracer v.1.4. *Encycl. Atmos. Sci.* **2007**, *141*, 2297–2305.
73. Bouckaert, R.R.; Heled, J.; Kuhnert, D.; Vaughan, T.G.; Wu, C.H.; Xie, D.; Suchard, M.A.; Rambaut, A.; Drummond, A.J. BEAST 2: A software platform for bayesian evolutionary analysis. *PLoS Computat. Biol.* **2014**, *10*, 38–49. [CrossRef] [PubMed]
74. Yu, Y.; Harris, A.J.; Blair, C.; He, X. RASP (Reconstruct Ancestral State in Phylogenies): A tool for historical biogeography. *Mol. Phylogenet. Evol.* **2015**, *87*, 46–49. [CrossRef] [PubMed]

10

Two Coiled-Coil Proteins, WEB1 and PMI2, Suppress the Signaling Pathway of Chloroplast Accumulation Response that is Mediated by Two Phototropin-Interacting Proteins, RPT2 and NCH1, in Seed Plants

Noriyuki Suetsugu [1],* and Masamitsu Wada [2]

[1]   Graduate School of Biostudies, Kyoto University, Kyoto 606-8502, Japan
[2]   School of Science and Engineering, Tokyo Metropolitan University, Tokyo 192-0397, Japan; masamitsu.wada@gmail.com
*   Correspondence: suetsugu@lif.kyoto-u.ac.jp

Abstract: Chloroplast movement is induced by blue light in a broad range of plant species. Weak light induces the chloroplast accumulation response and strong light induces the chloroplast avoidance response. Both responses are essential for efficient photosynthesis and are mediated by phototropin blue-light receptors. J-DOMAIN PROTEIN REQUIRED FOR CHLOROPLAST ACCUMULATION RESPONSE 1 (JAC1) and two coiled-coil domain proteins WEAK CHLOROPLAST MOVEMENT UNDER BLUE LIGHT 1 (WEB1) and PLASTID MOVEMENT IMPAIRED 2 (PMI2) are required for phototropin-mediated chloroplast movement. Genetic analysis suggests that JAC1 is essential for the accumulation response and WEB1/PMI2 inhibit the accumulation response through the suppression of JAC1 activity under the strong light. We recently identified two phototropin-interacting proteins, ROOT PHOTOTROPISM 2 (RPT2) and NPH3/RPT2-like (NRL) PROTEIN FOR CHLOROPLAST MOVEMENT 1 (NCH1) as the signaling components involved in chloroplast accumulation response. However, the relationship between RPT2/NCH1, JAC1 and WEB1/PMI2 remained to be determined. Here, we performed genetic analysis between RPT2/NCH1, JAC1, and WEB1/PMI2 to elucidate the signal transduction pathway.

Keywords: Arabidopsis; blue light; *Marchantia*; organelle movement; phototropin

1. Introduction

Phototropins (phot) are blue-light photoreceptor kinases that mediate phototropism, leaf flattening, stomatal opening, and chloroplast movement including low light-induced chloroplast accumulation response and strong light-induced chloroplast avoidance response (herein referred to as the accumulation and avoidance response, respectively). These responses contribute to optimal photosynthetic light utilization at the organ/tissue, cellular, and organelle level [1,2]. Most land plants have two or more phototropin genes and functional differences exist between them. There are two phototropins in *Arabidopsis thaliana*, phot1 and phot2. Phototropism, leaf flattening, stomatal opening, and the accumulation response are mediated by both phot1 and phot2; however, phot1 plays a greater role in these responses, especially at lower blue-light intensities [1,2]. In contrast, the avoidance response is mediated primarily by phot2 [1,2]. Clear functional divergence of phototropins in the accumulation and avoidance response is observed in the moss *Physcomitrella patens* and the fern *Adiantum capillus-veneris* [3,4]. However, in the liverwort *Marchantia polymorpha*, which is a basal land plant, a single phototropin mediates both the accumulation and avoidance responses [5].

Thus, phototropins can intrinsically mediate all phototropin-mediated responses and functional diversification of phototropins seems to have occurred during land plant evolution concomitant with phototropin gene duplication [6].

By regulating various signaling components, such as phototropin-interacting proteins, phototropins can regulate multiple, diverse responses. For example, BLUE LIGHT SIGNALING1 kinase is a direct phototropin substrate and specifically mediates stomatal opening [7]. In addition, the two phototropin-interacting bric á brac, tramtrack and broad complex/pox virus and zinc finger (BTB/POZ) domain proteins NONPHOTOTROPIC HYPOCOTYL 3 (NPH3) and ROOT PHOTOTROPISM 2 (RPT2), which belong to the NPH3/RPT2-like (NRL) protein family, mediate phototropism and leaf flattening [8–11]. Recently, we identified a phototropin-interacting NRL protein, NRL PROTEIN FOR CHLOROPLAST MOVEMENT 1 (NCH1), and found that NCH1 specifically mediates the accumulation response [12]. NCH1 is highly similar to RPT2 than NPH3 and contains four conserved regions including a BTB/POZ domain (Figure 1a). Furthermore, functional redundancy was found between NCH1 and RPT2 for the accumulation response, but not the avoidance response [12]. These results indicate that phototropism, leaf flattening, and the accumulation response are dependent on these NRL proteins, while stomatal opening and the avoidance response are independent of these NRL proteins. In *M. polymorpha*, the RPT2/NCH1 ortholog MpNCH1 specifically mediates the accumulation response, indicating that phototropin-regulated chloroplast movement is conserved in land plants [12].

Similar to NPH3 and RPT2, NCH1 is localized on the plasma membrane and interacts with phototropins [12], but the downstream function of NCH1 as well as other NRL proteins remained to be determined. Here, we performed genetic analysis of *RPT2* and *NCH1* using triple or quadruple mutant plants between *rpt2nch1* and other mutants that were implicated in the signal transduction of chloroplast movement.

2. Results and Discussion

J-domain protein required for chloroplast accumulation response 1 (*jac1*) and *rpt2nch1* plants are both defective in the accumulation response [12,13]. JAC1 protein is a C-terminal J-domain protein similar to clathrin uncoating factor auxilin (Figure 1a) [13]. To analyze chloroplast movement in *jac1* and *rpt2nch1* in detail, we performed analysis of the light-induced changes in leaf transmittance, reflective of light-induced chloroplast movements (Figure 1b) [14]. In response to 3 μmol m^{-2} s^{-1} of blue light, which induces the accumulation response in wild type, a clear avoidance response is induced in *rpt2nch1* but not in *jac1* (Figure 1c,d) [12]. Therefore, RPT2/NCH1 could suppress the induction of the avoidance response to facilitate efficient induction of the accumulation response under low light conditions (Figure 2). Compared to wild type and *jac1*, a faster avoidance response was induced by 20 μmol m^{-2} s^{-1} of blue light in *rpt2nch1* (Figure 1c,d; one-way ANOVA followed by Tukey–Kramer multiple comparison post hoc test, $p < 0.01$ for wild type or *jac1* vs. *rpt2nch1*), although similar transmittance changes (defined as "amplitude" in Figure 1e) were observed for both *rpt2nch1* and *jac1* following 40 min of 20 μmol m^{-2} s^{-1} blue-light irradiation (Figure 1c,e; one-way ANOVA followed by Tukey–Kramer multiple comparison post hoc test, $p > 0.05$ for *jac1* vs. *rpt2nch1*). Following subsequent application of 50 μmol m^{-2} s^{-1} blue light, only a slight additional avoidance response was observed in *rpt2nch1* which contrasted with the stronger avoidance response induced in wild type (Figure 1c) [12]. This result could be attributed to prior movement of the majority of chloroplasts to the side walls in *rpt2nch1* during the former irradiation period. Interestingly, avoidance responses of similar magnitudes were induced in *jac1* in response to both 20 and 50 μmol m^{-2} s^{-1} of blue light and a decreased rate of avoidance response induction was not observed during strong light irradiation (Figure 1c). The changes observed in leaf transmittance for *rpt2nch1jac1* were intermediate between those observed for *rpt2nch1* and *jac1*.

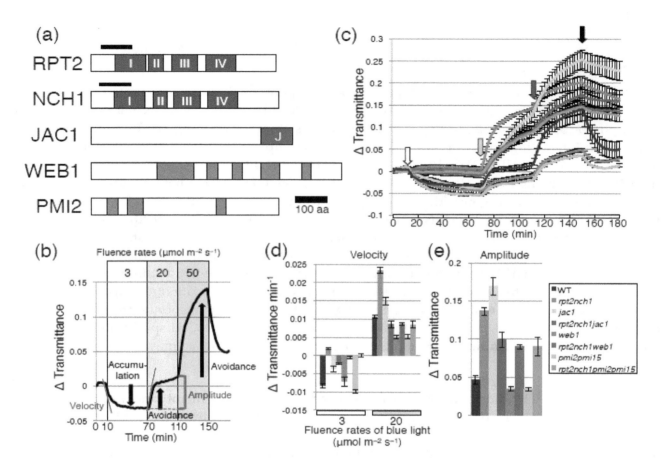

Figure 1. (**a**) Protein structure of ROOT PHOTOTROPISM 2 (RPT2), NRL PROTEIN FOR CHLOROPLAST MOVEMENT 1 (NCH1), J-DOMAIN PROTEIN REQUIRED FOR CHLOROPLAST ACCUMULATION RESPONSE 1 (JAC1), WEAK CHLOROPLAST MOVEMENT UNDER BLUE LIGHT 1 (WEB1), and PLASTID MOVEMENT IMPAIRED 2 (PMI2). **Blue** boxes indicate the four conserved regions of NPH3/RPT2-like (NRL) proteins. The position of the BTB/POZ domain is indicated by a **black** bar. **Red** box is a J-domain. **Green** boxes indicate the coiled-coil domains; (**b**) Measurement of light-induced changes in leaf transmittance as a result of chloroplast photorelocation movements. The depicted trace represents typical data collected for wild type under the various light irradiation conditions (indicated by color boxes). There is a decrease in leaf transmittance in response to 3 μmol m^{-2} s^{-1} of blue light, indicating that the accumulation response is induced (**downward arrow**). Conversely, there is an increase in leaf transmittance in response to 20 and 50 μmol m^{-2} s^{-1} of blue light, indicating that the avoidance response is induced (**upward arrows**). **Red lines** mark the initial linear fragments of leaf transmittance rate change during the first 2–6 min of the irradiation period, indicating the velocity. A **red parenthesis** marks the difference between the transmittance level observed following 60 min of 3 μmol m^{-2} s^{-1} blue-light irradiation and the transmittance level observed a following further 40 min of 20 μmol m^{-2} s^{-1} blue-light irradiation, indicating the amplitude of the avoidance response caused by 20 μmol m^{-2} s^{-1} blue-light irradiation; (**c–e**) Distinct chloroplast movements observed between *rpt2nch1* and *jac1*; (**c**) Light-induced changes in leaf transmittance of the indicated lines were measured using a custom-made plate reader system [14]. The samples were sequentially irradiated with 3, 20 and 50 μmol m^{-2} s^{-1} of continuous blue light. The beginning of each irradiation period is indicated by **white**, **cyan** and **blue** arrows, respectively. The light was extinguished after 150 min (**black arrow**); (**d**) The velocity of light-induced transmittance changes. (**e**) The amplitude of the avoidance response caused by 20 μmol m^{-2} s^{-1} blue-light irradiation. Data for wild type, *rpt2nch1*, *jac1* and *rpt2nch1jac1* from Suetsugu et al. (2016) [12] were used for comparison, because data for *web1*, *rpt2nch1web1*, *pmi2pmi15* and *rpt2nch1pmi2pmi15* were acquired in the same experiments using the same plate. Data are presented as means of three independent experiments and the error bars indicate standard errors. WT, wild type.

Previously, we showed that *JAC1* mutation suppresses the defective avoidance response in *weak chloroplast movement under blue light 1* (*web1*) and *plastid movement impaired 2* (*pmi2*) [15]. WEB1 and PMI2 are related coiled-coil domain proteins that interact with each other (Figure 1a) [15]. Although the low light-induced accumulation response was normal in *web1* and *pmi2pmi15*, both mutant plants exhibited attenuated avoidance response under the strong light conditions (Figure 1c to e) [15]. The *jac1web1* and *jac1pmi2pmi15* showed nearly the same phenotypes as *jac1* single mutants [15]. Importantly, the weak avoidance response phenotype observed in *web1* and *pmi2pmi15* was completely suppressed in *jac1web1* and *jac1pmi2pmi15*, respectively [15]. Therefore, we hypothesized that WEB1 and PMI2 suppress JAC1 function under strong light, preventing the induction of the JAC1-dependent accumulation response and leading to efficient induction of the avoidance response (Figure 2). The weak avoidance response phenotype observed in *web1* and *pmi2pmi15* was absent in *rpt2nch1web1* and *rpt2nch1pmi2pmi15*, similar to *jac1web1* and *jac1pmi2pmi15* (Figure 1c–e). Mutation of *JAC1* suppressed *web1* and *pmi2pmi15* phenotypes, because *jac1web1* and *jac1pmi2pmi15* phenotypes are indistinguishable from *jac1* [15]. Although mutation of *RPT2* and *NCH1* largely suppressed the weak avoidance response phenotypes observed in *web1* and *pmi2pmi15*, the velocity and amplitude of the avoidance response in these mutants did not match those in *rpt2nch1* (Figure 1c–e; one-way ANOVA followed by Tukey–Kramer multiple comparison post hoc test, $p < 0.01$ for *rpt2nch1web1* or *rpt2nch1pmi2pmi15* vs. *rpt2nch1* in velocity and $p < 0.05$ for *rpt2nch1web1* or *rpt2nch1pmi2pmi15* vs. *rpt2nch1* in amplitude). The phenotypes of *rpt2nch1web1* and *rpt2nch1pmi2pmi15* were very similar to *rpt2nch1jac1* in that no detectable chloroplast movement was observed under 3 µmol m^{-2} s^{-1} of blue light and their avoidance response phenotypes were similar to *jac1* (Figure 1c,d; one-way ANOVA followed by Tukey–Kramer multiple comparison post hoc test, $p > 0.05$ for *rpt2nch1web1* or *rpt2nch1pmi2pmi15* vs. *rpt2nch1jac1*). Collectively, our results indicate that RPT2 and NCH1 are essential for the accumulation response and regulate JAC1-dependent and -independent pathways and that WEB1/PMI2 represses the signaling pathway for the accumulation response under the strong light conditions (Figure 2). However, how WEB1 and PMI2 suppress the accumulation response pathway remained to be determined. Interaction of WEB1 and/or PMI2 with JAC1 has never been detected [15]. At the least, the amounts of phototropins were normal in *web1* and *pmi2pmi15* mutant plants [15]. Further analysis of the relationship between WEB1/PMI2, RPT2/NCH1 and JAC1 is required.

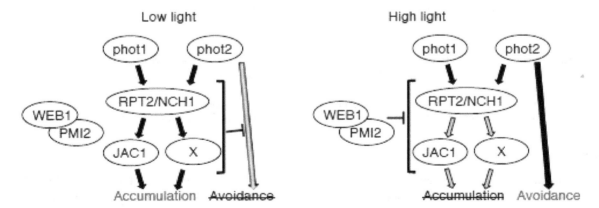

Figure 2. Working model of chloroplast photorelocation movements. The photoreceptors phot1 and phot2 mediate the accumulation response under a low light condition through RPT2 and NCH1. RPT2 and NCH1 might regulate both JAC1-dependent and -independent (X) pathways. The signaling pathway by RPT2/NCH1 and JAC1 suppresses that of the avoidance response under a low light condition. Under the high light condition, the WEB1/PMI2 complex suppresses the signaling pathway for the accumulation response that is regulated by RPT2/NCH1 and JAC1 through an unknown mechanism, resulting in the efficient induction of the avoidance response mediated by phot2. **Gray** arrows indicate the suppressed signaling pathways. **Black** arrows indicate the activated signaling pathways.

RPT2 and NCH1 are localized on the plasma membrane and interact with phototropins, indicating that RPT2 and NCH1 are the initial downstream signaling components involved in the phototropin-mediated accumulation response (Figure 2) [12]. Notably, RPT2 and NCH1 are conserved in land plants, but JAC1, WEB1 and PMI2 orthologs are found only in seed plants [16]. Thus, to maximize light utilization through chloroplast movement, land plants have evolved a sophisticated mechanism of controlling chloroplast movement by increasing the molecular components involved in blue-light signaling.

3. Materials and Methods

3.1. Arabidopsis Lines and the Growth Condition

The wild-type and mutant lines are a Columbia *gl1* background. Seeds were sown on 0.8% agar medium containing 1/3 strength Murashige & Skoog salt and 1% sucrose, and grown under white light at ca. ~100 µmol m^{-2} s^{-1} (16 h)/dark (8 h) cycle at 23 °C in an incubator. *rpt2-4nch1-1* [12], *jac1-1* [13], *web1-2* [15] and *pmi2-2pmi15-1* [15] were described previously. For *PMI2* mutant plants, *pmi2pmi15* was used, because PMI15 is closely related to PMI2 and *pmi15* exhibits a very weak defect in chloroplast movement [17]. *rpt2-4*, *nch1-1*, *pmi2-2* and *pmi15-1* are T-DNA knockout lines [12,17]. *jac1-1* carries a missense mutation [13] and *web1-2* carries a deletion of one nucleotide [15]. Western blot analysis showed that JAC1 and WEB1 proteins were not detected in *jac1-1* and *web1-2*, respectively [13,15]. Double, triple and quadruple mutants were generated by genetic crossing.

3.2. Analyses of Chloroplast Photorelocation Movements

Chloroplast photorelocation movements were analyzed by the measurement of light-induced changes in leaf transmittance as described previously [14]. Third leaves that were detached from 16-day-old seedlings were placed on 1% (*w/v*) gellan gum in a 96-well plate and then dark-adapted at least for 1 h before transmittance measurement.

3.3. Statistical Analysis

Statistical analyses were performed by one-way ANOVA followed by Tukey–Kramer multiple comparison post hoc test.

Acknowledgments: This work was supported in part by the Grant-in-Aid for Scientific Research Grants (26840097 and 15KK0254 to Noriyuki Suetsugu; 20227001, 23120523, 25120721, and 25251033 to Masamitsu Wada).

Author Contributions: Noriyuki Suetsugu conceived, designed and performed the experiments; Noriyuki Suetsugu and Masamitsu Wada analyzed the data, contributed reagents/materials/analysis tools and wrote the paper.

References

1. Christie, J.M. Phototropin blue-light receptors. *Annu. Rev. Plant Biol.* **2007**, *58*, 21–45. [CrossRef] [PubMed]
2. Suetsugu, N.; Wada, M. Evolution of three LOV blue light receptor families in green plants and photosynthetic stramenopiles: Phototropin, ZTL/FKF1/LKP2 and aureochrome. *Plant Cell Physiol.* **2013**, *54*, 8–23. [CrossRef] [PubMed]
3. Kasahara, M.; Kagawa, T.; Sato, Y.; Kiyosue, T.; Wada, M. Phototropins mediate blue and red light-induced chloroplast movements in *Physcomitrella patens*. *Plant Physiol.* **2004**, *135*, 1388–1397. [CrossRef] [PubMed]
4. Kagawa, T.; Kasahara, M.; Abe, T.; Yoshida, S.; Wada, M. Function analysis of phototropin2 using fern mutants deficient in blue light-induced chloroplast avoidance movement. *Plant Cell Physiol.* **2004**, *45*, 416–426. [CrossRef] [PubMed]
5. Komatsu, A.; Terai, M.; Ishizaki, K.; Suetsugu, N.; Tsuboi, H.; Nishihama, R.; Yamato, K.T.; Wada, M.; Kohchi, T. Phototropin encoded by a single-copy gene mediates chloroplast photorelocation movements in the liverwort *Marchantia polymorpha*. *Plant Physiol.* **2014**, *166*, 411–427. [CrossRef] [PubMed]

6.   Li, F.W.; Rothfels, C.J.; Melkonian, M.; Villarreal, J.C.; Stevenson, D.W.; Graham, S.W.; Wong, G.K.S.; Mathews, S.; Pryer, K.M. The origin and evolution of phototropins. *Front. Plant Sci.* **2015**, *6*, 637. [CrossRef] [PubMed]

7.   Takemiya, A.; Sugiyama, N.; Fujimoto, H.; Tsutsumi, T.; Yamauchi, S.; Hiyama, A.; Tada, Y.; Christie, J.M.; Shimazaki, K. Phosphorylation of BLUS1 kinase by phototropins is a primary step in stomatal opening. *Nat. Commun.* **2013**, *4*, 2094. [CrossRef] [PubMed]

8.   Motchoulski, A.; Liscum, E. Arabidopsis NPH3: A NPH1 photoreceptor-interacting protein essential for phototropism. *Science* **1999**, *286*, 961–964. [CrossRef] [PubMed]

9.   Sakai, T.; Wada, T.; Ishiguro, S.; Okada, K. RPT2: A signal transducer of the phototropic response in Arabidopsis. *Plant Cell* **2000**, *12*, 225–236. [CrossRef] [PubMed]

10.   Inoue, S.; Kinoshita, T.; Takemiya, A.; Doi, M.; Shimazaki, K. Leaf positioning of Arabidopsis in response to blue light. *Mol. Plant* **2008**, *1*, 15–26. [CrossRef] [PubMed]

11.   Harada, A.; Takemiya, A.; Inoue, S.; Sakai, T.; Shimazaki, K. Role of RPT2 in leaf positioning and flattening and a possible inhibition of phot2 signaling by phot1. *Plant Cell Physiol.* **2013**, *54*, 36–47. [CrossRef] [PubMed]

12.   Suetsugu, N.; Takemiya, A.; Kong, S.G.; Higa, T.; Komatsu, A.; Shimazaki, K.; Kohchi, T.; Wada, M. RPT2/NCH1 subfamily of NPH3-like proteins is essential for the chloroplast accumulation response in land plants. *Proc. Natl. Acad. Sci. USA* **2016**, *113*, 10424–10429. [CrossRef] [PubMed]

13.   Suetsugu, N.; Kagawa, T.; Wada, M. An auxilin-like J-domain protein, JAC1, regulates phototropin-mediated chloroplast movement in Arabidopsis. *Plant Physiol.* **2005**, *139*, 151–162. [CrossRef] [PubMed]

14.   Wada, M.; Kong, S.G. Analysis of chloroplast movement and relocation in Arabidopsis. *Methods Mol. Biol.* **2011**, *774*, 87–102. [CrossRef]

15.   Kodama, Y.; Suetsugu, N.; Kong, S.G.; Wada, M. Two interacting coiled-coil proteins, WEB1 and PMI2, maintain the chloroplast photorelocation movement velocity in Arabidopsis. *Proc. Natl. Acad. Sci. USA* **2010**, *107*, 19591–19596. [CrossRef] [PubMed]

16.   Suetsugu, N.; Wada, M. Evolution of the cp-actin-based motility system of chloroplasts in green plants. *Front. Plant Sci.* **2016**, *7*, 561. [CrossRef] [PubMed]

17.   Luesse, D.R.; DeBlasio, S.L.; Hangarter, R.P. Plastid movement impaired 2, a new gene involved in normal blue-light-induced chloroplast movements in Arabidopsis. *Plant Physiol.* **2006**, *141*, 1328–1337. [CrossRef] [PubMed]

11

Exploring the History of Chloroplast Capture in *Arabis* using Whole Chloroplast Genome Sequencing

Akira Kawabe *, Hiroaki Nukii and Hazuka Y. Furihata

Faculty of Life Sciences, Kyoto Sangyo University, Kyoto, Kyoto 603-8555, Japan;
g1447799@cc.kyoto-su.ac.jp (H.N.); hazuka.furi@cc.kyoto-su.ac.jp (H.Y.F.)
* Correspondence: akiraka@cc.kyoto-su.ac.jp

Abstract: Chloroplast capture occurs when the chloroplast of one plant species is introgressed into another plant species. The phylogenies of nuclear and chloroplast markers from East Asian *Arabis* species are incongruent, which indicates hybrid origin and shows chloroplast capture. In the present study, the complete chloroplast genomes of *A. hirsuta*, *A. nipponica*, and *A. flagellosa* were sequenced in order to analyze their divergence and their relationships. The chloroplast genomes of *A. nipponica* and *A. flagellosa* were similar, which indicates chloroplast replacement. If hybridization causing chloroplast capture occurred once, divergence between recipient species would be lower than between donor species. However, the chloroplast genomes of species with possible hybrid origins, *A. nipponica* and *A. stelleri*, differ at similar levels to possible maternal donor species *A. flagellosa*, which suggests that multiple hybridization events have occurred in their respective histories. The mitochondrial genomes exhibited similar patterns, while *A. nipponica* and *A. flagellosa* were more similar to each other than to *A. hirsuta*. This suggests that the two organellar genomes were co-transferred during the hybridization history of the East Asian *Arabis* species.

Keywords: *Arabis*; chloroplast capture; Brassicaceae

1. Introduction

The genus *Arabis* includes about 70 species that are distributed throughout the northern hemisphere. The genus previously included many more species, but a large number of these were reclassified into other genera, including *Arabidopsis*, *Turritis*, and *Boechera*, *Crucihimalaya*, *Scapiarabis*, and *Sinoarabis* [1–6]. Because of their highly variable morphology and life histories, *Arabis* species have been used for ecological and evolutionary studies of morphologic and phenotypic traits [7–11]. The whole genome of *Arabis alpina* has been sequenced, providing genomic information for evolutionary analyses [12,13].

Molecular phylogenetic studies of *Arabis* species have been conducted to determine species classification and also correlation to morphological evolution of *Arabis* species [10,14,15]. Despite having similar morphologies, *A. hirsuta* from Europe, North America, and East Asia have been placed in different phylogenetic positions and are now considered distinct species. For example, East Asian *A. hirsuta*, which was previously classified as *A. hirsuta* var. *nipponica*, is now designated as *A. nipponica* [16]. Meanwhile, nuclear ITS sequences indicated that *A. nipponica*, *A. stelleri*, and *A. takeshimana* were closely related to European *A. hirsuta*. However, chloroplast *trnLF* sequences indicated that the species were closely related to East Asian *Arabis* species [14,16]. Such incongruent nuclear and organellar phylogenies have been reported from in other plant species and this is generally known as "chloroplast capture" [17,18], which is a process that involves hybridization and many successive backcrosses [17]. When chloroplast capture happens, the chloroplast genome of a species is replaced by another species' chloroplast genome. *A. nipponica* may have originated from the hybridization of *A. hirsuta* or *A. sagittata* and East Asian *Arabis* species (similar to *A. serrata*, *A. paniculata*, and *A. flagellosa*), which act as paternal and maternal parents, respectively [14,16]. However, the evolutionary history and hybridization processes of *A. nipponica* and other East Asian *Arabis* species still need to be clarified. Because these conclusions for incongruence between nuclear and chloroplast phylogenies came from analyzing a small number of short sequences, hybridized species, the divergence level, and the classification of species are somewhat ambiguous. In the present study, the whole chloroplast genomes of three *Arabis* species were sequenced in order to analyze their divergence and evolutionary history. The whole chloroplast genome sequences also provide a basis for future marker development.

2. Results

2.1. Chloroplast Genome Structure of Arabis Species

The structures of the whole chloroplast genomes are summarized in Table 1, which also includes previously reported *Arabis* chloroplast genomes and the chloroplast genome of the closely related species *Draba nemorosa*. The chloroplast genome structure identified in the present study is shown as a circular map (see Figure 1). The complete chloroplast genomes of the *Arabis* species had total lengths of 152,866–153,758 base pairs, which included 82,338 to 82,811 base pair long single copy (LSC) regions and 17,938 to 18,156 base pair short single copy (SSC) regions, which were separated by a pair of 26,421 to 26,933 base pair inverted repeat (IR) regions. The structure and length are conserved, and are similar to other Brassicaceae species' chloroplast genome sequences [19–22]. The complete genomes contain 86 protein-coding genes, 37 tRNA genes, and eight rRNA genes. Of these, seven protein-coding genes, seven tRNA genes, and four rRNA genes were located in the IR regions, and were therefore duplicated. The *rps16* gene became a pseudogene in *A. flagellosa*, *A. hirsuta*, and *A. nipponica* strain Midori, which was previously reported as a related species [23]. In addition, the *rps16* sequences of *D. nemorosa*, *A. stelleri*, *A. flagellosa*, *A. hirsuta*, and *A. nipponica* shared a 10 base pair deletion in the first exon, while *A. stelleri*, *A. flagellosa*, *A. hirsuta*, and *A. nipponica* shared a 1 base pair deletion in the second exon and *D. nemorosa* lacked the second exon entirely. The *rps16* sequence of *A. alpina* also lacked part of the second exon and had mutations in the start and stop codons. Therefore, different patterns of *rps16* pseudogenization were observed in *A. alpina* and the other *Arabis* species, as was previously suggested [23]. The *A. alpina* lineage had acquired independent dysfunctional mutation(s). The patterns observed for the European *A. hirsuta* revealed that the pseudogenization of *rps16* in the other *Arabis* species might not have occurred independently but, instead, occurred before the divergence of *D. nemorosa* and other Arabis species after splitting from *A. alpina*.

Table 1. Summary of chloroplast genome structure in *Arabis* species.

Species	Strain	Nucleotide Length (bp)				GC Contents (%)				NCBI #	Reference
		Entire	LSC	SSC	IR	Entire	LSC	SSC	IR		
Draba nemorosa	JO21	153289	82457	18126	26353	36.47	34.27	29.3	42.39	AP009373 (NC009272)	
Arabis alpina		152866	82338	17938	26933	36.45	34.21	29.31	42.39	HF934132 (NC023367)	[25]
Arabis hirsuta	Brno	153758	82710	18156	26446	36.4	34.15	29.16	42.41	LC361350	this study
Arabis flagellosa	Kifune	153673	82775	18052	26423	36.4	34.13	29.22	42.41	LC361351	this study
Arabis stelleri		153683	82807	18030	26423	36.39	34.11	29.22	42.42	KY126841	[23]
Arabis nipponica	JO23	153689	82811	18036	26421	36.4	34.1	29.31	42.42	AP009369 (NC009268)	
Arabis nipponica	Midori	153668	82772	18052	26422	36.39	34.1	29.24	42.42	LC361349	this study

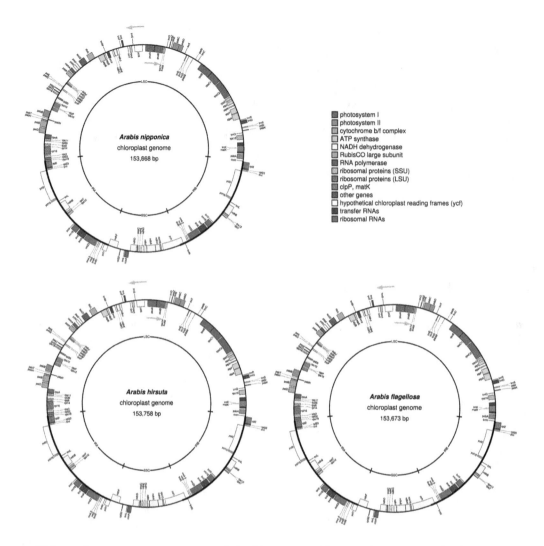

Figure 1. Chloroplast genome structure of *Arabis* species. Genes shown outside the map circles are transcribed clockwise, while those drawn inside are transcribed counterclockwise. Genes from different functional groups are color-coded according to the key at the top right. The positions of long single copy (LSC), short single copy (SSC), and two inverted repeat (IR: IRA and IRB) regions are shown in the inner circles.

2.2. Chloroplast Genome Divergence

Phylogenetic trees were generated by using whole chloroplast genome sequences and concatenated coding sequence (CDS) regions (see Figure 2). The inclusion of other Brassicaceae members revealed that *D. nemorosa* should be placed within *Arabis*, as previously reported [24]. In both trees, the two *A. nipponica* strains were grouped with *A. flagellosa* and *A. stelleri*. Although several nodes were supported by high bootstrap probabilities, the nearly identical sequences of the four East Asian *Arabis* species made them indistinguishable.

The divergence among the *Arabis* chloroplast genomes was shown using a VISTA plot (see Figure 3) and this was summarized in Table 2. The genome sequences of the two Japanese *A. nipponica* strains differed by only 55 nucleotide substitutions (0.036% per site), while those of *A. hirsuta* and *A. nipponica* differed by about 3500 sites (2.4% per site). The chloroplast genomes of *A. nipponica* and the other two East Asian *Arabis* species were also very similar (~100 nucleotide differences, <0.1% per site). Additionally, the 35 CDS regions, 29 tRNA genes, and four rRNA genes of the four East Asian *Arabis* species were identical, with three, 27, and four, respectively, also found to be identical in *A. hirsuta*. The levels of divergence between the East Asian *Arabis* species were similar to previously reported levels of variation within the local *A. alpina* population, in which 130 SNPs were identified among 24 individuals (Waterson's θ = 0.02%) [25]. If the hybridization event had facilitated chloroplast capture, the divergence between the *A. stelleri* and *A. nipponica* chloroplast genomes should have been less than their divergence from *A. flagellosa*. However, the divergence between the potential hybrid-origin species (*A. stelleri* and *A. nipponica*: 0.068 to 0.085) was similar to their divergence from *A. flagellosa* (0.056 to 0.086). Although the level of divergence was too low to make reliable comparisons, it is possible that *A. stelleri* and *A. nipponica* originated from independent hybridization events or the introgression process may still be ongoing.

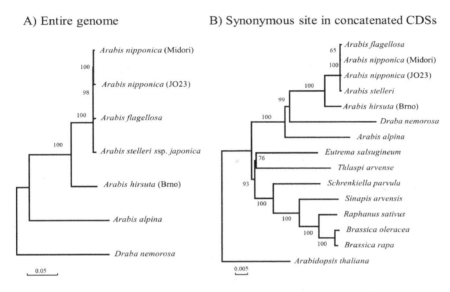

Figure 2. Chloroplast genome-based phylogenetic trees of *Arabis* species. The neighbor-joining trees were constructed using both (**A**) whole chloroplast genomes and (**B**) synonymous divergence from concatenated CDS. Numbers beside the nodes indicate bootstrap probabilities (%). Scale bars are shown at the bottom left of each tree.

Table 2. Divergence between species.

Compared Species			# of Differences	Divergence (%: Ks with JC Correction)
Draba nemorosa	vs.	Arabis alpina	4475	2.976
Draba nemorosa	vs.	Arabis hirsuta	4219	2.801
Draba nemorosa	vs.	Arabis flagellosa	4262	2.765
Draba nemorosa	vs.	Arabis stelleri	4171	2.771
Draba nemorosa	vs.	Arabis nipponica (JO23)	4150	2.757
Draba nemorosa	vs.	Arabis nipponica (Midori)	4131	2.745
Arabis alpina	vs.	Arabis hirsuta	3566	2.366
Arabis alpina	vs.	Arabis flagellosa	3571	2.371
Arabis alpina	vs.	Arabis stelleri	3565	2.366
Arabis alpina	vs.	Arabis nipponica (JO23)	3564	2.366
Arabis alpina	vs.	Arabis nipponica (Midori)	3547	2.355
Arabis hirsuta	vs.	Arabis flagellosa	1245	0.815
Arabis hirsuta	vs.	Arabis stelleri	1253	0.82
Arabis hirsuta	vs.	Arabis nipponica (JO23)	1234	0.808
Arabis hirsuta	vs.	Arabis nipponica (Midori)	1214	0.795
Arabis flagellosa	vs.	Arabis stelleri	132	0.086
Arabis flagellosa	vs.	Arabis nipponica (JO23)	111	0.072
Arabis flagellosa	vs.	Arabis nipponica (Midori)	86	0.056
Arabis stelleri	vs.	Arabis nipponica (JO23)	130	0.085
Arabis stelleri	vs.	Arabis nipponica (Midori)	104	0.068
Arabis nipponica (JO23)	vs.	Arabis nipponica (Midori)	55	0.036

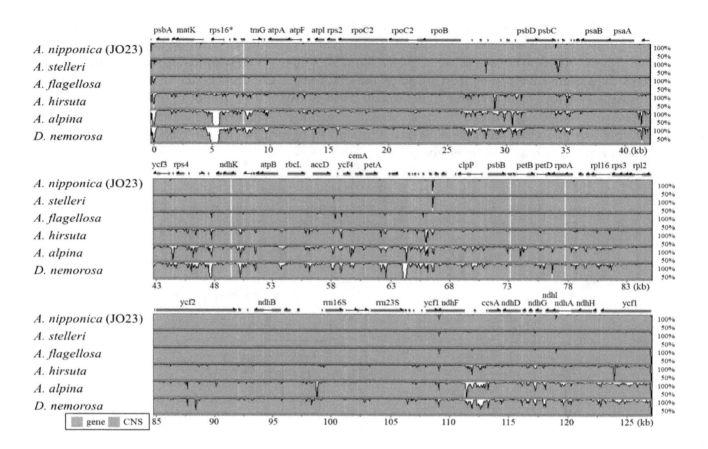

Figure 3. Alignment of the seven chloroplast genomes. VISTA-based identity plots of chloroplast genomes from six *Arabis* species and *Draba nemorosa* are compared to *A. nipponica* strain Midori. Arrows above the alignment indicate genes and their orientation. The names of genes ≥500 bp in length are also shown. A 70% identity cut-off was used for making the plots, and the Y-axis represents percent identity (50–100%), while the X-axis represents the location in the chloroplast genome. The blue and pink regions indicate genes and conserved noncoding sequences, respectively.

2.3. Distribution of Simple Sequence Repeats in the Chloroplast Genomes

Because the extremely low divergence among the East Asian *Arabis* species made it difficult to resolve their evolutionary relationships, other highly variable markers were needed. Therefore, simple sequence repeat (SSR) regions throughout the chloroplast genome were assessed for their ability to provide high-resolution species definition. A total of 74 mono-nucleotide, 22 di-nucleotide, and two tri-nucleotide repeat regions of ≥10 base pairs in length were identified (see Table 3).

However, these repeat regions were still unable to completely resolve the relationships of the East Asian *Arabis* species. Fifty of the 98 SSRs exhibited no variation among the East Asian *Arabis* species, while only 29 SSRs exhibited species-specific variation, including nine in *A. flagellosa*, 15 in *A. stelleri*, four in *A. nipponica* strain JO23, and one in *A. nipponica* strain Midori. Five of the SSRs were shared by the two *A. nipponica* strains, which suggests that they were also species-specific. Although the two *A. nipponica* strains were similar to each other, *A. flagellosa*, *A. stelleri*, and *A. nipponica* differ to a similar degree in terms of of variable SSRs, which suggests that the occurrence of chloroplast capture would be independent or still ongoing. This was suggested by the patterns of nucleotide substitutions.

Table 3. Simple sequence repeats (SSRs) in *Arabis* chloroplast genome.

Position in *A. nipponica* (Midori) Genome		UNIT	*A. nipponica*		*A. stelleri*	*A. flagellosa*	*A. hirsuta*	*A. alpina*
from	to		Midori	JO23				
287	318	AT	16	15	15	12	13 with 2 mutations	29 bp with several mutations
1922	1932	A	11	11	9	12	11	9
3929	3938	T	10	9	10	9	7	7
4258	4270	T	13	18	18	17	13	13
7713	7727	T	15	15	15	15	12	11
7729	7738	A	10	10	10	9	10	10
8203	8216	TA	7	6	7	7	6	6
8273	8282	TA	5	5	5	5	5	6
8289	8302	AT	7	7	6	7	8	6
8321	8330	TA	5	5	4	5	5	deletion
9677	9690	T	14	14	T4GT10	15	14	14
9982	9991	TA	5	5	5	5	5	5
11,660	11,669	A	10	10	9	10	10	7
12,406	12,414	T	9	9	10	10	T3AT6	T3AT6
13,010	13,018	T	9	9	10	9	T7AT2	T10AT2
13,810	13,821	ATT	4	4	4	4	4	ATTATATTCTT
14,101	14,110	A	10	10	14	10	12	10
18,027	18,037	T	CDS 11	11	11	11	11	11
19,398	19,408	TA	5	5	5	5	5	5
22,549	22,558	T	10	10	11	10	9	15
25,777	25,786	T	CDS 10	10	10	10	10	10
27,601	27,611	G	11	11	11	15	12	9
28,808	28,817	T	10	10	9	10	10	T5CT3G2
30,293	30,310	A	18	17	17	18	12	A4CA5
30,737	30,751	T	15	15	14	15	15	5
30,830	30,839	A	10	9	11	10	8	6
30,918	30,929	TA	6	6	6	6	6	4
31,260	31,269	AT	5	5	5	5	5	3
35,309	35,316	G	8	11	10	11	7	10
35,516	35,525	AT	5	5	5	5	5	5
35,538	35,555	AT	9	9	9	9	3	deletion
41,508	41,522	T	15	13	11	13	18nt	101nt
41,768	41,778	A	11	12	12	11	A12GA4	11
43,656	43,665	A	10	10	10	11	A8TA2	A9TA2TA2
43,887	43,895	T	9	15	T4AT4AT4	T4AT4	4	4
45,038	45,046	T	9	9	9	10	8	7
45,771	45,788	T	18	18	18	18	13	9
46,034	46,057	A	24	24	24	24	16	11
46,116	46,133	AT	9	9	9	8	9	7
46,135	46,144	TA	5	5	5	5	5	3
46,782	46,791	T	10	11	10	11	14	10
47,368	47,378	A	11	11	12	12	10	13
47,586	47,595	T	10	10	10	10	13	TCT8
47,624	47,633	A	10	10	11	11	8	7
49,061	49,070	T	10	10	11	10	8	T3AT10
49,631	49,640	T	10	10	10	10	8	8
50,329	50,340	A	12	12	12	12	11	11
51,202	51,211	TA	5	5	5	5	19nt	deletion
51,215	51,230	T	16	17	17	16	13	13
53,088	53,097	T	CDS 10	10	10	10	10	12
53,592	53,601	C	10	11	9	12	9	9
55,477	55,490	T	14	14	14	14	complement A11	complement A13
55,891	55,906	T	16	16	16	16	13	15
56,476	56,485	T	10	10	10	10	10	A2T8
58,301	58,310	T	10	10	10	10	10	6
59,338	59,348	T	11	10	9	11	11	4
61,731	61,739	C	9	13	9	12	8	C3AC3
62,108	62,117	TA	5	5	5	5	6	4
62,161	62,182	T	22	22	22	31	2nt shorter A5TA3	2nt shorter A5TA3
62,202	62,210	A	9	10	9	9	A5TA3	A5TA3
63,523	63,538	T	16	15	15	T5GT10	16	7
64,629	64,639	T	11	11	11	11	11	T6GT3G
65,636	65,645	C	10	13	11	13	8	C2TCTGC7
66,253	66,262	AT	5	5	5	5	4	7
66,851	66,864	A	14	14	14	19	17	12
68,965	68,977	T	13	13	13	13	11	11
69,965	69,975	T	11	11	12	11	11	8
75,328	75,340	A	13	14	14	13	19	14

Table 3. *Cont.*

Position in A. nipponica (Midori) Genome		UNIT	A. nipponica		A. stelleri	A. flagellosa	A. hirsuta	A. alpina
76,614	76,626	T	13	13	13	13	13	13
78,154	78,162	TTG	3	5	3	3	4	2
80,484	80,493	A	10	11	10	10	10	9
81,019	81,035	T	17	17	17	17	17	17
81,178	81,191	T	14	14	14	14	18	8
82,568	82,578	A	11	10	9	10	9	10
83,489	83,498	TA	5	5	5	5	5	4
93,127	93,136	TA	5	5	5	5	5	4
97,975	97,984	A	10	10	10	10	12	9
98,781	98,791	T	11	11	11	11	10	14
107,287	107,295	AT	5	5	5	5	5	7
107,313	107,324	T	12	11	13	13	T2(AT)4T7	14
111,481	111,490	TA	5	5	5	5	TA2TGTA	4
111,589	111,598	AT	5	5	5	5	5	10
111,665	111,672	T	8	8	10	8	7	10
111,801	111,810	A	A7CA2	A7CA2	10	A7CA2	A7CA2	A7TAC
112,472	112,481	A	10	10	10	10	11	10
116,836	116,845	T	10	9	10	11	T7AT3	10
123,173	123,184	T	12	12	12	12	12	12
123,285	123,383	T	10	10	10	10	10	10
123,884	123,893	T	10	10	10	10	10	10
123,975	123,987	A	13	13	13	13	13	13
124,356	124,365	TA	5	5	5	5	5	5
124,874	124,886	T	13	13	13	13	13	13
125,029	125,041	A	13	13	13	13	13	13
126,052	125,385	T	15	15	15	15	15	17
126,087	126,097	T	11	11	11	11	11	11
126,117	126,128	A	12	12	12	12	12	12
126,952	126,962	T	11	11	11	11	T8CT2	T8CT2
127,241	127,252	A	12	12	12	12	6	6

2.4. Mitochondrial Genome Analysis

Chloroplast capture could have originated from hybridization events that also affected other cytoplasmic genomes. Due to this, variation in the mitochondrial genome sequences was analyzed. Mapping next-generation sequencing (NGS) reads to the *Eruca vesicaria* mitochondrial genome revealed that 29 sites with five or more mapped reads varied among the *A. nipponica* strain Midori, *A. flagellosa*, and *A. hirsuta* (see Table 4). Twenty-eight of the sites were conserved among *A. nipponica* and *A. flagellosa*. One site was specific to *A. nipponica* and provided 100% support for the relationship between *A. nipponica* and *A. flagellosa*. Even though reliability decreased, 123 of 125 sites with two or more reads (98.4%) also supported the similarity of the *A. nipponica* and *A. flagellosa* mitochondrial genomes. These findings suggest that the hybridization history of the species affects both the chloroplast and the mitochondrial genomes similarly.

Table 4. Nucleotide variation in the mitochondrial genome of *Arabis* species.

		Number of Mapped Reads			
		5 and More	4 and More	3 and More	2 and More
Number of variable sites	Total	29	46	74	129
Specific to	A. nipponica	1	1	4	12
	A. flagellosa	0	0	0	3
	A. hirsuta	14	25	35	62
Shared with	A. flagellosa and A. nipponica	14	19	31	46
	A. nipponica and A. hirsuta	0	0	1	1
	A. flagellosa and A. hirsuta	0	0	1	1
	other type	0	1	2	4

3. Discussion

Chloroplast capture results in the incongruence of chloroplast and nuclear phylogenies, which has been reported in many plant taxa and is considered common among plants [17,18,26–37]. Furthermore, it is possible that the introgression of chloroplast genomes occurs more frequently than that of nuclear genomes as a result of uniparental inheritance, lack of recombination, and low selective constraint [38–40]. Chloroplast capture could occur by using several factors including sampling error, convergence, evolutionary rate heterogeneity, wrong lineage sorting, and hybridization/introgression [17]. Introgression-induced chloroplast capture occurred through hybridization between distant but compatible species, which was followed by backcrossing with pollen donor species [41,42].

East Asian *Arabis* species have previously been reported to show evidence of chloroplast capture [14,16]. More specifically, detailed phylogenetic analyses of nuclear and chloroplast marker genes has suggested that *A. nipponica*, *A. stelleri*, and *A. takeshimana* originated from the hybridization of *A. hirsuta* (or *A. sagittata*) and East Asian *Arabis* species (close to *A. serrata*, *A. paniculata*, and *A. flagellosa*), which act as paternal and maternal parents, respectively [14,16]. In the present study, comparing the whole chloroplast genomes of four plants from three East Asian *Arabis* species (two *A. nipponica*, one each of *A. stelleri*, and *A. flagellosa*) revealed genome-wide similarities that indicated chloroplast capture by *A. nipponica* and *A. stelleri*. The study also compared the species' partial mitochondrial genomes, which indicated a closer relationship between *A. nipponica* and *A. flagellosa* than between the former and European *A. hirsuta*. This suggested that *A. nipponica* also has a history of mitochondrial capture. This is not surprising, because hybridization and backcrossing could have similar effects on both organellar genomes. Also, cyto-nuclear incompatibility caused by a mitochondrial genome could lead cytoplasmic replacement to exhibit chloroplast capture [17,41,42]. The pattern of variation in the mitochondrial genomes suggested that both the chloroplast and mitochondrial genomes were co-transmitted during the evolutionary history of East Asian *Arabis* species. Future research should focus on the process of chloroplast (organellar) capture. Simple backcrossing could show the mechanisms of cytoplasm replacement and could produce results in as few as a hundred generations under certain conditions [42]. In the present study, the divergence between the genomes of hybrid-origin species and putative pollen-donor species was similar to the divergence observed within species, which suggests that the hybridization event was relatively recent. Nuclear genome markers are needed to estimate the proportion of parental genome fragments in the current nuclear genome of *A. nipponica*.

4. Materials and Methods

4.1. Plant Materials

Arabis nipponica (*A. hirsuta* var. *nipponica*, sampled from Midori, Gifu Prefecture, Japan), *A. flagellosa* (sampled from Kifune, Kyoto Prefecture, Japan), and *A. hirsuta* (strain Brno from Ulm Botanical Garden, Germany) were used in the present study.

4.2. DNA Isolation, NGS Sequencing, and Genome Assembly

Chloroplasts were isolated from *A. hirsuta* and *A. nipponica* as described in Okegawa and Motohashi [43]. DNA was isolated from the chloroplasts using the DNeasy Plant Mini Kit (Qiagen, Valencia, CA, USA), while the total DNA was isolated from leaves of *A. flagellosa*. NGS libraries were constructed using the Nextera DNA Sample Preparation Kit (Illumina, San Diego, CA, USA) and sequenced as single-ended reads using the NextSeq500 platform (Illumina). About 2 Gb (1.4 Gb, 12 M clean reads) of sequences were obtained for *A. flagellosa* (43 Mb mapped reads, 282.69× coverage). Additionally, 400 Mb (300 Mb, 2.5 M clean reads) were obtained for both *A. hirsuta* (64 Mb mapped reads, 417.17× coverage) and *A. nipponica* (72 Mb mapped reads, 455.87× coverage). The generated reads were assembled using velvet 1.2.10 [44] and assembled into complete chloroplast genomes by mapping to previously published whole chloroplast genome sequences. Sequence gaps were resolved using Sanger sequencing. Genes were annotated using DOGMA [45] and BLAST. The newly constructed chloroplast genomes were deposited in the DDBJ database under the accession numbers LC361349-51. Finally, the circular chloroplast genome maps were drawn using OGDRAW [46].

4.3. Molecular Evolutionary Analyses

The whole chloroplast genome sequences of *A. nipponica* (strain JO23: AP009369), *A. stelleri* (KY126841) [23], *A. alpina* (HF934132) [25], and *D. nemorosa* (strain JO21: AP009373) in the GenBank were also used. Whole chloroplast sequences were aligned in order to construct neighbor-joining trees with Jukes and Cantor distances. The sequences of 77 known functional genes were linked in a series after excluding initiation and stop codons and were then used for phylogenetic analyses along with sequences from the related clade species *Brassica oleracea* (KR233156) [47], *B. rapa* (DQ231548), *Eutrema salsugineum* (KR584659) [48], *Raphanus sativus* (KJ716483) [49], *Scherenkiella parvula* (KT222186) [48], *Sinapis arvensis* (KU050690), and *Thlaspi arvense* (KX886351) [21] using *A. thaliana* (AP000423) [50] as an outgroup. The synonymous divergence of the concatenated CDS was estimated using the Nei and Gojobori method. All phylogenetic analyses were performed using MEGA 7.0 [51]. Levels of divergence throughout the chloroplast genome were visualized using mVISTA [52] with Shuffle-LAGAN alignment [53].

4.4. Mapping NGS Reads to Mitochondrial Genome Sequences

Because the chloroplast isolation method used in the present study did not completely exclude mitochondria, about 1% of the sequence reads were derived from mitochondrial genomes. Although this proportion is too low to be useful for assembling whole mitochondrial genomes, the reads were nevertheless mapped to the mitochondrial genome of *Eruca vesicaria* (KF442616) [54] in order to measure mitochondrial genome divergence. Regions with at least five mapped reads were used for the analysis.

Acknowledgments: This study was supported in part by JSPS KAKENHI Grant Number 17K19361 and Grants-in-Aid from MEXT-Supported Program for the Strategic Research Foundation at Private Universities (S1511023) to Akira Kawabe.

Author Contributions: Akira Kawabe designed the study. Hiroaki Nukii and Hazuka Y. Furihata performed the experiments. Akira Kawabe and Hazuka Y. Furihata analyzed the data. Akira Kawabe wrote the manuscript. All authors read and approved the final manuscript.

References

1.   Al-Shehbaz, I.A. Transfer of most North American species of Arabis to Boechera (Brassicaceae). *Novon* **2003**, *13*, 381–391. [CrossRef]

2.   O'Kane, S.L.; Al-Shehbaz, I.A. Phylogenetic Position and Generic Limits of Arabidopsis (Brassicaceae) Based on Sequences of Nuclear Ribosomal DNA. *Ann. Mo. Bot. Gard.* **2003**, *90*, 603–612. [CrossRef]

3.   Al-Shehbaz, I.A.; Beilstein, M.A.; Kellogg, E.A. Systematics and phylogeny of the Brassicaceae (Cruciferae): An overview. *Plant Syst. Evol.* **2006**, *259*, 89–120. [CrossRef]

4.   Al-Shehbaz, I.A.; German, D.A.; Karl, R.; Ingrid, J.T.; Koch, M.A. Nomenclatural adjustments in the tribe Arabideae (Brassicaceae). *Plant Div. Evol.* **2011**, *129*, 71–76. [CrossRef]

5.   Koch, M.A.; Karl, R.; German, D.A.; Al-Shehbaz, I.A. Systematics, taxonomy and biogeography of three new Asian genera of Brassicaceae tribe Arabideae: An ancient distribution circle around the Asian high mountains. *Taxon* **2012**, *61*, 955–969.

6.   Kiefer, M.; Schmickl, R.; German, D.A.; Mandáková, T.; Lysak, M.A.; Al-Shehbaz, I.A.; Franzke, A.; Mummenhoff, K.; Stamatakis, A.; Koch, M.A. BrassiBase: Introduction to a novel knowledge database on Brassicaceae evolution. *Plant Cell Physiol.* **2014**, *55*, e3. [CrossRef] [PubMed]

7.   Ansell, S.W.; Grundmann, M.; Russell, S.J.; Schneider, H.; Vogel, J.C. Genetic discontinuity, breeding-system change and population history of *Arabis alpina* in the Italian Peninsula and adjacent Alps. *Mol Ecol.* **2008**, *17*, 2245–2257. [CrossRef] [PubMed]

8.   Bergonzi, S.; Albani, M.C.; ver Loren van Themaat, E.; Nordström, K.J.; Wang, R.; Schneeberger, K.;

Moerland, P.D.; Coupland, G. Mechanisms of age-dependent response to winter temperature in perennial flowering of *Arabis alpina*. *Science* **2013**, *340*, 1094–1097. [CrossRef] [PubMed]

9. Karl, R.; Koch, M.A. A world-wide perspective on crucifer speciation and evolution: Phylogenetics, biogeography and trait evolution in tribe Arabideae. *Ann. Bot.* **2013**, *112*, 983–1001. [CrossRef] [PubMed]

10. Toräng, P.; Vikström, L.; Wunder, J.; Wötzel, S.; Coupland, G.; Ågren, J. Evolution of the selfing syndrome: Anther orientation and herkogamy together determine reproductive assurance in a self-compatible plant. *Evolution* **2017**, *71*, 2206–2218. [CrossRef] [PubMed]

11. Heidel, A.J.; Kiefer, C.; Coupland, G.; Rose, L.E. Pinpointing genes underlying annual/perennial transitions with comparative genomics. *BMC Genom.* **2016**, *17*, 921. [CrossRef] [PubMed]

12. Willing, E.M.; Rawat, V.; Mandáková, T.; Maumus, F.; James, G.V.; Nordström, K.J.; Becker, C.; Warthmann, N.; Chica, C.; Szarzynska, B.; et al. Genome expansion of *Arabis alpina* linked with retrotransposition and reduced symmetric DNA methylation. *Nat. Plants* **2015**, *1*, 14023. [CrossRef] [PubMed]

13. Jiao, W.B.; Accinelli, G.G.; Hartwig, B.; Kiefer, C.; Baker, D.; Severing, E.; Willing, E.M.; Piednoel, M.; Woetzel, S.; Madrid-Herrero, E.; et al. Improving and correcting the contiguity of long-read genome assemblies of three plant species using optical mapping and chromosome conformation capture data. *Genome Res.* **2017**, *27*, 778–786. [CrossRef] [PubMed]

14. Koch, M.A.; Karl, R.; Kiefer, C.; Al-Shehbaz, I.A. Colonizing the American continent: Systematics of the genus Arabis in North America (Brassicaceae). *Am. J. Bot.* **2010**, *97*, 1040–1057. [CrossRef] [PubMed]

15. Karl, R.; Kiefer, C.; Ansell, S.W.; Koch, M.A. Systematics and evolution of Arctic-Alpine *Arabis alpina* (Brassicaceae) and its closest relatives in the eastern Mediterranean. *Am. J. Bot.* **2012**, *99*, 778–794. [CrossRef] [PubMed]

16. Karl, R.; Koch, M.A. Phylogenetic signatures of adaptation: The Arabis hirsuta species aggregate (Brassicaceae) revisited. *Perspect. Plant Ecol. Evol. Syst.* **2014**, *16*, 247–264. [CrossRef]

17. Rieseberg, L.H.; Soltis, D.E. Phylogenetic consequences of cytoplasmic gene flow in plants. *Evolut. Trends Plants* **1991**, *5*, 65–84.

18. Soltis, D.E.; Kuzoff, R.K. Discordance between nuclear and chloroplast phylogenies in the Heuchera group (Saxifragaceae). *Evolution* **1995**, *49*, 727–742. [CrossRef] [PubMed]

19. Ruhfel, B.R.; Gitzendanner, M.A.; Soltis, P.S.; Soltis, D.E.; Burleigh, J.G. From algae to angiosperms-inferring the phylogeny of green plants (Viridiplantae) from 360 plastid genomes. *BMC Evol. Biol.* **2014**, *14*, 23. [CrossRef] [PubMed]

20. Hohmann, N.; Wolf, E.M.; Lysak, M.A.; Koch, M.A. A Time-calibrated road map of brassicaceae species radiation and evolutionary history. *Plant Cell* **2015**, *27*, 2770–2784. [CrossRef] [PubMed]

21. Guo, X.; Liu, J.; Hao, G.; Zhang, L.; Mao, K.; Wang, X.; Zhang, D.; Ma, T.; Hu, Q.; Al-Shehbaz, I.A.; Koch, M.A. Plastome phylogeny and early diversification of Brassicaceae. *BMC Genom.* **2017**, *18*, 176. [CrossRef] [PubMed]

22. Mandáková, T.; Hloušková, P.; German, D.A.; Lysak, M.A. Monophyletic origin and evolution of the largest crucifer genomes. *Plant Physiol.* **2017**, *174*, 2062–2071. [CrossRef] [PubMed]

23. Raman, G.; Park, V.; Kwak, M.; Lee, B.; Park, S. Characterization of the complete chloroplast genome of Arabis stellari and comparisons with related species. *PLoS ONE* **2017**, *12*, e0183197. [CrossRef] [PubMed]

24. Jordon-Thaden, I.; Hase, I.; Al-Shehbaz, I.A.; Koch, M.A. Molecular phylogeny and systematics of the genus Draba (Brassicaceae) and identification of its most closely related genera. *Mol. Phylogenet. Evol.* **2010**, *55*, 524–540. [CrossRef] [PubMed]

25. Melodelima, C.; Lobréaux, S. Complete *Arabis alpina* chloroplast genome sequence and insight into its polymorphism. *Meta Gene* **2013**, *1*, 65–75. [CrossRef] [PubMed]

26. Acosta, M.C.; Premoli, A.C. Evidence of chloroplast capture in South American *Nothofagus* (subgenus *Nothofagus*, Nothofagaceae). *Mol. Phylogenet. Evol.* **2010**, *54*, 235–242. [CrossRef] [PubMed]

27. Dorado, O.; Rieseberg, L.H.; Arias, D.M. Chloroplast DNA introgression in southern California sunflowers. *Evolution* **1992**, *46*, 566–572. [CrossRef] [PubMed]

28. Fehrer, J.; Gemeinholzer, B.; Chrtek, J.; Bräutigam, S. Incongruent plastid and nuclear DNA phylogenies reveal ancient intergeneric hybridization in *Pilosella* hawkweeds (*Hieracium*, Cichorieae, Asteraceae). *Mol. Phylogenet. Evol.* **2007**, *42*, 347–361. [CrossRef] [PubMed]

29. Gurushidze, M.; Fritsch, R.M.; Blattner, F.R. Species-level phylogeny of *Allium* subgenus *Melanocrommyum*: Incomplete lineage sorting, hybridization and *trnF* gene duplication. *Taxon* **2010**, *59*, 829–840.

30. Liston, A.; Kadereit, J.W. Chloroplast DNA evidence for introgression and long distance dispersal in the desert annual *Senecio flavus* (*Asteraceae*). *Plant Syst. Evol.* **1995**, *197*, 33–41. [CrossRef]

31. Mir, C.; Jarne, P.; Sarda, V.; Bonin, A.; Lumaret, R. Contrasting nuclear and cytoplasmic exchanges between phylogenetically distant oak species (*Quercus suber* L. and *Q. ilex* L.) in Southern France: Inferring crosses and dynamics. *Plant Biol.* **2009**, *11*, 213–226. [CrossRef] [PubMed]

32. Okuyama, Y.; Fujii, N.; Wakabayashi, M.; Kawakita, A.; Ito, M.; Watanabe, M.; Murakami, N.; Kato, M. Nonuniform concerted evolution and chloroplast capture: Heterogeneity of observed introgression patterns in three molecular data partition phylogenies of Asian *Mitella* (Saxifragaceae). *Mol. Biol. Evol.* **2005**, *22*, 285–296. [CrossRef] [PubMed]

33. Rieseberg, L.H.; Choi, H.C.; Ham, F. Differential cytoplasmic versus nuclear introgression in *Helianthus*. *J. Hered.* **1991**, *82*, 489–493. [CrossRef]

34. Schilling, E.E.; Panero, J.K. Phylogenetic reticulation in subtribe *Helianthinae*. *Am. J. Bot.* **1996**, *83*, 939–948. [CrossRef]

35. Wolfe, A.D.; Elisens, W.J. Evidence of chloroplast capture and pollen-mediated gene flow in Penstemon sect. Peltanthera (Scrophulariaceae). *Syst. Bot.* **1995**, *20*, 395–412. [CrossRef]

36. Yi, T.S.; Jin, G.H.; Wen, J. Chloroplast capture and intra-and inter-continental biogeographic diversification in the Asian–New World disjunct plant genus Osmorhiza (Apiaceae). *Mol. Phylogenet. Evol.* **2015**, *85*, 10–21. [CrossRef] [PubMed]

37. Yuan, Y.W.; Olmstead, R.G. A species-level phylogenetic study of the Verbena complex (Verbenaceae) indicates two independent intergeneric chloroplast transfers. *Mol. Phylogenet. Evol.* **2008**, *48*, 23–33. [CrossRef] [PubMed]

38. Avise, J.C. *Molecular Markers, Natural History and Evolution*, 2nd ed.; Sinauer: Sunderland, MA, USA, 2004.

39. Rieseberg, L.H.; Wendel, J. Introgression and its consequences in plants. In *Hybrid Zones and the Evolutionary Process*; Harrison, R., Ed.; Oxford University Press: New York, NY, USA, 1993; pp. 70–109.

40. Martinsen, G.D.; Whitham, T.G.; Turek, R.J.; Keim, P. Hybrid populations selectively filter gene introgression between species. *Evolution* **2001**, *55*, 1325–1335. [CrossRef] [PubMed]

41. Rieseberg, L.H. The role of hybridization in evolution: Old wine in new skins. *Am. J. Bot.* **1995**, *82*, 944–953. [CrossRef]

42. Tsitrone, A.; Kirkpatrick, M.; Levin, D.A. A model for chloroplast capture. *Evolution* **2003**, *57*, 1776–1782. [CrossRef] [PubMed]

43. Okegawa, Y.; Motohashi, K. Chloroplastic thioredoxin m functions as a major regulator of Calvin cycle enzymes during photosynthesis in vivo. *Plant J.* **2015**, *84*, 900–913. [CrossRef] [PubMed]

44. Zerbino, D.R.; Birney, E. Velvet: Algorithms for de novo short read assembly using de Bruijn graphs. *Genome Res.* **2008**, *18*, 821–829. [CrossRef] [PubMed]

45. Wyman, S.K.; Jansen, R.K.; Boore, J.L. Automatic annotation of organellar genomes with DOGMA. *Bioinformatics* **2004**, *20*, 3252–3255. [CrossRef] [PubMed]

46. Lohse, M.; Drechsel, O.; Bock, R. OrganellarGenomeDRAW (OGDRAW)—A tool for the easy generation of high-quality custom graphical maps of plastid and mitochondrial genomes. *Curr. Genet.* **2007**, *52*, 267–274. [CrossRef] [PubMed]

47. Seol, Y.J.; Kim, K.; Kang, S.H.; Perumal, S.; Lee, J.; Kim, C.K. The complete chloroplast genome of two *Brassica* species, *Brassica nigra* and *B. oleracea*. *Mitochondrial DNA Part A* **2017**, *28*, 167–168. [CrossRef] [PubMed]

48. He, Q.; Hao, G.; Wang, X.; Bi, H.; Li, Y.; Guo, X.; Ma, T. The complete chloroplast genome of *Schrenkiella parvula* (Brassicaceae). *Mitochondrial DNA Part A* **2016**, *27*, 3527–3528. [CrossRef] [PubMed]

49. Jeong, Y.M.; Chung, W.H.; Mun, J.H.; Kim, N.; Yu, H.J. De novo assembly and characterization of the complete chloroplast genome of radish (*Raphanus sativus* L.). *Gene* **2014**, *551*, 39–48. [CrossRef] [PubMed]

50. Sato, S.; Nakamura, Y.; Kaneko, T.; Asamizu, E.; Tabata, S. Complete structure of the chloroplast genome of *Arabidopsis thaliana*. *DNA Res.* **1999**, *6*, 283–290. [CrossRef] [PubMed]

51. Kumar, S.; Stecher, G.; Tamura, K. MEGA7: Molecular Evolutionary Genetics Analysis Version 7.0 for Bigger Datasets. *Mol. Biol. Evol.* **2016**, *33*, 1870–1874. [CrossRef] [PubMed]

52. Frazer, K.A.; Pachter, L.; Poliakov, A.; Rubin, E.M.; Dubchak, I. VISTA: Computational tools for comparative genomics. *Nucleic Acids Res.* **2004**, *32*, W273–W279. [CrossRef] [PubMed]

53. Brudno, M.; Malde, S.; Poliakov, A.; Do, C.B.; Couronne, O.; Dubchak, I.; Batzoglou, S. Glocal Alignment: Finding rearrangements during alignment. *Bioinformatics* **2003**, *19*, i54–i62. [CrossRef] [PubMed]

54. Wang, Y.; Chu, P.; Yang, Q.; Chang, S.; Chen, J.; Hu, M.; Guan, R. Complete mitochondrial genome of *Eruca sativa* Mill. (Garden rocket). *PLoS ONE* **2014**, *9*, e105748. [CrossRef] [PubMed]

12

Comparative Plastid Genomes of *Primula* Species: Sequence Divergence and Phylogenetic Relationships

Ting Ren [1], Yanci Yang [1], Tao Zhou [2] and Zhan-Lin Liu [1,*]

[1] Key Laboratory of Resource Biology and Biotechnology in Western China (Ministry of Education), College of Life Sciences, Northwest University, Xi'an 710069, China; renting92@stumail.nwu.edu.cn (T.R.); yycjyl1@gmail.com (Y.Y.)

[2] School of Pharmacy, Xi'an Jiaotong University, Xi'an 710061, China; zhoutao196@mail.xjtu.edu.cn

* Correspondence: liuzl@nwu.edu.cn

Abstract: Compared to traditional DNA markers, genome-scale datasets can provide mass information to effectively address historically difficult phylogenies. *Primula* is the largest genus in the family Primulaceae, with members distributed mainly throughout temperate and arctic areas of the Northern Hemisphere. The phylogenetic relationships among *Primula* taxa still maintain unresolved, mainly due to intra- and interspecific morphological variation, which was caused by frequent hybridization and introgression. In this study, we sequenced and assembled four complete plastid genomes (*Primula handeliana*, *Primula woodwardii*, *Primula knuthiana*, and *Androsace laxa*) by Illumina paired-end sequencing. A total of 10 *Primula* species (including 7 published plastid genomes) were analyzed to investigate the plastid genome sequence divergence and their inferences for the phylogeny of *Primula*. The 10 *Primula* plastid genomes were similar in terms of their gene content and order, GC content, and codon usage, but slightly different in the number of the repeat. Moderate sequence divergence was observed among *Primula* plastid genomes. Phylogenetic analysis strongly supported that *Primula* was monophyletic and more closely related to *Androsace* in the Primulaceae family. The phylogenetic relationships among the 10 *Primula* species showed that the placement of *P. knuthiana*–*P. veris* clade was uncertain in the phylogenetic tree. This study indicated that plastid genome data were highly effective to investigate the phylogeny.

Keywords: plastid genome; phylogenetic relationship; *Primula*; repeat; sequence divergence

1. Introduction

Primula is the largest genus in the family Primulaceae with approximately 500 species [1,2], where they are especially rich in the temperate and arctic areas of the Northern Hemisphere, with only a few outliers found in the Southern Hemisphere. China is the center of *Primula* diversity and speciation with over 300 species [1,3]. Many *Primula* species are grown widely as ornamental and landscape plants because of their attractive flowers and long flowering period. Therefore, *Primula* is reputed to be one of the great garden plant genera throughout the world [2,3].

As a typical cross-pollinated plant with heterostyly, *Primula* has been a particular focus of many botanists, and various studies are involved in hybridization [4], pollination biology [5,6], and distyly [7,8]. According to morphological traits, the taxonomic study of *Primula* has been revised for several times. Smith and Fletcher (1947) firstly proposed an infrageneric system with a total of 31 sections [9]. Considering some putative reticulate evolutionary relationships, Wendelbo (1961) posed a revised system with seven subgenera [10]. Richards (1993) later amended Wendelbo's version and classified six subgenera [11]. Hu and Kelso (1996) delimited the Chinese *Primula* species into 24 sections [1]. Recently, numerous molecular phylogenetic works of the genus *Primula* have also been conducted by using plastid and/or nuclear gene fragments [12–14]. These studies have greatly advanced our understanding of the evolutionary history of *Primula* species. However, the phylogenetic relationships within the genus *Primula* are still uncertain, mainly due to intra- and interspecific morphological variation, which was caused by frequent hybridization and introgression [1,2,14]. Further research has been hindered by the insufficient information of the traditional DNA markers, such as one or few chloroplast gene fragments, and by the complex evolutionary relationships in *Primula*. Therefore, more sequence resources and genome data are required in order to obtain a better understanding of the phylogeny of the genus *Primula*.

In general, the plastid genome in angiosperms is a typical quadripartite structure, where the size ranges from 115 to 165 kb, with two copies of inverted repeat (IR) regions separated by a large single copy (LSC) region and a small single copy (SSC) region [15]. Approximately 110–130 distinct genes are located along the plastid genome [16]. Most of these are protein-coding genes, the remainder being transfer RNA (*tRNA*) or ribosomal RNA (*rRNA*) genes [16]. Due to its particular advantages—such as small size, uniparental inheritance, low substitution rates, and high conservation in terms of the gene content and genome structure [17,18]—the plastid genome is considered a very promising tool for phylogenetic studies [19,20]. Significant advances in next-generation sequencing technology made it fairly inexpensive and convenient to obtain plastid genome sequences [21,22]. As a result, phylogenomic analyses have also been greatly facilitated. For example, the plastid phylogenomic analyses supported Tofieldiaceae as the most basal lineage within Alismatales [23]. The relationships between wild and domestic *Citrus* species could also be resolved with 34 plastid genomes [24]. Similarly, 142 plastid genomes were used to successfully infer deep phylogenetic relationships and the diversification history of Rosaceae [25]. These studies strongly indicate that plastid phylogenomics is helpful in determining the phylogenetic positions of various questionable lineages of angiosperms.

In the present study, we analyzed the complete plastid genomes of 10 *Primula* species including 7 published plastid genomes and 3 new data (*Primula handeliana*, *Primula woodwardii*, and *Primula knuthiana*) by using Illumina sequencing technology. Our primary aims were to: (1) compare the complete plastid genomes of 10 *Primula* species; (2) document that the extent of sequence divergence among the *Primula* plastid genomes; and (3) increase more sequence resources and genome information for investigating the phylogeny in genus *Primula*. The complete plastid genome of *Androsace laxa* from a closely related genus was used as the outgroup in the phylogenomic analysis of genus *Primula*. This study will not only contribute to further studies on the phylogeny, taxonomy, and evolutionary history of the genus *Primula*, but also provide insight into the plastid genome evolution of *Primula*.

2. Results

2.1. Genome Features

The sizes of the plastid genomes of the 10 *Primula* species ranged from 150,856 bp to 153,757 bp, where they had a typical quadripartite structure, including a LSC region (82,048–84,479 bp) and a SSC region (17,568–17,896 bp) separated by a pair of IR regions (25,182–25,855 bp) (Table 1). In the 10 *Primula* plastid genomes, gene content was similar and gene order was identical. The *Primula* plastid genomes contained about 130–132 genes, including 85–86 protein-coding genes, 37 *tRNA* genes, and 8 *rRNA* genes (Tables 1 and S4). The *accD* gene was a pseudogene in *P. sinensis*, whereas it was missing in *P. persimilis* and *P. kwangtungensis*. The *P. poissonii* plastid genome contained a pseudogene (*infA*). Among these genes, 15 genes harbored a single intron (*trnA-UGC, trnG-UCC, trnI-GAU, trnK-UUU, trnL-UAA, trnV-UAC, atpF, ndhA, ndhB, petB, petD, rpoC1, rpl2, rpl16,* and *rps16*) and three genes (*pafI, clpP,* and *rps12*) harbored two introns. Seven *tRNA* genes, seven protein-coding genes, and all four *rRNA* genes were completely duplicated in the IR regions (Table 1). *trnk-UUU* had the largest intron (2487–2568 bp) containing the *matK* gene. The GC contents of the LSC, SSC, and IR regions, as well as those of the whole plastid genomes, were nearly identical in the 10 *Primula* plastid genomes (Table 1). The complete plastid genome of *A. laxa* was 151,942 bp in length and contained 132 genes (Table 1). The overall GC content of the *A. laxa* plastid genome was 37.3%, and the corresponding values for the LSC, SSC, and IR regions were 35.2, 30.9, and 42.7%, respectively (Table 1).

Table 1. Plastid genomic characteristics of the 10 *Primula* species and *A. laxa*.

Taxa	A. laxa *	P. handeliana *	P. woodwardii *	P. knuthiana *	P. poissonii	P. sinensis	P. veris
Assembly reads	16,137,534	12,884,542	25,149,710	15,928,364	/	/	/
Mean coverage	293.4×	482.4×	508.3×	405.3×	/	/	/
GenBank numbers	MG181220	MG181221	MG181222	MG181223	NC_024543	NC_030609	NC_031428
Total genome size (bp)	151,942	151,081	151,666	152,502	151,664	150,859	150,856
LSC (bp)	83,078	82,785	83,325	83,446	83,444	82,064	82,048
IRs (bp)	25,970	25,200	25,290	25,604	25,199	25,535	25,524
SSC (bp)	16,924	17,896	17,761	17,848	17,822	17,725	17,760
Total GC content (%)	37.3	37	37	37	37	37.2	37.1
LSC (%)	35.2	34.9	34.9	34.9	34.9	35.2	35.1
IRs (%)	42.7	42.9	42.8	42.7	42.9	42.8	42.7
SSC (%)	30.9	30.2	30.2	30.3	30.1	30.5	30.2
Total number of genes	132	131	131	131	132	131	131
Protein-coding	87 (7)	86 (7)	86 (7)	86 (7)	86 (7)	85 (7)	86 (7)
tRNA	37 (7)	37 (7)	37 (7)	37 (7)	37 (7)	37 (7)	37 (7)
rRNA	8 (4)	8 (4)	8 (4)	8 (4)	8 (4)	8 (4)	8 (4)
Pseudogenes	/	/	/	/	infA	accD	/

Taxa	P. kwangtungensis	P. chrysochlora	P. stenodonta	P. persimilis
Raw Base (G)	/	/	/	/
Mean coverage	/	/	/	/
GenBank numbers	NC_034371	KX668178	KX668176	KX641757
Total genome size (bp)	153,757	151,944	150,785	152,756
LSC (bp)	84,479	83,953	82,682	83,537
IRs (bp)	25,855	25,460	25,182	25,753
SSC (bp)	17,568	17,801	17,739	17,713
Total GC content (%)	37.1	37	37.1	37.2
LSC (%)	35	35	35	35.2
IRs (%)	42.7	42.8	43	42.8
SSC (%)	30.4	30.2	30.2	30.6
Total number of genes	130	131	131	130
Protein-coding	85 (7)	86 (7)	86 (7)	85 (7)
tRNA	37 (7)	37 (7)	37 (7)	37 (7)
rRNA	8 (4)	8 (4)	8 (4)	8 (4)
Pseudogenes	/	/	/	/

*, The four newly generated plastid genomes. LSC, large single copy region, IR, inverted repeat regions, and SSC, small single copy region.

2.2. Codon Usage Analysis

Codon usage plays a crucial role in evolution of plastid genome. Here, we first analyzed codons of the protein-coding genes in the 10 *Primula* plastid genomes. The number of encoded codons ranged from 25,781 (*P. sinensis*) to 26,505 (*P. knuthiana*) (Table S5). Detailed codon analysis showed that the

10 *Primula* species had similar codon usage and relative synonymous codon usage (RSCU) values (Table S5). Leucine and Cysteine were the highest (2743–2823 codons) and lowest (280–298 codons) frequent amino acids in these species, respectively (Table S5). RSCU > 1 denotes that the codon is biased and used more frequently, RSCU = 1 shows that the codon has no bias, and RSCU < 1 indicates that the codon is used less frequently. All 10 *Primula* plastid genomes had 30 biased codons with RSCU > 1 (Table S5). The biased codons had higher representation rates for A or T at the third codon position in a similar manner to the majority of angiosperm plastid genomes. Except for TTG, all of the types of biased codons (RSCU > 1) ended with A or T. The GC% was quite different at the three codon positions (Table S6). The average values of GC% for the first, second, and third codon positions of 10 *Primula* species were 45.3, 37.9, and 29.2%, respectively (Table S6). The observation of GC% level also indicated that plastid genome in *Primula* was a strong bias toward A or T at the third codon position.

2.3. Analysis of Repeat Elements

Three categories of repeats (dispersed, palindromic, and tandem repeats) were identified in the 10 *Primula* plastid genomes. We detected 326 repeats in total comprising 144 dispersed, 123 palindromic, and 59 tandem repeats (Figure 1A and Table S7). Among them, repeats of *P. sinensis* (45) were the greatest and that of *P. woodwardii* (26) were the lowest (Figure 1A and Table S7). The majority of the repeats (95.4%) ranged in size from 14 to 62 bp (Figure 1B and Table S7). Repeats located in intergenic spacer (IGS) and intron regions comprised 44.2% (144 repeats) of the total repeats and 47.8% (156 repeats) were located in *ycf2* gene, whereas only a minority were located in other coding DNA sequence (CDS) regions, such as *psaB, trnS-GCU, ycf1, rpoB, ndhF,* etc. (Table S7).

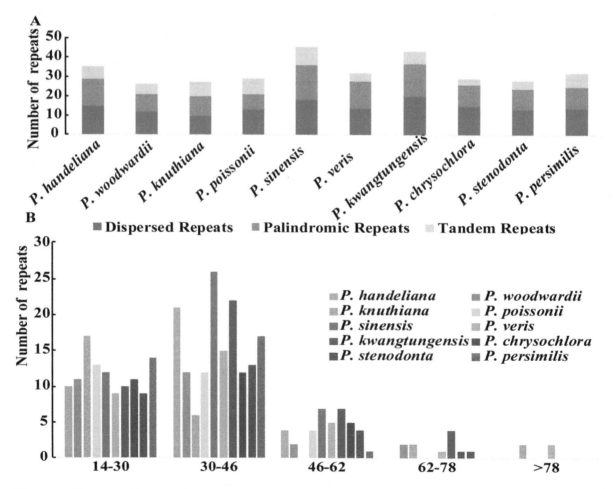

Figure 1. The type of repeated sequences in the 10 *Primula* plastid genomes. (**A**) Number of three repeat types; (**B**) number of repeat sequences by length.

A total of 496 simple sequence repeats (SSRs) measuring at least 10 bp in length were also analyzed (Figure 2A and Table S8). Among these SSRs, the mononucleotide, dinucleotide, trinucleotide, tetranucleotide, pentanucleotide, and hexanucleotide SSRs were all detected. The mononucleotide SSR were the richest with a proportion of 76.6%, followed by dinucleotide SSR (11.7%), tetranucleotide SSR (7.7%), and trinucleotide SSR (2.8%) (Figure 2A and Table S8). We only detected six pentanucleotide and hexanucleotide SSRs in the 10 *Primula* cp genomes (Figure 2A and Table S8). Unsurprisingly, the mononucleotide A/T SSR occupied the highest portion (368; 74.2%) (Table S8). The number of mononucleotide A/T SSR was significantly higher than that of the mononucleotide G/C SSR (Figure 2B and Table S8). Furthermore, most of the SSRs were found in IGS regions (56.9%), followed by CDS regions (25%), and intron regions (17.9%) (Table S8). SSRs located in the CDS region were mainly found in the *ycf1* gene.

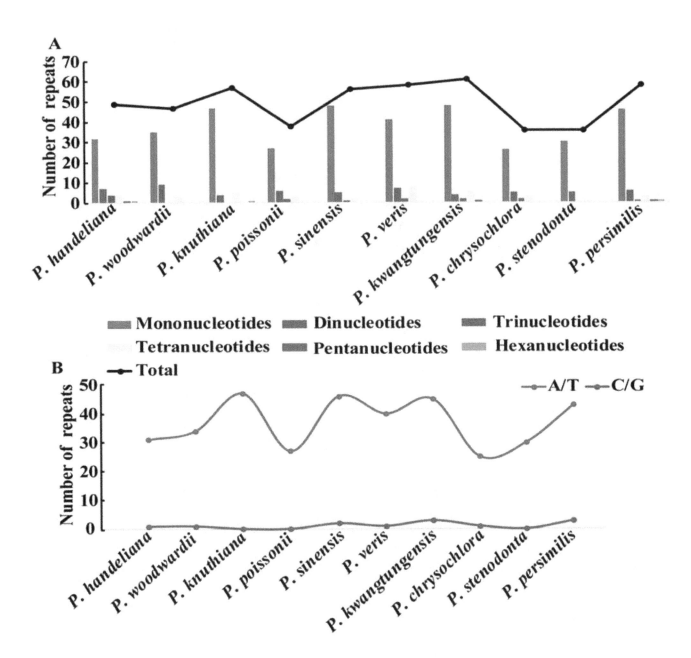

Figure 2. Simple sequence repeats (SSRs) in the 10 *Primula* plastid genomes. (**A**) Number of SSR types; (**B**) number of mononucleotide A/T and G/C SSRs.

2.4. IR/SC Boundary and Genome Rearrangement

The IR/SC boundary contents of 10 *Primula* plastid genomes were compared (Figure 3). The gene content and gene order were conserved at the IR/SC boundary, but the *Primula* plastid genomes exhibited more obvious differences. In the *P. kwangtungensis* plastid genome, the *rps19* gene was located entirely in the LSC, whereas IRb extended in a variable manner 7–175 bp into the *rps19* gene in all the other species. In the *P. chrysochlora* plastid genome, IRb even crossed completely into the *rps19* gene. IRb extended 7–74 bp in a variable manner into the *ndhF* genes, except in the *P. handeliana* and *P. poissonii* plastid genomes. In all of the *Primula* plastid genomes, IRa extended into the *ycf1* genes, where the smallest and largest extensions occurred in the *P. handeliana* (888 bp) and *P. kwangtungensis* (1048 bp) plastid genomes. The whole-genome alignment of the 10 *Primula* plastid genomes showed no rearrangement events in *Primula* (Figure S1).

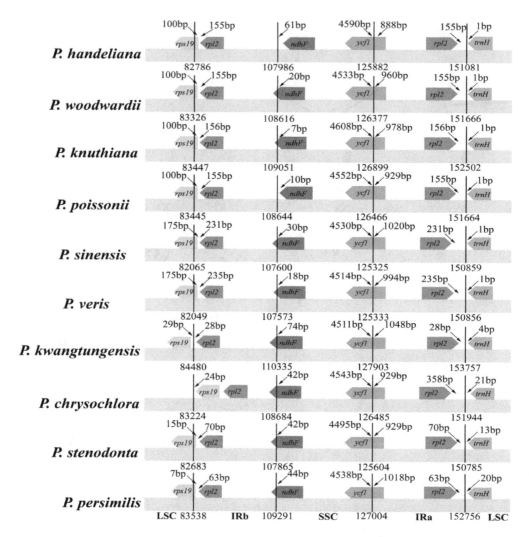

Figure 3. Comparison of the LSC, IR, and SSC border regions among the 10 *Primula* plastid genomes. Number above the gene features means the distance between the ends of genes and the borders sites. These features are not to scale.

2.5. Sequence Divergence

To investigate the levels of sequence divergence, the 10 *Primula* plastid genomes were plotted using mVISTA with *P. poissonii* as the reference (Figure 4). The *Primula* plastid genomes exhibited moderate sequence divergence (Figure 4). As expected, coding and IR regions exhibited more sequence conservation than non-coding and SC regions, respectively (Figure 4). We then calculated the percentage of variable characters for each coding region and non-coding regions with an alignment length of more than 200 bp (Table S9). The average percentage of variation in non-coding regions is 0.38, which was significantly higher than that in the coding regions (0.088 on average; Table S9). The *accD* gene contained various indels and it was a pseudogene in *P. sinensis* and missing in *P. persimilis* and *P. kwangtungensis*, which may have caused the most divergent coding region. In addition, 15 genes had a percentage of variation greater than 0.10 (Table S9), i.e., *ycf1* (0.23), *matK* (0.18), *ycf15* (0.17), *ndhF* (0.17), *rpl33* (0.16), *rpl22* (0.16), *rps16* (0.15), *rps8* (0.12), *ccsA* (0.12), *rps15* (0.12), *rpoC2* (0.11), *psbH* (0.11), *ndhD* (0.11), *rpoA* (0.10), and *ndhA* (0.10). Among the 16 genes with higher percentages of variation, 15 genes were found in SC regions and only one gene in IR regions (Table S9). The average percentages of variations in the LSC, SSC, and IR regions were 0.42, 0.43 and 0.15 in the non-coding regions, while the corresponding values in the coding regions were 0.09, 0.11, and 0.04, respectively (Table S9). All of the results demonstrated that the IR regions were more conserved than the SC regions. The overall sequence divergence based on the p-distance among the 10 *Primula* species was 0.028143 (Table S10). The pairwise p-distance between the 10 species ranged from 0.005857 to 0.041629 (Table S10). These results suggested that moderate sequence divergence has occurred within the genus *Primula*.

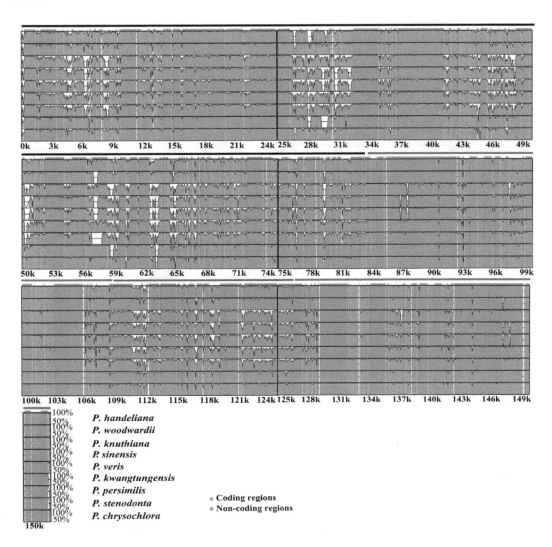

Figure 4. Sequence identity plot of the 10 *Primula* plastid genomes, with *Primula poissonii* as a reference. The *y*-axis represents % identity ranging from 50% to 100%. Coding and non-coding regions are marked in purple and pink, respectively. The red, black, and gray lines show the IRs, LSC, and SSC regions, respectively.

2.6. Phylogenomic Analysis

To investigate the phylogenetic position of *Primula*, three datasets (76 shared protein-coding genes, codon positions 1 + 2, and codon position 3) were used to conduct the BI and ML analyses (Figures 5 and S2). The selected models of each dataset were shown in Table 2. Support values were generally high for almost all relationships inferred from 76 shared protein-coding genes (the support values had a range of 78/0.91–100/1) (Figure 5). All phylogenetic trees clearly identified that *Primula* was monophyletic and more closely related to *Androsace* with high support values (Figures 5 and S2).

Table 2. Datasets and selected model in ML and BI analysis

Datasets	Best Fit Model	Model in ML	Model in BI
76 shared protein-coding genes	TVM + I + G	GTR + G	TVM + I + G
Codon positions 1 + 2	TVM + I + G	GTR + G	TVM + I + G
Codon position 3	GTR + I + G	GTR + G	GTR + I + G
Whole plastid genomes	TVM + I + G	GTR + G	TVM + I + G
Protein-coding regions	TVM + I + G	GTR + G	TVM + I + G
Introns & intergenic spacers	TVM + I + G	GTR + G	TVM + I + G
IRs	TVM + I + G	GTR + G	TVM + I + G
LSC	GTR + I + G	GTR + G	GTR + I + G
SSC	TVM + I + G	GTR + G	TVM + I + G

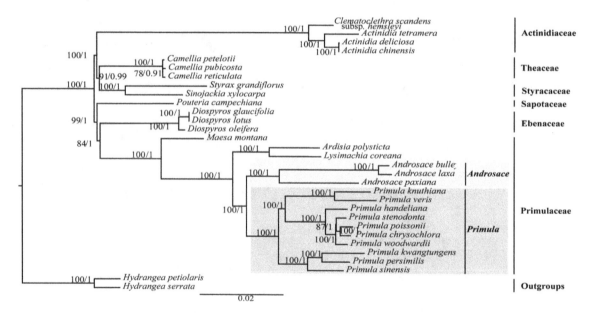

Figure 5. Phylogenetic relationship of the 31 species inferred from ML and BI analyses based on 76·shared protein-coding genes. The numbers near each node are bootstrap support values and posterior probability. *Hydrangea petiolaris* and *Hydrangea serrata* were used as the outgroups.

We then constructed six datasets (whole plastid genome, protein-coding regions, LSC, SSC, IRs, and introns & intergenic spacers) to analyze the phylogenetic relationships among the members of the genus *Primula*. The plastid genome of *A. laxa* was used as the outgroup. The selected models for each dataset used in BI and ML analyses were displayed in Table 2. The different datasets generally produced congruent phylogenetic trees (two topological structures) with moderate to high support values (Figure 6). All of the phylogenetic trees showed that *P. stenodonta*, *P. poissonii*, and *P. chrysochlora* formed a monophyletic group, where they belong to Sect. *Proliferae*. Although *P. woodwardii* and *P. handeliana* belong to Sect. *Crystallophlomis*, they were not monophyletic. *P. kwangtungensis*, *P. persimilis*, and *P. sinensis* belong to different sections, but they clustered together in the phylogenetic trees. In addition, *P. knuthiana* was more closely related to *P. veris* than other *Primula* species, but their placements varied in topological structure.

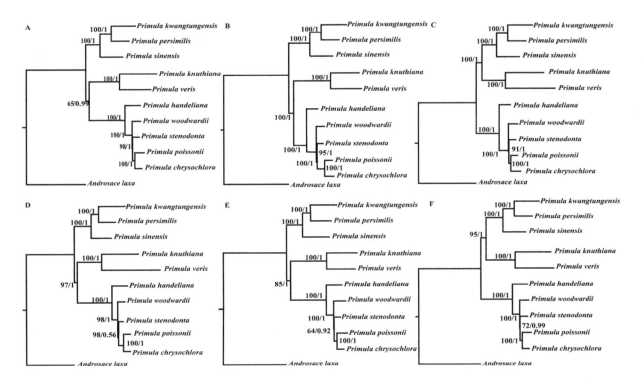

Figure 6. Phylogenetic relationships of the 10 *Primula* species and *A. laxa* inferred from ML and BI analyses. (**A**) Whole plastid genomes; (**B**) protein-coding regions; (**C**) introns and intergenic spacer regions; (**D**) IR regions; (**E**) SSC regions; and (**F**) LSC regions. The numbers near each node are bootstrap support values and posterior probability.

3. Discussion

3.1. Evolution of the Plastid Genome

Most angiosperm plastid genomes are highly conserved in terms of their gene content and order, but gene loss (deletion or production of pseudogenes) has occurred in several angiosperm lineages [26,27]. In our study, the *accD* gene was found in seven *Primula* plastid genomes, while it was a pseudogene in *P. sinensis* and was missing in *P. persimilis* and *P. kwangtungensis*. The *accD* gene encodes the acetyl-CoA carboxylase subunit D, which has been lost either partially or completely from some members of the Poales and Acoraceae [28]. The *infA* gene was a pseudogene in *P. poissonii* plastid genome, but it has been entirely lost from the other *Primula* plastid genome. The *infA* gene encodes translation initiation factor 1, which assists with the assembly of the translation initiation complex [18]. Similar events have also occurred in other angiosperm plastid genomes, such as those of *Hagenia abyssinica* [29] and *Morella rubra* [30], although the plastid genome of *A. laxa* contains the *infA* gene. The photosystem assembly factors (*ycf3* and *ycf4*) that act on photosystem I complex [31,32] should be renamed as *pafI* and *pafII* (respectively) according to recent studies [18]. Here, we use the new names of the two genes in both *Primula* and *A. laxa* plastid genomes.

IRs are the most conserved regions in the plastid genomes, where the contraction and expansion of the IR regions have occurred frequently. Our results indicated more obvious differences at the IR/SC boundaries. Particularly, in the *P. kwangtungensis* plastid genome, the *rps19* gene was located entirely in the LSC. By contrast, IRb extended into the *rps19* gene and it even completely crossed the *rps19* gene in the *P. chrysochlora* plastid genome. In addition, IRa extended into the *ycf1* genes where the smallest and largest extensions occurred in *P. handeliana* (888 bp) and *P. kwangtungensis* (1048 bp). The expansions of IRs into the *rps19* gene and *ycf1* gene have been also observed in *Cardiocrinum* [33] and *Amana* [34]. IR regions contraction and expansion events are relatively common evolutionary phenomena in plants [35]. Moreover, IR region loss was observed in some species [36,37].

Large and complex repeat sequences may play important roles in the arrangement and recombination of the plastid genome [38,39]. In all, 326 repeats were detected in the 10 *Primula* plastid genomes. Compared with other angiosperm species [40], this number is relatively small. Most of repeats ranged in size from 14 to 62 bp and almost all were not large repeats (>100 bp), which were in a similar manner to those reported in other plants [41–43]. *Pelargonium, Trifolium,* and *Trachelium*, the most highly rearranged plastid genomes contain a high frequency of large repeats (>100 bp) [44]. Our study revealed that no rearrangement events occurred in *Primula*, we thus deduced that may be mainly ascribed to no large repeats in these 10 *Primula* plastid genomes. Repeats located in *ycf2* gene occupied 47.8% of the total repeats. The *ycf2* gene is the largest gene in the *Primula* plastid genomes with over 6000 bp in length, and is completely duplicated in the IR regions. This phenomenon has also been reported in *Cardiocrinum* [33]. SSRs are highly polymorphic, and thus they are employed as molecular markers for population genetics and phylogenetic investigations [45,46]. Notably, the majority of the SSRs in the 10 *Primula* plastid genomes were the mononucleotide A/T SSRs (74.2%), which supports previous reports that SSRs in the plastid genome generally comprise short polyadenine (polyA) or polythymine (polyT) repeats [47,48]. Most of the SSRs were found in IGS regions (56.9%), followed by CDS regions (25%) and introns (17.9%). The CDS region with the highest number of SSRs was *ycf1*, as found in other species, such as *Cardiocrinum* [33] and *Vigna radiata* [49]. In the 10 *Primula* plastid genomes, the *ycf1* gene usually spanned the small single copy (SSC) and the inverted repeat a (IRa) region. It is very interesting that all but two of the SSRs in the *ycf1* gene are distributed in the SSC region. It is possible because the section of *ycf1* gene in the IRa region is shorter (less than one kilobase long) than these in SSC region (more than four kilobase long) [50]. The cpSSRs reported here would be potential molecular markers for future studies of *Primula* species.

According to the results obtained using mVISTA, the *Primula* plastid genomes exhibited moderate sequence divergence, especially in the non-coding regions. Our study showed that the coding regions were more conserved than the non-coding regions, as found in many plants [41–43]. Besides, the IR regions were more conserved than the SC regions as previous studies [51]. This fact that the two IR regions were less variable was attributed to the conservation of the ribosomal RNA genes, which comprised about one-third of the IR region in the plastid genomes [17]. The p-distance results also confirmed that moderate sequence divergence exists within the genus *Primula*. Compared with related herbaceous plants, trees, and shrubs generally have relatively long generation times and low rates of molecular evolution [52]. Herbs have shorter generation times and show much higher rates of molecular change and variance in rates [52]. The genetic diversity of heterotypic flower plants is higher than that of self-pollinated plants, indicating that genetic variation is easy to occur in interspecific and intraspecific species of heterotypic flower plants [53]. Therefore, the moderate sequence divergence probably be related to biology characteristics of these *Primula* species, such as perennial herbs, shorter generation times, cross-pollination, distyly, etc.

3.2. Phylogenetic Relationships

Plastid genomes have been successfully used to resolve the phylogenetic relationships in plant groups [23,25,54]. In this study, we used two methods (ML and BI) to construct the phylogenetic trees. We used three datasets to investigate the phylogenetic position of *Primula*. All of the phylogenetic trees indicated that *Primula* was monophyletic and more closely related to *Androsace* in Primulaceae family. Besides, in the genus *Primula*, all of the phylogenetic trees showed that Sect. *Proliferae* (*P. chrysochlora*, *P. poissonii*, and *P. stenodonta*) formed a monophyletic group and *P. chrysochlora* was closely related to *P. poissonii* [55]. Both *P. woodwardii* and *P. handeliana* belong to Sect. *Crystallophlomis*, but they did not have the closest relationship. The phylogenetic trees indicated that *P. woodwardii* and Sect. *Proliferae* were sister groups, then they clustered with *P. handeliana* in the same clade. Section *Crystallophlomis* and *Proliferae* were clustered into one clade in this study, which was also supported by karyotype study [56], but was inconsistent with the morphological work [9]. The placement of *P. knuthiana-P. veris* clade was uncertain in the phylogenetic tree. This was partly due to the rapid evolution of genus *Primula* [14,57]. The lack of samples might also affect the results of the phylogenetic analysis. In fact, for this large genus, our study could not fully clarify the relationships among *Primula* species due to the limited taxa sampled. Hence, more species and comprehensive analyses should be included in the future phylogenetic studies of *Primula* species. All in all, our analysis based on plastid genomes provides a valuable resource that should facilitate future phylogeny, taxonomy, and evolutionary history studies of this genus.

4. Materials and Methods

4.1. Plant Materials and DNA Extraction

The four plant materials (*Primula handeliana*, *Primula woodwardii*, *Primula knuthiana*, and *Androsace laxa*) used in this study were sampled from Taibai Mountain (Shaanxi, China; 107.77 °E, 33.95 °N). Total genomic DNA was extracted from silica-dried leaves with a modified CTAB method [58] by Biomarker Technologies Inc., Beijing, China. Voucher specimens were deposited in the Key Laboratory of Resource Biology and Biotechnology, Northwest University. All of the newly generated complete plastid genome sequences were deposited in GenBank (https://www.ncbi.nlm.nih.gov) (Table 1). The complete plastid genomes of *Primula poissonii* (NC_024543) [59], *Primula sinensis* (NC_030609) [57], *Primula veris* (NC_031428) [60], *Primula kwangtungensis* (NC_034371) [61], *Primula chrysochlora* (KX668178) [55], *Primula stenodonta* (KX668176) [62], and *Primula persimilis* (KX641757) [63] were recovered in order to conduct follow-up analysis (Table S1).

4.2. Illumina Sequencing, Assembly, and Annotation

Whole-genome sequencing was performed using the 150 bp pair-end sequencing method with the Illumina Hiseq 2500 Platform by Biomarker Technologies Inc. (Beijing, China). First, the raw Illumina reads were quality trimmed using the NGSQC Toolkit_v2.3.3 [64] with the default cutoff values. The clean reads were then subjected to reference-guided assembly with the MIRA v4.0.2 program [65] (parameters: job = genome, mapping, accurate; technology = solexa; segment_placement = FR). We used *Primula poissonii* (NC_024543) and *Androsace bulleyana* (KU513438) as reference genomes to assemble the *Primula* species and *A. laxa*, respectively. The resultant contigs were further assembled using a baiting and iteration method based on MITObim v1.8 [66] with default parameters. In addition, we also used the SPAdes v3.6.2 [67] (k = 33, 55, 77) to assemble the resultant clean reads of four species. We performed de novo assembly in order to verify the validity and accuracy of assembly results. Finally, a few gaps containing some ambiguous bases "N" and low-coverage regions in the assembled plastid genomes were confirmed by PCR-based Sanger sequencing. The primer pairs were designed online with the Primer3 program [68] and listed in the Supplementary Table S2. All of the genes were annotated using Dual Organellar Genome Annotator (DOGMA) software [69] with the default parameters. We then corrected the annotations with the GENEIOUS R8.0.2 program (Biomatters Ltd., Auckland, New Zealand) based on comparisons with related species. Codon usage and relative synonymous codon usage (RSCU) [70] value were estimated for all exons in the protein-coding genes with the CodonW v1.4.2 program [71].

4.3. Identification of Repeat Sequences

We used the online REPuter program [72] to identify dispersed and palindromic repeats with a minimum repeat size of 30 bp and two repeats comprising not less than 90% (Hamming distance = 3). Tandem repeats were detected using the Tandem Repeat Finder program [73] by setting two, seven, and seven as the alignment parameters for match, mismatch, and indels, respectively. The minimum alignment score and maximum period size were 80 and 500, respectively. Simple sequence repeats (SSRs) were detected using the Perl script MISA (http://pgrc.ipk-gatersleben.de/misa/) by setting the minimum number of repeats to 10, 5, 4, 3, 3, and 3 for mono-, di-, tri-, tetra-, penta-, and hexanucleotide SSRs, respectively.

4.4. Whole Plastid Genomes Comparison

Whole genome alignment with 10 *Primula* plastid genomes was run in MAUVE [74] under default settings to test rearrangement events across genomes.

4.5. Sequence Divergence Analysis

The mVISTA program [75] was used to compare the 10 *Primula* plastid genomes with *P. poissonii* as the reference. The percentages of variable characters in each coding region and non-coding region with an aligned length of more than 200 bp were calculated as described in a previous study of Poaceae species [76]. The average genetic divergences of these *Primula* plastid genomes were estimated using p-distance with MEGA6 [77]. Substitution included transition and transversion. Gaps and missing data were completely deleted.

4.6. Phylogenomic Analysis

To investigate the phylogenetic position of *Primula*, we used 31 complete plastid genomes (Table S3). Among them, 29 were from Ericales, and two *Hydrangea* species (*Hydrangea serrata* and *Hydrangea petiolaris*) were used as the outgroups. 76 shared protein-coding genes, codon positions 1 + 2, and codon position 3, were used to conduct the phylogenetic analysis.

Then, six datasets, including the whole plastid genomes, protein-coding regions, LSC, SSC, IRs, and introns & intergenic spacers were used to conduct the phylogenetic analysis among genus *Primula* with *A. laxa* as the outgroup.

All of the datasets were aligned with MAFFT [78] using the default settings. In order to examine the phylogenetic utility of different datasets, phylogenetic analyses were conducted using maximum likelihood (ML) and Bayesian inference (BI) methods. The ML analysis was conducted using RAxMLv7.2.8 [79] with 1000 bootstrap replicates. The GTRGAMMA model was used in all of the ML analyses, as suggested in the RAxML manual. For the BI analysis, the best substitution model was determined according to Akaike's information criterion (AIC) with Modeltest v3.7 [80]. The BI analysis was performed using MrBayes v3.1.2 [81]. The Markov chain Monte Carlo (MCMC) algorithm was run for two million generations and the trees were sampled very 100 generations. Convergence was determined by examining the average standard deviation of the split frequencies (<0.01). The first 25% of the trees were discarded as a burn-in and the remaining trees were used to generate the consensus tree.

Acknowledgments: This work was financially supported by the National Natural Science Foundation of China (31670219, 31370353).

Author Contributions: Zhan-Lin Liu conceived and designed the work. Ting Ren, Yanci Yang, and Tao Zhou performed the experiments and analyzed the data. Ting Ren wrote the manuscript. Yanci Yang and Zhan-Lin Liu revised the manuscript. All authors gave final approval of the paper.

References

1. Hu, C.M.; Kelso, S. *Flora of China*; Science Press: Beijing, China, 1996; Volume 15, pp. 99–185.
2. Richards, A.J. *Primula*, 2nd ed.; B. T. Batsford Ltd.: London, UK, 2002.
3. Yan, H.F.; He, C.H.; Peng, C.I.; Hu, C.M.; Hao, G. Circumscription of *Primula* subgenus *Auganthus* (Primulaceae) based on chloroplast DNA sequences. *J. Syst. Evol.* **2010**, *48*, 123–132. [CrossRef]
4. Woodell, S. Natural hybridization between the cowsip (*Primula veris* L.) and the primrose (*P. vulgaris* Huds.) in Britain. *Watsonia* **1965**, *6*, 190–202.
5. Ornduff, R. Pollen flow in a population of *Primula vulgaris* Huds. *Bot. J. Linn. Soc.* **1979**, *78*, 1–10. [CrossRef]
6. Shen, L.L. Research advances on the pollination biology of *Primula*. *J. Anhui. Agric. Sci.* **2010**, *38*, 5574–5585.
7. Li, J.H.; Webster, M.A.; Smith, M.C.; Gilmartin, P.M. Floral heteromorphy in *Primula vulgaris*: Progress towards isolation and characterization of the *S. locus*. *Ann. Bot.* **2011**, *108*, 715–726. [CrossRef] [PubMed]
8. Nowak, M.D.; Russo, G.; Schlapbach, R.; Huu, C.N.; Lenhard, M.; Conti, E. The draft genome of *Primula veris* yields insights into the molecular basis of heterostyly. *Genome Biol.* **2015**, *16*, 12. [CrossRef] [PubMed]

9. Smith, W.W.; Fletcher, H.R. XVII.–The genus *Primula*: Sections *Obconica, Sinenses, Reinii, Pinnatae, Malacoides, Bullatae, Carolinella, Grandis* and *Denticulata*. *Trans. R. Soc. Edinb.* **1947**, *61*, 415–478. [CrossRef]

10. Wendelbo, P. Studies in Primulaceae. II. An account of *Primula* subgenus *Sphondylia* (*Syn. Sect. Floribundae*) with a review of the subdivisions of the genus. *Matematisk-Naturvitenskapelig Ser.* **1961**, *11*, 1–46.

11. Richards, A.J. *Primula*; B. T. Batsford Ltd.: London, UK, 1993.

12. Conti, E.; Suring, E.; Boyd, D.; Jorgensen, J.; Grant, J.; Kelso, S. Phylogenetic relationships and character evolution in *Primula* L.: The usefulness of ITS sequence data. *Plant Biosyst.* **2000**, *134*, 385–392. [CrossRef]

13. Mast, A.R.; Kelso, S.; Richards, A.J.; Lang, D.J.; Feller, D.M.; Conti, E. Phylogenetic relationships in *Primula* L. and related genera (Primulaceae) based on noncoding chloroplast DNA. *Int. J. Plant Sci.* **2001**, *162*, 1381–1400. [CrossRef]

14. Yan, H.F.; Liu, Y.J.; Xie, X.F.; Zhang, C.Y.; Hu, C.M.; Hao, G.; Ge, X.J. DNA barcoding evaluation and its taxonomic implications in the species-rich genus *Primula* L. in China. *PLoS ONE* **2015**, *10*, e0122903. [CrossRef] [PubMed]

15. Ravi, V.; Khurana, J.P.; Tyagi, A.K.; Khurana, P. An update on chloroplast genomes. *Plant Syst. Evol.* **2008**, *271*, 101–122. [CrossRef]

16. Jansen, R.K.; Raubeson, L.A.; Boore, J.L.; Chumley, T.W.; Haberle, R.C.; Wyman, S.K. Methods for obtaining and analyzing whole chloroplast genome sequences. *Methods Enzymol.* **2005**, *395*, 348–384. [PubMed]

17. Palmer, J.D. Comparative organization of chloroplast genomes. *Annu. Rev. Genet.* **1985**, *19*, 325–354. [CrossRef] [PubMed]

18. Wicke, S.; Schneeweiss, G.M.; dePamphilis, C.W.; Müller, K.F.; Quandt, D. The evolution of the plastid chromosome in land plants: Gene content, gene order, gene function. *Plant Mol. Biol.* **2011**, *76*, 273–297. [CrossRef] [PubMed]

19. Jansen, R.K.; Cai, Z.Q.; Raubeson, L.A.; Daniell, H.; dePamphilis, C.W.; Leebens-Mack, J.; Müller, K.F.; Guisinger-Bellian, M.; Haberle, C.R.; Hansen, A.K.; et al. Analysis of 81 genes from 64 plastid genomes resolves relationships in angiosperms and identifies genome-scale evolutionary patterns. *Proc. Natl. Acad. Sci. USA* **2007**, *104*, 19369–19374. [CrossRef] [PubMed]

20. Moore, M.J.; Bell, C.D.; Soltis, P.S.; Soltis, D.E. Using plastid genome-scale data to resolve enigmatic relationships among basal angiosperms. *Proc. Natl. Acad. Sci. USA* **2007**, *104*, 19363–19368. [CrossRef] [PubMed]

21. Cronn, R.; Liston, A.; Parks, M.; Gernandt, D.S.; Shen, R.; Mockler, T. Multiplex sequencing of plant chloroplast genomes using Solexa sequencing-by-synthesis technology. *Nucleic Acids Res.* **2008**, *36*, e122. [CrossRef] [PubMed]

22. Mardis, E.R. The impact of next-generation sequencing technology on genetics. *Trends Genet.* **2008**, *24*, 133–141. [CrossRef] [PubMed]

23. Luo, Y.; Ma, P.F.; Li, H.T.; Yang, J.B.; Wang, H.; Li, D.Z. Plastid phylogenomic analyses resolve Tofieldiaceae as the root of the early diverging monocot order *Alismatales*. *Genome Biol. Evol.* **2016**, *8*, 932–945. [CrossRef] [PubMed]

24. Carbonell-Caballero, J.; Alonso, R.; Ibañez, V.; Terol, J.; Talon, M.; Dopazo, J. A phylogenetic analysis of 34 chloroplast genomes elucidates the relationships between wild and domestic species within the genus *Citrus*. *Mol. Biol. Evol.* **2015**, *32*, 2015–2035. [CrossRef] [PubMed]

25. Zhang, S.D.; Jin, J.J.; Chen, S.Y.; Chase, M.W.; Soltis, D.E.; Li, H.T.; Yang, J.B.; Li, D.Z.; Yi, T.S. Diversification of Rosaceae since the Late Cretaceous based on plastid phylogenomics. *New Phytol.* **2017**, *214*, 1355–1367. [CrossRef] [PubMed]

26. Braukmann, T.; Kuzmina, M.; Stefanović, S. Plastid genome evolution across the genus *Cuscuta* (Convolvulaceae): Two clades within subgenus *Grammica* exhibit extensive gene loss. *J. Exp. Bot.* **2013**, *64*, 977–989. [CrossRef] [PubMed]

27. Logacheva, M.D.; Schelkunov, M.I.; Nuraliev, M.S.; Samigullin, T.H.; Penin, A.A. The plastid genome of mycoheterotrophic monocot *Petrosavia stellaris* exhibits both gene losses and multiple rearrangements. *Genome Biol. Evol.* **2014**, *6*, 238–246. [CrossRef] [PubMed]

28. Katayama, H.; Ogihara, Y. Phylogenetic affinities of the grasses to other monocots as revealed by molecular analysis of chloroplast DNA. *Curr. Genet.* **1996**, *29*, 572–581. [CrossRef] [PubMed]

29. Gichira, A.W.; Li, Z.Z.; Saina, J.K.; Long, Z.C.; Hu, G.W.; Gituru, R.W.; Wang, Q.F.; Chen, J.M. The complete chloroplast genome sequence of an endemic monotypic genus *Hagenia* (Rosaceae): Structural comparative analysis, gene content and microsatellite detection. *PeerJ* **2017**, *5*, e2846. [CrossRef] [PubMed]

30. Liu, L.X.; Li, R.; Worth, J.R.; Li, X.; Li, P.; Cameron, K.M.; Fu, C.X. The complete chloroplast genome of Chinese bayberry (*Morella rubra*, Myricaceae): Implications for understanding the evolution of Fagales. *Front. Plant Sci.* **2017**, *8*. [CrossRef] [PubMed]

31. Naver, H.; Boudreau, E.; Rochaix, J.D. Functional studies of Ycf3: Its role in assembly of photosystem I and interactions with some of its subunits. *Plant Cell* **2001**, *13*, 2731–2745. [CrossRef] [PubMed]

32. Ozawa, S.I.; Nield, J.; Terao, A.; Stauber, E.J.; Hippler, M.; Koike, H.; Rochaix, J.D.; Takahashi, Y. Biochemical and structural studies of the large Ycf4-photosystem I assembly complex of the green alga *Chlamydomonas reinhardtii*. *Plant Cell* **2009**, *21*, 2424–2442. [CrossRef] [PubMed]

33. Lu, R.S.; Li, P.; Qiu, Y.X. The complete chloroplast genomes of three *Cardiocrinum* (Liliaceae) species: Comparative genomic and phylogenetic analyses. *Front. Plant Sci.* **2017**, *7*, 2054. [CrossRef] [PubMed]

34. Li, P.; Lu, R.S.; Xu, W.Q.; Ohitoma, T.; Cai, M.Q.; Qiu, Y.X.; Cameron, M.K.; Fu, C.X. Comparative genomics and phylogenomics of East Asian tulips (*Amana*, Liliaceae). *Front. Plant Sci.* **2017**, *8*, 451. [CrossRef] [PubMed]

35. Kim, K.J.; Lee, H.L. Complete chloroplast genome sequences from Korean Ginseng (*Panax schinseng* Nees) and comparative analysis of sequence evolution among 17 vascular plants. *DNA Res.* **2004**, *11*, 247–261. [CrossRef] [PubMed]

36. Perry, A.S.; Wolfe, K.H. Nucleotide substitution rates in legume chloroplast DNA depend on the presence of the inverted repeat. *J. Mol. Evol.* **2002**, *55*, 501–508. [CrossRef] [PubMed]

37. Yi, X.; Gao, L.; Wang, B.; Su, Y.J.; Wang, T. The complete chloroplast genome sequence *of Cephalotaxus oliveri* (Cephalotaxaceae): Evolutionary comparison of *Cephalotaxus* chloroplast DNAs and insights into the loss of inverted repeat copies in Gymnosperms. *Genome Biol. Evol.* **2013**, *5*, 688–698. [CrossRef] [PubMed]

38. Ogihara, Y.; Terachi, T.; Sasakuma, T. Intramolecular recombination of chloroplast genome mediated by short direct-repeat sequences in wheat species. *Proc. Natl. Acad. Sci. USA* **1988**, *85*, 8573–8577. [CrossRef] [PubMed]

39. Weng, M.L.; Blazier, J.C.; Govindu, M.; Jansen, R.K. Reconstruction of the ancestral plastid genome in Geraniaceae reveals a correlation between genome rearrangements, repeats and nucleotide substitution rates. *Mol. Biol. Evol.* **2013**, *31*, 645–659. [CrossRef] [PubMed]

40. Zhang, X.; Zhou, T.; Kanwal, N.; Zhao, Y.M.; Bai, G.Q.; Zhao, G.F. Completion of eight *Gynostemma* BL. (Cucurbitaceae) chloroplast genomes: Characterization, comparative analysis, and phylogenetic relationships. *Front. Plant Sci.* **2017**, *8*, 1583. [CrossRef] [PubMed]

41. Hu, Y.H.; Woeste, K.E.; Zhao, P. Completion of the chloroplast genomes of five Chinese *Juglans* and their contribution to chloroplast phylogeny. *Front. Plant Sci.* **2016**, *7*, 1955. [CrossRef] [PubMed]

42. Yang, Y.C.; Zhou, T.; Duan, D.; Yang, J.; Feng, L.; Zhao, G.F. Comparative analysis of the complete chloroplast genomes of five *Quercus* species. *Front. Plant Sci.* **2016**, *7*, 959. [CrossRef] [PubMed]

43. Zhou, T.; Chen, C.; Wei, Y.; Chang, Y.X.; Bai, G.Q.; Li, Z.H.; Kanwal, N.; Zhao, G.F. Comparative transcriptome and chloroplast genome analyses of two related *Dipteronia* species. *Front. Plant Sci.* **2016**, *7*, 1512. [CrossRef] [PubMed]

44. Guisinger, M.M.; Kuehl, J.V.; Boore, J.L.; Jansen, R.K. Extreme reconfiguration of plastid genomes in the angiosperm family Geraniaceae: Rearrangements, repeats, and codon usage. *Mol. Biol. Evol.* **2011**, *28*, 583–600. [CrossRef] [PubMed]

45. Powell, W.; Morgante, M.; Andre, C.; McNicol, J.W.; Machray, G.C.; Doyle, J.J. Hypervariable microsatellites provide a general source of polymorphic DNA markers for the chloroplast genome. *Curr. Biol.* **1995**, *5*, 1023–1029. [CrossRef]

46. He, S.L.; Wang, Y.S.; Volis, S.; Li, D.Z.; Yi, T.S. Genetic diversity and population structure: Implications for conservation of wild soybean (*Glycine soja* Sieb. et Zucc) based on nuclear and chloroplast microsatellite variation. *Int. J. Mol. Sci.* **2012**, *13*, 12608–12628. [CrossRef] [PubMed]

47. Kuang, D.Y.; Wu, H.; Wang, Y.L.; Gao, L.M.; Zhang, S.Z.; Lu, L. Complete chloroplast genome sequence of *Magnolia kwangsiensis* (Magnoliaceae): Implication for DNA barcoding and population genetics. *Genome* **2011**, *54*, 663–673. [CrossRef] [PubMed]

48. Martin, G.; Baurens, F.C.; Cardi, C.; Aury, J.M.; D'Hont, A. The complete chloroplast genome of banana (*Musa acuminata*, Zingiberales): Insight into plastid monocotyledon evolution. *PLoS ONE* **2013**, *8*, e67350. [CrossRef] [PubMed]

49. Tangphatsornruang, S.; Sangsrakru, D.; Chanprasert, J.; Uthaipaisanwong, P.; Yoocha, T.; Jomchai, N.; Tragoonrung, S. The chloroplast genome sequence of mungbean (*Vigna radiata*) determined by high-throughput

pyrosequencing: Structural organization and phylogenetic relationships. *DNA Res.* **2009**, *17*, 11–22. [CrossRef] [PubMed]

50. Dong, W.P.; Xu, C.; Li, C.H.; Sun, J.H.; Zuo, Y.J.; Shi, S.; Cheng, T.; Guo, J.J.; Zhou, S.L. *ycf1*, the most promising plastid DNA barcode of land plants. *Sci. Rep.* **2015**, *5*. [CrossRef] [PubMed]

51. Zhu, A.D.; Guo, W.H.; Gupta, S.; Fan, W.S.; Mower, J.P. Evolutionary dynamics of the plastid inverted repeat: The effects of expansion, contraction, and loss on substitution rates. *New Phytol.* **2016**, *209*, 1747–1756. [CrossRef] [PubMed]

52. Smith, S.A.; Donoghue, M.J. Rates of molecular evolution are linked to life history in flowering plants. *Science* **2008**, *322*, 86–89. [CrossRef] [PubMed]

53. Weller, S.G.; Sakai, A.K.; Straub, C. Allozyme diversity and genetic identity in *Schiedea* and *Alsinidendron* (Caryophyllaceae: Alsinoideae) in the Hawaiian Islands. *Evolution* **1996**, *50*, 23–34. [CrossRef] [PubMed]

54. Ma, P.F.; Zhang, Y.X.; Zeng, C.X.; Guo, Z.H.; Li, D.Z. Chloroplast phylogenomic analyses resolve deep-level relationships of an intractable bamboo tribe *Arundinarieae* (*Poaceae*). *Syst. Biol.* **2014**, *63*, 933–950. [CrossRef] [PubMed]

55. Zhang, C.Y.; Liu, T.J.; Xu, Y.; Yan, H.F. Characterization of the whole chloroplast genome of a rare candelabra primrose *Primula chrysochlora* (Primulaceae). *Conserv. Genet. Resour.* **2017**, *9*, 361–363. [CrossRef]

56. Bruun, H.G. Cytological Studies in Primula with Special Reference to the Relation between the Karyology and Taxonomy of the Genus. Ph.D. Thesis, Acta Universitatis Upsaliensis, Uppsala, Sweden, 1932.

57. Liu, T.J.; Zhang, C.Y.; Yan, H.F.; Zhang, L.; Ge, X.J.; Hao, G. Complete plastid genome sequence of *Primula sinensis* (Primulaceae): Structure comparison, sequence variation and evidence for *accD* transfer to nucleus. *PeerJ* **2016**, *4*, e2101. [CrossRef] [PubMed]

58. Doyle, J.J. A rapid DNA isolation procedure for small quantities of fresh leaf tissue. *Phytochem. Bull.* **1987**, *19*, 11–15.

59. Yang, J.B.; Li, D.Z.; Li, H.T. Highly effective sequencing whole chloroplast genomes of angiosperms by nine novel universal primer pairs. *Mol. Ecol. Resour.* **2014**, *14*, 1024–1031. [CrossRef] [PubMed]

60. Zhou, T.; Zhao, J.X.; Chen, C.; Meng, X.; Zhao, G.F. Characterization of the complete chloroplast genome sequence of *Primula veris* (Ericales: Primulaceae). *Conserv. Genet. Resour.* **2016**, *8*, 455–458. [CrossRef]

61. Zhang, C.Y.; Liu, T.J.; Xu, Y.; Yan, H.F.; Hao, G.; Ge, X.J. Characterization of the whole chloroplast genome of an endangered species *Primula kwangtungensis* (Primulaceae). *Conserv. Genet. Resour.* **2017**, *9*, 87–89. [CrossRef]

62. Zhang, C.Y.; Liu, T.J.; Yan, H.F.; Ge, X.J.; Hao, G. The complete chloroplast genome of a rare candelabra primrose *Primula stenodonta* (Primulaceae). *Conserv. Genet. Resour.* **2017**, *9*, 123–125. [CrossRef]

63. Zhang, C.Y.; Liu, T.J.; Yan, H.F.; Xu, Y. The complete chloroplast genome of *Primula persimilis* (Primulaceae). *Conserv. Genet. Resour.* **2017**, *9*, 189–191. [CrossRef]

64. Patel, R.K.; Jain, M. NGS QC Toolkit: A toolkit for quality control of next generation sequencing data. *PLoS ONE* **2012**, *7*, e30619. [CrossRef] [PubMed]

65. Chevreux, B.; Pfisterer, T.; Drescher, B.; Driesel, A.J.; Müller, W.E.; Wetter, T.; Suhai, S. Using the miraEST assembler for reliable and automated mRNA transcript assembly and SNP detection in sequenced ESTs. *Genome Res.* **2004**, *14*, 1147–1159. [CrossRef] [PubMed]

66. Hahn, C.; Bachmann, L.; Chevreux, B. Reconstructing mitochondrial genomes directly from genomic next-generation sequencing reads-a baiting and iterative mapping approach. *Nucleic Acids Res.* **2013**, *41*, e129. [CrossRef] [PubMed]

67. Bankevich, A.; Nurk, S.; Antipov, D.; Gurevich, A.A.; Dvorkin, M.; Kulikov, A.S.; Lesin, V.M.; Nikolenko, S.I.; Pham, S.; Prjibelski, A.D.; et al. SPAdes: A new genome assembly algorithm and its applications to single-cell sequencing. *J. Comput. Biol.* **2012**, *19*, 455–477. [CrossRef] [PubMed]

68. Untergrasser, A.; Cutcutache, I.; Koressaar, T.; Ye, J.; Faircloth, B.C.; Remm, M.; Rozen, S.G. Primer3-new capabilities and interfaces. *Nucleic Acids Res.* **2012**, *40*, e115. [CrossRef] [PubMed]

69. Wyman, S.K.; Jansen, R.K.; Boore, J.L. Automatic annotation of organellar genomes with DOGMA. *Bioinformatics* **2004**, *20*, 3252–3255. [CrossRef] [PubMed]

70. Sharp, P.M.; Li, W.H. The codon adaptation index-a measure of directional synonymous codon usage bias, and its potential applications. *Nucleic Acids Res.* **1987**, *15*, 1281–1295. [CrossRef] [PubMed]

71. Peden, J.F. Analysis of codon usage. Ph.D. Thesis, University of Nottingham, University of Nottingham, UK, 1999.

72. Kurtz, S.; Choudhuri, J.V.; Ohlebusch, E.; Schleiermacher, C.; Stoye, J.; Giegerich, R. REPuter: The manifold applications of repeat analysis on a genomic scale. *Nucleic Acids Res.* **2001**, *29*, 4633–4642. [CrossRef] [PubMed]

73. Benson, G. Tandem repeats finder: A program to analyze DNA sequences. *Nucleic Acids Res.* **1999**, *27*, 573–580. [CrossRef] [PubMed]

74. Darling, A.C.E.; Mau, B.; Blattner, F.R.; Perna, N.T. Mauve: Multiple alignment of conserved genomic sequence with rearrangements. *Genome Res.* **2004**, *14*, 1394–1403. [CrossRef] [PubMed]

75. Frazer, K.A.; Pachter, L.; Poliakov, A.; Rubin, E.M.; Dubchak, I. VISTA: Computational tools for comparative genomics. *Nucleic Acids Res.* **2004**, *32*, W273–W279. [CrossRef] [PubMed]

76. Zhang, Y.J.; Ma, P.F.; Li, D.Z. High-throughput sequencing of six bamboo chloroplast genomes: Phylogenetic implications for temperate woody bamboos (*Poaceae*: *Bambusoideae*). *PLoS ONE* **2011**, *6*, e20596. [CrossRef] [PubMed]

77. Tamura, K.; Stecher, G.; Peterson, D.; Filipski, A.; Kumar, S. MEGA6: Molecular evolutionary genetics analysis version 6.0. *Mol. Biol. Evol.* **2013**, *30*, 2725–2729. [CrossRef] [PubMed]

78. Katoh, K.; Standley, D.M. MAFFT multiple sequence alignment software version 7: Improvements in performance and usability. *Mol. Biol. Evol.* **2013**, *30*, 772–780. [CrossRef] [PubMed]

79. Stamatakis, A. RAxML-VI-HPC: Maximum likelihood-based phylogenetic analysis with thousands of taxa and mixed models. *Bioinformatics* **2006**, *22*, 2688–2690. [CrossRef] [PubMed]

80. Posada, D.; Crandall, K.A. Modeltest: Testing the model of DNA substitution. *Bioinformatics* **1998**, *14*, 817–818. [CrossRef] [PubMed]

81. Ronquist, F.; Huelsenbeck, J.P. MrBayes 3: Bayesian phylogenetic inference under mixed models. *Bioinformatics* **2003**, *19*, 1572–1574. [CrossRef] [PubMed]

13

Metabolic Reprogramming in Chloroplasts under Heat Stress in Plants

Qing-Long Wang [1], Juan-Hua Chen [1], Ning-Yu He [1] and Fang-Qing Guo [1,2,*]

[1] The National Key Laboratory of Plant Molecular Genetics, Institute of Plant Physiology & Ecology, Chinese Academy of Sciences, 300 Fenglin Road, Shanghai 200032, China; qlwang@sibs.ac.cn (Q.-L.W.); jhchen02@sibs.ac.cn (J.-H.C.); nyhe@sibs.ac.cn (N.-Y.H.)

[2] CAS Center for Excellence in Molecular Plant Sciences, Institute of Plant Physiology & Ecology, Chinese Academy of Sciences, 300 Fenglin Road, Shanghai 200032, China

* Correspondence: fqguo@sibs.ac.cn

Abstract: Increases in ambient temperatures have been a severe threat to crop production in many countries around the world under climate change. Chloroplasts serve as metabolic centers and play a key role in physiological adaptive processes to heat stress. In addition to expressing heat shock proteins that protect proteins from heat-induced damage, metabolic reprogramming occurs during adaptive physiological processes in chloroplasts. Heat stress leads to inhibition of plant photosynthetic activity by damaging key components functioning in a variety of metabolic processes, with concomitant reductions in biomass production and crop yield. In this review article, we will focus on events through extensive and transient metabolic reprogramming in response to heat stress, which included chlorophyll breakdown, generation of reactive oxygen species (ROS), antioxidant defense, protein turnover, and metabolic alterations with carbon assimilation. Such diverse metabolic reprogramming in chloroplasts is required for systemic acquired acclimation to heat stress in plants.

Keywords: heat stress; metabolic reprogramming; chloroplasts; chlorophyll breakdown; reactive oxygen species (ROS); antioxidant defense; protein turnover; Photosystem II (PSII) core subunit; PSII repair cycle; carbon assimilation

1. Introduction

Photosynthesis is thought to be the most important photo-chemical reaction, during which sunlight is trapped and converted into biological energy in plants. In general, leaves function as a highly specialized organ that are basically appointed in the photosynthetic process in higher plants. Leaf photosynthesis is substantially affected, often lethally, by high temperature stress, usually 10–15 °C above an optimum temperature for plant growth as plants are not capable of moving to more favorable environments [1–3]. Housed in chloroplasts, the photosynthetic apparatus is susceptible to be damaged by heat stress and the chloroplasts have been demonstrated to play an essential role in activation of cellular heat stress signaling [4–6]. Given that Photosystem II (PSII) is the most susceptible target within the chloroplast thylakoid membrane protein complexes for heat stress, heat stress commonly causes severe thermal damages to PSII, dramatically affecting photosynthetic electron transfer and ATP synthesis [1–3,7,8]. Under heat stress, heat stress-induced damages lead to alterations of photochemical reactions in thylakoid lamellae of chloroplast, reflected as a significant reduction in the ratio of variable fluorescence to maximum fluorescence (Fv/Fm) [1,2,9–11]. Exposure to high temperature stress causes oxidative stress in plants, particularly leading to the dissociation of oxygen evolving complex (OEC) in PSII, which further results in inhibition of the electron transportation from OEC to the acceptor side of PSII [1,2,12,13]. Heat stress causes cleavage of the reaction center-binding protein D1 of PSII in spinach thylakoids and induces dissociation of a manganese (Mn)-stabilizing 33-kDa proteins from PSII reaction center complex [14]. As a photosynthetic carbon fixation cycle, the Calvin–Benson cycle is responsible for the fixation of CO_2 into carbohydrates, as well as assimilation, transport, and utilization of photoassimilates as the organic products of photosynthesis. In addition to the disruption of OEC in PSII, heat stress also results in dysfunction in the system of carbon assimilation metabolism in the stroma of chloroplast [2,8]. It has been observed that the disruption of electron transport and inactivation of the oxygen evolving enzymes of PSII dramatically inhibit the rate of ribulose-1,5-bisphosphate (RuBP) regeneration [10,15]. Heat stress-induced inhibition in the activity of ribulose-1,5-bisphosphate carboxylase/oxygenase (Rubisco) mainly results from inactivation of Rubisco activase that is extremely sensitive to heat stress because the enzyme Rubisco of higher plants is heat-stable [8,15]. In addition to the early effects on photochemical reactions and carbon assimilation, alterations in the microscopic ultrastructures of chloroplast and the integrity of thylakoid membranes were reported to be severely disrupted, including membrane destacking and reorganization when subjected to heat stress [2,16–19].

In photosynthetic organisms, chloroplasts respond to a variety of environmental stresses, including heat stress, with adjustments for major metabolic processes to optimize carbon fixation and growth requirements [2,3,6,8,20]. As one of the subcellular energy organelles, chloroplast in plant cells conducts major metabolic reprogramming processes including chlorophyll breakdown, generation and scavenging of reactive oxygen species (ROS), protein turnover, and metabolic alterations of carbon assimilation in response to heat stress (Figure 1). Here we review studies that have investigated the effects of heat stress on photosynthesis and its associated metabolic adaptations for optimizing plant growth and development under stress conditions. We will assess the role of chloroplast from an organellar perspective to begin building insights into better understanding the importance and significance of metabolic reprogramming within this organelle during high temperature stresses.

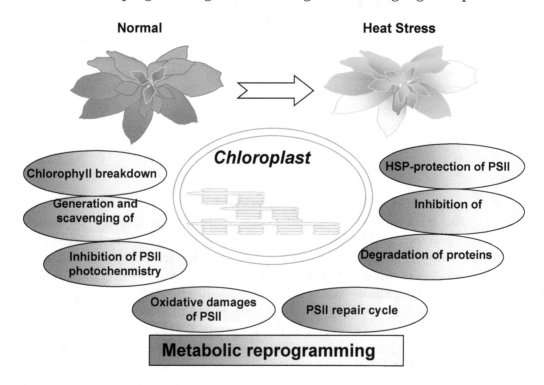

Figure 1. Extensive and transient metabolic reprogramming in chloroplasts under heat stress. Major events of metabolic reprogramming in response to heat stress include chlorophyll breakdown, generation of reactive oxygen species (ROS), antioxidant defense, protein turnover, and metabolic alterations with carbon assimilation. With respect to the systemic acquired acclimation to heat stress in plants, diverse metabolic reprogramming in chloroplasts is required for optimizing plant growth and development during high temperature stresses.

2. Chlorophyll Breakdown under Heat Stress

Chlorophyll (Chl) functions in harvesting light energy and driving electron transfer during the initial and indispensable processes of photosynthesis, and is a pigment consisting of two moieties: a chlorin ring containing a magnesium ion at its center and a long hydrophobic phytol chain, joined by an ester bond. It should be noted that when the photosynthetic apparatus is overexcited, and oxygen receives the absorbed energy from Chl, Chl acts as a harmful molecule, negatively affecting plant cells including most porphyrins [21,22]. Importantly, the breakdown of Chl is a physiological process as a prerequisite for protecting plant cells from hazardous effects of phototoxic pigments in association with recycling nitrogen sources from Chl-binding proteins in chloroplasts when leaves are senescing [22–25].

Recent studies have provided advances in better understanding of the pathway of Chl catabolism in higher plants. In brief, the removal of the phytol residue and the central Mg by chlorophyllase occurs at the initial reaction in the Chl breakdown pathway. Chlorophyllase genes, termed *CLHs*, were reported to be involved in Chl breakdown [26], but recent studies questioned the involvement of CLHs in Chl degradation in vivo during leaf senescence [23,27]. In 2009, functional analysis of pheophytinase (PPH) supports its role in porphyrin-phytol hydrolysis involved in senescence-related chlorophyll breakdown in vivo [28]. Several reports have revealed that the resulting pheophorbide (pheide) a is converted into a primary fluorescent chlorophyll catabolite (pFCC) in the next-step reactions, and two enzymes are involved in the reactions, including pheide a oxygenase (PAO) and red chl catabolite reductase (RCCR) [29–32]. *PAO* encodes a Rieske-type iron-sulfur oxygenase that is bound to chloroplast envelop, also named as *ACCELERATED CELL DEATH (ACD1)* [30]. PAO functions as a key enzyme that catalyzing the cleavage of the porphyrin ring in Chl breakdown pathway, and the red chlorophyll catabolite resulted from reactions can be further catalyzed into pFCC by RCCR. Next, a primary active transport system is responsible for exporting the resulting primary fluorescent catabolite pFCCs from the plastid and importing into the vacuole [33,34]. Based on the genetic analysis of the stay-green mutants, *NON-YELLOW COLORING1 (NYC1)* and *NYC1-LIKE (NOL)* were cloned and found to selectively retain photosystem II (PSII) light-harvesting complex subunits [35,36]. The first half of chlorophyll *b* can be catalyzed into chlorophyll *a* by chlorophyll *b* reductase that is composed of two subunits NYC1 and NOL [36]. In addition, leaves of the loss-of-function mutant *pao* showed a stay-green phenotype under dark-induction conditions and a light-dependent lesion mimic phenotype was also observed because of the increase in levels of phototoxic pheide *a* [31,37,38]. Interestingly, the *Bf993* mutant of *Festuca pratensis* is classified into the third type of stay-green mutants in which the stay-green gene named as *senescence-induced degradation (SID)* is defective [39–41]. *SGR (STAY-GREEN)*, designated as the orthologous gene of *SID*, has been cloned and characterized in a lot of plant species, including *Arabidopsis* [42], rice [43–45], pea [45,46], bell pepper [47,48], and tomato [47]. Based on the accumulated evidence, the direct interacting relationship has been characterized to exist between SGR and a subset of the protein components of the light harvesting chlorophyll *a/b*-protein complex II (LHCPII), suggesting that SGR is likely to play a role in making pigment-protein complexes unstable as a prerequisite for the enzymes in chlorophyll breakdown pathway to reach their substrate when leaves are in senescing processes [22,44,49].

A visible sign of leaf senescence and fruit ripening, loss of green color is resulted from massive Chl breakdown into nonphototoxic breakdown products in combination with carotenoid retention or anthocyanin accumulation [22,23,50]. Under normal growth conditions, chlorophyll maintains at a steady level due to the balance of its biosynthesis and degradation without a visible change in chlorophyll content [23,25]. In contrast, chlorophylls undergo turnover or breakdown when the partial or complete dismantling of the photosynthetic machinery occurs in response to environmental stresses, including heat stress. Heat stress symptoms in plants are typically characterized by leaf senescence or chlorosis due to a decline in chlorophyll content [2,51,52]. Heat-induced leaf chlorosis has been observed in a variety of plant species, including *Arabidopsis* [4,5,53], Soybean (*Glycine max* L. Merr.) [54], sorghum (*Sorghum bicolor*) [55], wheat (*Triticum aestivum*) [56,57] and creeping bentgrass (*Agrostis stolonifera*) [58]. However, the underlying mechanisms by which leaf senescence is regulated during heat stress remain elusive. Further studies are needed for answering the question of whether heat-induced chlorophyll loss is caused by heat-induced inhibition of chlorophyll synthesis and/or heat-enhanced chlorophyll degradation in plant leaves.

In *Arabidopsis*, chlorophyll a (Chl *a*) content was gradually reduced and chlorophyllase (Chlase) activity substantially increased during the high temperature treatment [53]. Upon heat treatment, no significant difference was detected in the enzyme activity of a key chlorophyll-synthesizing enzyme, porphobilinogen deaminase across all the lines of bentgrass (*Agrostis* spp.). However, the activities of chlorophyll-degrading enzymes, including chlorophyllase and chlorophyll-degrading peroxidase, increased significantly after heat stress whereas pheophytinase activity was unchanged [52]. Interestingly, lower activities of chlorophyll-degrading enzymes were detected in heat-tolerant transgenic lines in which the expression of isopentenyl transferase (*ipt*) gene is driven by a senescence-activated promoter (*SAG12*) or heat shock promoter (*HSP18.2*) in modulation of cytokinin biosynthesis when compared with the WT under heat stress [52]. The authors suggested that the enhanced degradation of chlorophyll under heat stress could result in a severe loss of the chlorophyll content in heat-challenged bentgrass. Studies on genetic variations contrasting in heat tolerance in the strength of leaf senescence under heat stress in hybrids of colonial (*Agrostis capillaris*) x creeping bentgrass (*Agrostis stolonifera*) supported that heat-induced loss of chlorophyll in leaves was caused by the rapid breakdown of chlorophyll, as manifested by the high-level activation of genes encoding chlorophyllase and pheophytinase, and the activity of pheophytinase (PPH) [59]. Recently, the map-based cloning of a semidominant, heat-sensitive, missense allele (*cld1-1*) led to identify a putative hydrolase, named as CHLOROPHYLL DEPHYTYLASE1 (CLD1) that is capable of dephytylating chlorophyll [60]. Their findings suggest that CLD1 is conserved in oxygenic photosynthetic organisms and plays a key role as the long-sought enzyme in making the phytol chain removed from chlorophyll in its degradation process at the steady-state level in chloroplasts.

Unlike studies on hormonal regulation of leaf senescence, less attention has been paid to effects of hormones on heat-induced chlorophyll loss. Exogenous application of a synthetic form of cytokinin, zeatin riboside (ZR) helped maintain higher leaf chlorophyll content creeping bentgrass exposed to heat stress by slowing down the action of protease and by induction or upregulation of heat-shock proteins [61]. The ethylene-inhibiting compound 1-methylcyclopropene (1-MCP) treatment can delay leaf senescence in cotton plants under high temperature by reducing lipid peroxidation, membrane leakage, soluble sugar content, and increasing chlorophyll content [62]. It was reported that the novel ethylene antagonist, 3-cyclopropyl-1-enyl-propanoic acid sodium salt (CPAS), increases grain yield in wheat by delaying leaf senescence under extreme weather conditions [63].

Plant chlorophyll retention-staygreen is considered a valuable trait under heat stress. Accumulating data indicate that understanding the physiological and molecular mechanisms of "STAY-GREEN" trait or delayed leaf senescence is required for regulating photosynthetic capability and may be also a key to break the plateau of productivity associated with adaptation to high temperature [64]. Staygreen traits are associated with heat tolerance in bread wheat and quantitative trait loci (QTL) for staygreen and related traits were identified across the genome co-located with agronomic and physiological traits associated to plant performance under heat stress, confirming that the staygreen phenotype is a useful trait for productivity enhancement in hot-irrigated environments [65]. In soybean, high temperatures and drought stress can lead to chlorophyll retention in mature seeds, termed as "green seed problem", which is usually related to lower oil and bad seed quality, thus inducing a yield loss of soybean seeds. A "mild" stay-green phenotype was observed in a susceptible soybean cultivar when subjected to combined abiotic stresses of heat and drought stress and also the transcript levels of the *STAY-GREEN 1* and *STAY-GREEN 2* (*D1, D2*), *PHEOPHORBIDASE 2* (*PPH2*) and *NON-YELLOW COLORING 1* (*NYC1*) genes were downregulated in soybean seeds, indicating that the high-level transcriptional activation of these genes mentioned above in fully mature seeds is critical for a tolerant cultivar to cope with stresses and conduct a rapid and complete turnover of chlorophyll [66].

It is well-established that the major responses of crop plants to heat stress are classified into several aspects including the enhancement of leaf senescence, reduction of photosynthesis, deactivation of photosynthetic enzymes, and generation of oxidative damages to the chloroplasts. With respect to crop yield, heat stress also reduces grain number and size by affecting grain setting, assimilate translocation and duration and growth rate of grains [57]. In wheat, delayed senescence, or stay-green, contributes to a long grain-filling period and stable yield under heat stress [67]. Waxy maize (*Zea mays* L. sinensis Kulesh) is frequently exposed to high temperatures during grain filling in southern China. The heat-sensitive waxy maize variety exhibited a significant increase in the translocation amount and rate of assimilates pre-pollination and the accelerated leaf senescence phenotype under heat stress [68]. Short heat waves during grain filling can reduce grain size and consequently yield in wheat (*Triticum aestivum* L.). The four susceptible varieties of wheat showed greater heat-triggered reductions in final grain weight, grain filling duration and chlorophyll contents in flag leaves under heat stress, suggesting that grain size effects of heat may be driven by premature senescence [69,70].

3. Generation and Homeostasis of ROS in Chloroplasts under Heat Stress

As a significant source of reactive oxygen species (ROS) in plant cells, the chloroplast produces a variety of ROS such as hydrogen peroxide (H_2O_2), superoxide, hydroxyl radicals ($\bullet OH$), and $^1O_2$ during photosynthesis [71]. The transfer of excitation energy in the PSII antenna complex and the electron transport in the PSII reaction center can be inhibited by a variety of abiotic stresses, resulting in the formation of ROS in algae and higher plants [72]. Given that ROS generation is an unavoidable consequence of aerobic metabolism, plants have evolved a large array of ROS-scavenging mechanisms [71]. ROS such as $^1O_2$ is formed by the excitation energy transfer, whereas superoxide anion radical ($O_2{}^{\bullet-}$), H_2O_2 and $\bullet OH$ are formed by the electron transport [73]. In chloroplasts, ROS are mainly generated in the reaction centers of PSI and PSII in the chloroplast thylakoid membranes [71]. Photoreduction of oxygen to superoxide occurs in PSI and in PSII, oxygen of the ground (triplet) state of oxygen ($^3O_2$) is excited to the excited singlet state of oxygen ($^1O_2$) by the P680 reaction center chlorophyll [71]. Under extreme conditions, these ROS synthesis rates can increase, leading to an oxidative stress in both organelle and whole-cell functions [72,74]. It is well established that the chloroplast is extremely sensitive to high temperature stress during photosynthesis [1,2,9,10]. Accumulated data indicate that oxidative bursts of superoxide and/or hydrogen peroxide can be induced rapidly in response to a variety of environmental stresses, including also heat stress in plants [75–77]. It was reported that ROS, produced in PSI, PSII as well as in the Calvin-Benson cycle, can cause irreversible oxidative damage to cells when plants were subjected to heat stress [71,72,78]. Under high temperature conditions, large amounts of ROS were generated in tobacco cells for initiating signaling events involved in PCD, which is consistent with the role of applying the antioxidants ascorbate or superoxide dismutase (SOD) to the cultures in supporting the survival of cells [79]. In the leaves of tobacco (*Nicotiana tabacum*) defective in ndhC-ndhK-ndhJ (Delta ndhCKJ), hydrogen peroxide was rapidly produced in response to heat treatment, implying a role of the NAD(P)H dehydrogenase-dependent pathway in repressing generation of reactive oxygen species in chloroplasts [80]. In exposure to moderate heat treatment conditions, the oxidative damages of the reaction center-binding D1 protein of photosystem II increased and showed a tight positive relationship with the accumulated levels of $^1O_2$ and $\bullet OH$ in spinach PSII membranes, implying that inhibition of a water-oxidizing manganese complex led to a rapid production of ROS through lipid peroxidation under heat stress [81]. In *Arabidopsis*, large amounts of chlorophyllide *a* caused a surge of phototoxic singlet oxygen in the chlorophyll synthase mutant (*chlg-1*) under heat stress, suggesting that chlorophyll synthase acts in maintenance of ROS homeostasis in response to heat stress [82].

It is well known that ROS burst can trigger the oxidative damage to pigments, proteins, and lipids in the thylakoid membrane [71,73,83]. ROS are primarily agents of damage, but this view is questioned by data proving their beneficial role and particularly their signaling function. The chloroplast harbors ROS-producing centers (triplet chlorophyll, ETC in PSI and PSII) and a diversified ROS-scavenging network (antioxidants, SOD, APX-glutathione cycle, and a thioredoxin system) to keep the equilibrium between ROS production and scavenging [71]. The non-enzymatic and enzymatic ROS scavenging systems are engaged in preventing harmful effects of ROS on the thylakoid membrane components to keep ROS level in chloroplasts under control [71,73,74,83–87]. The efficient enzymatic scavenging systems are composed of several key enzymes, including superoxide dismutase (SOD), catalase (CAT), ascorbate peroxidase (APX), glutathione reductase (GR), monodehydroascorbate reductase (MDHAR), dehydroascorbate reductase (DHAR), glutathione peroxidase (GPX) and glutathione-S-transferase (GST) and non-enzymatic systems contain antioxidants such as ascorbic acid (ASH), glutathione (GSH), phenolic compounds, alkaloids, non-protein amino acids and alpha-tocopherols [21,71,73,83,87–90]. These antioxidant defense systems work in concert to maintain homeostasis of ROS, protecting plant cells from oxidative damage by scavenging of ROS.

In addition to triggering ROS burst, heat stress also affects the scavenging systems in plants. Heat induced degradation of chloroplast Cu/Zn superoxide dismutase as shown by reduced protein levels and isozyme-specific SOD activity [91]. Loss of Cu/Zn SOD and induction of catalase activity would explain the altered balance between hydrogen peroxide and superoxide under stress. The authors proposed that degradation of PSII could thus be caused by the loss of components of chloroplast antioxidant defense systems and subsequent decreased function of PSII [91]. In a recent study aiming at identifying heat protective mechanisms promoted by CO_2 in coffee crop, the results likely favored that the maintenance of reactive oxygen species (ROS) at controlled levels contributed to mitigate of PSII photoinhibition under the high temperature stress [92]. Exogenous application of spermidine protected rice seedlings from heat-induced damage as marked by lower levels of malondialdehyde (MDA), H_2O_2, and proline content coupled with increased levels of Ascorbate (AsA), GSH, AsA and GSH redox status [93]. The authors conclude that heat exposure provoked an oxidative burden while enhancement of the antioxidative and glyoxalase systems by spermidine rendered rice seedlings more tolerant to heat stress. Under the later high temperature stress, heat priming contributed to a better redox homeostasis, as exemplified by the higher activities of superoxide dismutase (SOD) in chloroplasts and glutathione reductase (GR), and of peroxidase (POD) in mitochondria, which led to the lower superoxide radical production rate and malondialdehyde concentration in both chloroplasts and mitochondria [94]. These results suggested that heat priming effectively improved thermo-tolerance of wheat seedlings subjected to a later high temperature stress, which could be largely ascribed to the enhanced anti-oxidation at the subcellular level. In Kentucky bluegrass (*Poa pratensis*), higher activities of superoxide dismutase (SOD), catalase (CAT), peroxidase (POD), ascorbate peroxidase (APX), and glutathione reductase were detected in plants of heat-tolerant "Midnight" after long-term heat stress (21 and 28 d) in comparison with plants of heat-sensitive "Brilliant" [95]. Meanwhile, transcript levels of chloroplastic Cu/Zn SOD, Fe SOD, CAT, POD and cytosolic (cyt) APX were significantly higher in "Midnight" than in "Brilliant" under long-term heat stress. These results supported the hypothesis that enzymatic ROS scavenging systems could play predominant roles in antioxidant protection against oxidative damages from long-term heat stress. In tomato, studies on the role of a tomato (*Lycopersicon esculentum*) chloroplast-targeted DnaJ protein (LeCDJ1) showed that the sense transgenic tomato plants were more tolerant to heat stress due to the higher activities of ascorbate peroxidase (APX) and superoxide dismutase (SOD) [96]. The high temperature stress in southern China is one of the major factors leading to loss of the yield and quality of wucai (*Brassica campestris* L.). Comparative investigations on two cultivars of wucai (heat-sensitive and heat-tolerant) showed that greater severity of damage to the photosynthetic apparatus and membrane system was observed in heat-sensitive cultivar probably due to a high-level accumulation of ROS and malondialdehyde (MDA) [97]. In line with the studies mentioned above, plants of heat-tolerant Cucurbit species exhibited comparatively little oxidative damage, with the lowest hydrogen peroxide (H_2O_2), superoxide (O^{2-}) and malondialdehyde (MDA) compared with the thermolabile and moderately heat-tolerant interspecific inbred line [98]. The enzyme activities of superoxide dismutase (SOD), ascorbate peroxidase (APX), catalase (CAT) and peroxidase (POD) were found to be increased with heat stress in tolerant genotypes and the significant inductions of FeSOD, MnSOD, APX2, CAT1 and CAT3 isoforms in tolerant genotypes suggested their participation in heat tolerance [98].

Generally, the chloroplast possesses a variety of constitutively expressed antioxidant defense mechanisms to scavenge the various types of ROS generated during heat stress in preventing and minimizing oxidative damage to biological macromolecules (Figure 2). Therefore, the capacity of antioxidant defense is critical for heat stress adaptation and its strength is correlated with acquisition of thermotolerance with respect to buffering the effect of heat stress on the metabolic system [3,74,99,100]. On the other hand, ROS are produced in chloroplasts can function as plastid signals to inform the nucleus to activate the expression of genes encoding antioxidant enzyme and to adjust the stress-responsive machinery for more efficient adaptation to environmental stresses [6,47,72,74,100,101].

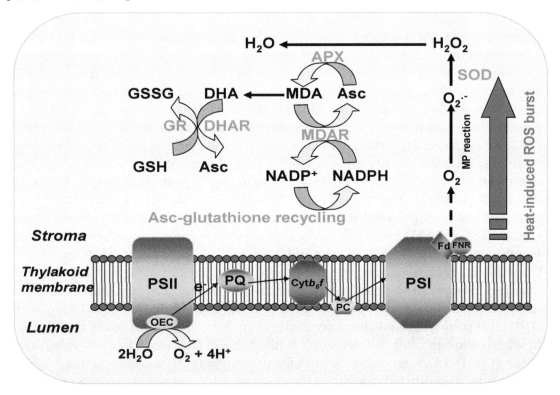

Figure 2. A representative scheme of reactive oxygen species (ROS) generation and scavenging in chloroplasts under heat stress. High temperature stress triggers oxidative bursts of superoxide and/or hydrogen peroxide in plants. The transfer of excitation energy in the photosystem II (PSII) antenna complex and the electron transport in the PSII reaction center can be inhibited by heat stress. It has been established that ROS are generated both on the electron acceptor and the electron donor side of PSII under heat stress during which electron transport from the manganese complex to plastoquinone (PQ) is limited. The leakage of electrons to molecular oxygen on the electron acceptor side of PSII forms $O_2^{\bullet-}$, inducing initiation of a cascade reaction leading to the formation of H_2O_2. A diversified ROS-scavenging network functions in concert in chloroplasts, mainly including antioxidants and APX-glutathione cycle, to keep the equilibrium between ROS production and scavenging. The efficient enzymatic scavenging systems are composed of several key enzymes, including superoxide dismutase (SOD), ascorbate peroxidase (APX), glutathione reductase (GR), monodehydroascorbate reductase (MDHAR), dehydroascorbate reductase (DHAR), glutathione peroxidase (GPX) and glutathione-S-transferase (GST) and non-enzymatic systems contain antioxidants such as ascorbic acid (Asc) and glutathione (GSH).

4. Turnover of PSII Core Subunits and PSII Protection under Heat Stress

The deleterious effects of high temperatures on proteins in chloroplasts include protein denaturation and aggregation [18,102,103]. ROS generation is one of the earliest responses of plant cells in response to heat stress and chloroplasts are the main targets of ROS-linked damage [71,102]. Given that chloroplasts are a major site of protein degradation, turnover of the damaged proteins is critical for plants to adapt to heat stress through the process of acclimation. As a multisubunit thylakoid membrane pigment-protein complex, PSII is vulnerable to light and heat damages that inhibit light-driven oxidation of water and reduction of plastoquinone [2,71]. PSII produces ROS, responsible for the frequent damage and turnover of this megacomplex that occur under physiological and stress conditions [71]. Accumulating data suggest that although more than 40 proteins are known to associate with PSII, the damage mainly targeted to one of its core proteins, the D1 protein under light and heat stress conditions [18,102,104,105]. Importantly, D1 protein has been demonstrated as a very susceptible target of $^1O_2$; on the other hand, it appears to function as a major scavenger of $^1O_2$ due to its close localization to the site of $^1O_2$ formation in the reaction center of PSII [106]. In addition to the D1 protein, β-carotene, plastoquinol, and a-tocopherol have also shown to play a role in scavenging $^1O_2$ and protect PSII against photo-oxidative damage [107,108]. Under heat stress conditions, two processes have been characterized in the thylakoids: dephosphorylation of the D1 protein in the stroma thylakoids, and aggregation of the phosphorylated D1 protein in the grana [102]. Heat stress also induced the release of the extrinsic Photosystem II Subunit O (PsbO), P and Q proteins from Photosystem II, which affected D1 degradation and aggregation significantly [102]. In spinach thylakoids, cleavage of the D1 protein was detected in response to moderate heat stress (40 °C, 30 min), producing an N-terminal 23-kDa fragment, a C-terminal 9-kDa fragment, and aggregation of the D1 protein [102,103]. ROS are known to specifically modify PSII proteins. Using high-resolution tandem mass spectrometry, oxidative modifications were identified on 36 amino acid residues on the lumenal side of PSII, in the core PSII proteins D1, D2, and CP43 of the cyanobacterium *Synechocystis* sp. PCC 6803, providing the compelling evidence to date of physiologically relevant oxidized residues in PSII [109]. Taken together, the oxidative damage of the D1 protein is caused by reactive oxygen species, mostly singlet oxygen, and also by endogenous cationic radicals generated by the photochemical reactions of photosystem II [71,73]. Under heat stress, the damage to the D1 protein by moderate heat stress is due to reactive oxygen species produced by lipid peroxidation near photosystem II [73,110]. Moreover, the damage of the D1 protein has been shown to be directly proportional to light intensity [111,112] or strength of heat stress [102]. D1 protein is well characterized as a protein of high turnover rate due to its rapid degradation when oxidized after its interaction with 1O2 and its replacement by newly synthesized D1 polypeptides [105]. It was reported that the D1 protein was shown to have the half-life of 2.4 h, the fourth fastest turn-over rate of barley (*Hordeum vulgare*) proteins when plants were growing under normal growth light intensity (500 μmol · m^{-2} · s^{-2}) [113]. Meanwhile, higher degradation rates of the D2, CP43 and PsbH subunits were also detected when compared with the other PSII subunits [113–115]. The maintenance of PSII activity is critical, but also a challenge for oxygenic photosynthetic organisms to survive under normal growth or stress conditions.

Thus, replacing the photo- or heat-damaged D1 with a newly synthesized copy is essential for maintaining PSII activity [105,116]. In chloroplasts, the damaged D1 is degraded by a concerted action of particular filamentation temperature sensitive H (FtsH) protease and Deg (for degradation of periplasmic proteins) isoforms during its rapid and specific turnover and replaced with a de novo synthesized one in a system which is termed the PSII repair cycle [105]. It has been a subject of extensive studies about the basic concept for replacement of the damaged D1 protein by a newly-synthetized copy in the PSII repair cycle [105,116–119]. Repair of damaged D1 protein in PSII includes five steps: (1) migration of damaged PSII core complex to the stroma thylakoid, (2) partial PSII disassembly of the PSII core monomer, (3) access of protease degrading damaged D1, (4) concomitant D1 synthesis, and (5) reassembly of PSII into grana thylakoid [105,116,120,121]. Deg /HtrA (for high temperature requirement A) proteases, a family of serine-type ATP-independent proteases, have been shown in higher plants to be involved in the degradation of the Photosystem II reaction center protein D1 [105,116,122]. In *Arabidopsis*, five DEGs (Deg1, Deg2, Deg5, Deg7, and Deg8) have shown to be peripherally attached to the thylakoid membrane of chloroplasts: Deg1, Deg5, and Deg8 are localized on the lumenal side, and Deg2 and Deg7 are localized on the stromal side [123–125]. DEG5 and DEG8 may have synergistic function in degradation of D1 protein in the repair cycle of PSII under heat stress based on functional analysis of *deg5*, *deg8* and the double mutant *deg5 deg8* of *Arabidopsis thaliana* [126]. Under photoinhibitory conditions, cooperative degradation of D1 by Deg and FtsH has been demonstrated in vivo, in which Deg cleavage assists FtsH processive degradation [127]. In Arabidopsis, FtsH11-encoded protease have shown to play a direct role in thermotolerance, a function previously reported for bacterial and yeast FtsH proteases [128]. It should be noted that photosynthetic capability and PSII quantum yield are greatly reduced in the leaves of FtsH11 mutants when exposed to the moderately high temperature whereas under high light conditions, FtsH11 mutants and wild-type plants showed no significant difference in photosynthesis capacity [128]. In general, several possible mechanisms have been proposed for activation of these proteases, which depend on oligomerization of the monomer subunits [129]. In line with the hypothesis that hexamers of the FtsH proteases are probably localized near the Photosystem II complexes at the grana, degradation of the D1 protein could take place in the grana rather than in the stroma thylakoids to circumvent long-distance migration of both the Photosystem II complexes containing the photodamaged D1 protein and the proteases [129]. Under high light conditions, the lumen-exposed loops of the D1 protein at specific sites were cleaved by all of three lumenal serine proteases, Deg1, Deg5 and Deg8, during PSII repair cycle [130–133]. It was reported that Deg5 and Deg8 interact to form an active protease complex under high light [133]. Interestingly, Deg1 is activated when the Deg1 monomers are transformed into a proteolytically active hexamer at acidic pH upon protonation of a histidine amino acid residue [134]. In addition to functioning as a protease, Deg1 also plays a novel role as a chaperone/assembly factor of PSII [135]. In addition, Deg1 has been shown to be responsible for the proteolytic activity against the PsbO protein in vitro [136]. During the PSII repair cycle, only the damaged reaction center protein D1 and occasionally also the D2, CP43 and PsbH subunits are replaced while the other protein components of the complex are recycled, indicating that many aspects of PSII repair cycle and de novo biogenesis are partially overlapping [113–115,137]. Originally, the vulnerability of PSII to high light or heat stress was taken as an inherent fault of the photosynthetic machinery. However, recent studies strongly support that the constant, yet highly regulated, photodamage and repair cycle of PSII are of a strong physiological basis. Collectively, the photodamage of PSII is likely to act as a PSI protection mechanism instead of being considered solely as an undesired consequence of the highly oxidizing chemistry of the water splitting PSII [105].

The synthesis of heat shock proteins (HSPs) is characterized as a major response of all organisms responding to heat stress. The HSPs act as chaperones by assisting in protein folding and preventing irreversible protein aggregation [138,139]. A chloroplast-localized sHSP, HSP21, has been identified in diverse higher plant species, including both dicots and monocots [139,140] and its precursor polypeptide is ~5 kD larger than the mature protein [141,142]. HSP21 is thought to protect photosynthetic electron transport, specifically that of Photosystem II, during heat stress [143–147] and oxidative stress [145,148]. In addition, HSP21 has been demonstrated to play a dual role in protecting PSII from oxidative stress and promoting color changes during fruit maturation whereas no protective effects for the transgene were detected on PSII thermotolerance [149]. Importantly, studies using the transgenic tomato plants overexpressing HSP21 have shown that this protein associates with proteins of Photosystem II and does not reactivate heat-denatured Photosystem II, but instead protects this complex from damage during heat stress [150]. Interestingly, around two thirds of chloroplast HSP21 proteins are translocated into the thylakoid membranes in response to heat treatment in plants, suggesting that the association with membranes should be considered to fully understand the role of sHsps in physiological adaptation processes under stress conditions [151]. Despite extensive studies on HSP21, the specific roles of HSP21 in protecting PSII from heat stress remain elusive.

Recently, HSP21 has been demonstrated to protect PSII from heat stress-induced damages by directly binding to D1 and D2 proteins, the core subunits of PSII. Importantly, heat-responsive transcriptional activation of *HSP21* is regulated by the chloroplast retrograde signaling pathway in which GUN5 acts as a determinant upstream signaling component in *Arabidopsis* [5]. Based on these findings, an auto-adaptation loop working module has emerged in which the GUN5-dependent plastid signal(s) is triggered in response to heat stress and in turn communicated into the nucleus to activate the heat-responsive expression of *HSP21* for optimizing particular demands of chloroplasts in making photosynthetic complexes stable during adaptation to heat stress in plants [5,6].

5. Effects of Heat Stress on Metabolic Flux through the Calvin-Benson-Bassham Cycle

In the Calvin-Benson-Bassham Cycle, Ribulose-1,5-bisphosphate carboxylase/oxygenase (Rubisco) plays a critical role in catalyzing the carboxylation of the 5-carbon sugar ribulose-1,5-bisphosphate (RuBP) when atmospheric CO_2 during is fixed during photosynthesis. Rubisco activase (RCA) regulates the activity of Rubisco by facilitating the dissociation of inhibitory sugar phosphates from the active site of Rubisco in an ATP-dependent manner [152]. Extensive evidence supports the conclusion that reduction of plant photosynthesis arises primarily from thermal inactivation of Rubisco activity due to the inhibition of RCA under moderately elevated temperatures [8,152–155]. In addition to Rubisco activation, electron transport activity, ATP synthesis, and RuBP regeneration are also inhibited by moderately heat stress [156–158]. As the temperature increases further above the thermal optimum, the physical integrity of electron transport components of the photosynthetic apparatus can be severely damaged, leading to the increased limitation in photosynthesis [15]. It has been the subject of extensive investigations of elucidating the biochemical basis for the decrease in Rubisco activation state under heat stress [8,15,153,154,159]. Initially, studies on thermal stability of purified RCA showed that heat treatment only slightly inhibited the activities of this enzyme [160,161] and later experiments confirmed that heat stress caused thermal denaturation of activase in wheat and cotton leaves [162]. In line with the thermal denaturation of RCA under heat stress, Feller et al. (1998) suggested that RCA exhibited the exceptional thermal lability in vivo and the thermal stabilities of activases were different in plants from contrasting thermal environments [15,159]. Thus, loss of activase activity during heat stress is caused by an exceptional sensitivity of the protein to thermal denaturation and is responsible, in part, for deactivation of Rubisco [163]. It has been assumed that the stability of RCA could be influenced by heat-induced changes either in redox state [11,164,165] or the concentrations of ions, nucleotides, or other chloroplast constituents in plants [154]. On the other hand, the thermotolerance of Rubisco activase has been proposed to be responsible for restricting the distribution of certain plant species [154] as demonstrated by the response of Rubisco activase activity to temperature for cotton, a warm-season species, and *Camelina sativa*, a cool-season species. With respect to the effects of high growth temperature on the relative contribution of diffusive and biochemical limitations to photosynthesis, our knowledge is limited although there is abundant evidence that photosynthesis can acclimate to temperature [166,167]. Accumulating data suggest that the biochemical mechanisms about the decrease in Rubisco activation can be attributed to: (1) more rapid de-activation of Rubisco caused by a faster rate of dead-end product formation; and (2) slower re-activation of Rubisco by activase [168]. In a word, the resulting consequence is that RCA becomes less effective in keeping Rubisco catalytically competent as temperature increases.

Inhibition of net photosynthesis by heat stress has been attributed to an inability of Rubisco activase to maintain Rubisco in an active form because of the low thermal stability of Rubisco's chaperone, activase. These results support a role for RCA in limiting photosynthesis at high temperature when the temperature exceeds the optimum range for plants. In cotton (*Gossypium hirsutum* L.), activase gene expression is influenced by post-transcriptional mechanisms that may contribute to acclimation of photosynthesis during extended periods of heat stress [169]. In wheat, northern blot analysis showed maximum accumulation of *TaRCA1* transcript in thermotolerant cv. during mealy-ripe stage, as compared to thermosusceptible ones [170]. To test the hypothesis that thermostable RCA can improve photosynthesis under elevated temperatures, gene shuffling technology was used to generate several *Arabidopsis thaliana* RCA1 (short isoform) variants exhibiting improved thermostability [171]. In line with the findings mentioned above, transgenic Arabidopsis lines expressing a thermostable chimeric activase showed higher rates of photosynthesis than the wild type after a short exposure to higher temperatures and they also recovered better, when they were returned to the normal temperature [172]. The results showed that photosynthesis and growth were improved under moderate heat stress in transgenic Arabidopsis expressing these thermotolerant RCA isoforms, providing evidence that manipulation of activase properties can improve C3 photosynthesis. In addition, the transcriptional level of wheat *RCA* (45–46 kDa) positively correlated with the yield of plants under heat-stress conditions in a very significant and linear manner [173]. At present, accumulating data indicate that RCA could affect plant productivity in relation to its endogenous levels under temperature stress conditions. Critically, RCA as the molecular chaperone plays a key role in constant engagement and remodeling of Rubisco to maintain metabolic flux through the Calvin-Benson-Bassham cycle as Rubisco is characterized as a dead-end inhibited complex in higher plants. In plants of the crassulacean acid metabolism (CAM), it has been assumed that possessing thermostable RCA is necessary for these plants to support the metabolic flux of Calvin-Benson-Bassham cycle when closure of stomata is a limitation factor during the day [174]. It is interesting that the CAM Rca isoforms (*Agave tequilana*) were found to be approximately 10 °C more thermostable when compared with the C3 isoforms of Rca isolated from rice (*Oryza sativa*) [174]. Interestingly, sequence analysis and immuno-blotting identified the beta-subunit of chaperonin-60 (cpn60 beta), the chloroplast GroEL homologue, as a protein that was bound to Rubisco activase from leaf extracts prepared from heat-stressed, but not control plants [175]. Rubisco requires RCA, an AAA+ ATPase that reactivates Rubisco by remodeling the conformation of inhibitor-bound sites. RCA is regulated by the ratio of ADP:ATP, with the precise response potentiated by redox regulation of the alpha-isoform [176]. Given that RCA uses the energy from ATP hydrolysis to restore catalytic competence to Rubisco, manipulation of RCA by redox regulation of the a-isoform might provide a strategy for enhancing photosynthetic performance in *Arabidopsis* [177]. In rice, heat stress significantly induced the expression of *RCA* large isoform (*RCAL*) as determined by both mRNA and protein levels and correlative analysis indicated that and RCA small isoform (RCAS) protein content was very tightly correlated to Rubisco initial activity and net photosynthetic rate under both heat stress and normal conditions [178]. In two *Populus* species adapted to contrasting thermal environments, the difference in the primary sequence of Rubisco activases between the species is more significant in the regions conferring ATPase activity and Rubisco recognition, suggesting that the genotypic distinctive characterizations in Rubisco activase are likely to underlie the specificities with respect to the heat-sensitive strength of Rubisco activase and photosynthesis under moderate high temperature conditions [179]. Recent studies on the effects of heat and drought on three major cereal crops, including rice, wheat, and maize, indicate that reductions in Rubisco activation might be not dependent on the amount of Rubisco and RCA, but could be resulted from the inhibition of RCA activity, as evidenced by the mutual reduction and positive relationship existed between the activation state of Rubisco and the rate of electron transport [153]. Critically, Rubisco activase acts as a key player in photosynthesis under heat stress conditions (non-stomatal limitation) [180]. When exposed to a moderate heat stress, Rca can be inhibited reversibly, but is irreversibly inhibited under a higher temperature and/or longer exposure due to heat stress-induced insolubilization and degradation of the Rca protein [180].

6. Conclusions and Perspectives

In many regions of the world, high temperature stress is one of the most important constraints to plant growth and productivity, especially for crop plants. The mechanism underlying the development of heat-tolerance for important agricultural crops as well as plant responses and adaptation to elevated temperatures needs to be better understood. Extensive studies have shown that metabolic regulation of adaptation processes during heat stress is not only an important developmental process, but also allows for flexibility of physiological responses to heat stress. In photosynthetic organisms, heat stress can affect photosynthesis through altered carbon assimilation metabolism in chloroplasts with remobilizing their starch reserve to release energy, sugars and derived metabolites in order to help mitigate the stress. This is thought to be an essential process for plant fitness with important implications for plant productivity under high temperature stress. One future challenges is to dissect the complex interaction networks between heat stress sensing, signal transduction and activations of key genes involved in metabolic reprogramming in coordination with developmental programmes. Accumulation and modification of metabolites in chloroplasts under heat stress may play a key role in the regulation of adaptation processes at cellular levels in plants, allowing plants to interact with their environment and to activate cellular heat stress responses at the optimal time in order to maintain photosynthesis. This kind of metabolic reprogramming is critical for plants to survive stress periods, and to prevent further damage to the whole plant.

The role of chloroplast in the metabolic regulation of heat stress responses has attracted increasing attention and extensive investigations from an organellar perspective have provided insights into better understanding the hypothesis stated that the heat stress-induced reprogramming, including decline in photosynthesis and alterations in photosynthetic metabolites which, in turn, could act as signal(s) or trigger the initial signal cascades to activate cellular heat stress responses. The present knowledge concerning the interplay between the chloroplast and nucleus in heat stress signal perception and activation of cellular heat stress responses is emerging, but more efforts are needed to reach a detailed overview. It can be predicted that uncovering the molecular mechanisms of heat sensing will pave the way to engineering plants capable of tolerating heat stress. It is well known that the ability of plants adapting to different climate regimes vary dramatically across and within species. Identification and functional analysis of the valuable heat-tolerant genetic resources will bring about a further significant improvement in manipulation of photosynthesis to increase crop yield based on a direct comparative analysis between the different manipulations with all the transgenic and wild-type plants grown and assessed in parallel under filed growth conditions. Thus, in-depth analyses of the interactions between the chloroplast and nucleus in heat stress responses are likely to be in focus during forthcoming years. On the other hand, Rubisco activase and enzymes functioning in the detoxification of reactive oxygen species are thought to be critical targets for breeding heat-tolerant crop plants with high yields under high temperature stress.

Acknowledgments: This study was supported by Chinese Academy of Sciences (Strategic Priority Research Program XDPB0404), the Ministry of Science and Technology of China (National Key R&D Program of China, 2016YFD0100405), and the National Natural Science Foundation of China (31770314 and 31570260).

Author Contributions: Fang-Qing Guo and Qing-Long Wang conceived and designed the outline and contents of this review article; Fang-Qing Guo and Qing-Long Wang wrote the paper with inputs from Juan-Hua Chen and Ning-Yu He.

Abbreviations

1-MCP	1-methylcyclopropene
$^1O_2$	singlet state of oxygen
$^3O_2$	triplet state of oxygen
ACD1	Accelerated cell death
APX	ascorbate peroxidase
AsA	Ascorbate
ASH	ascorbic acid
ATP	adenosine 5'-triphosphate
CAM	crassulacean acid metabolism
CAT	catalase
Chl	Chlorophyll
Chl a	Chlorophyll a

Chlase	chlorophyllase
CLD1	CHLOROPHYLL DEPHYTYLASE1
CLHs	Chlorophyllase genes
CPAS	3-cyclopropyl-1-enyl-propanoic acid sodium salt
cpn60 beta	beta-subunit of chaperonin-60
Deg/HtrA	high temperature requirement A
DHAR	dehydroascorbate reductase
FtsH	filamentation temperature sensitive H
Fv/Fm	variable fluorescence to maximum fluorescence
GPX	glutathione peroxidase
GR	glutathione reductase
GSH	glutathione
GST	glutathione-S-transferase
GUN5	genomes uncoupled 5
H_2O_2	hydrogen peroxide
HSPs	heat shock proteins
ipt	isopentenyl transferase
LeCDJ1	tomato (*Lycopersicon esculentum*) chloroplast-targeted DnaJ protein
LHCPII	light harvesting chlorophyll a/b-protein complex II
MDA	malondialdehyde
MDHAR	monodehydroascorbate reductase
NAD(P)H	nicotinamide adenine dinucleotide phosphate (NADP)
NOL	NYC1-LIKE
NYC1	NON-YELLOW COLORING1
$O_2^{\bullet-}$	superoxide anion radical
OEC	oxygen evolving complex
OH	hydroxyl radicals
PAO	pheide a oxygenase
pFCC	primary fluorescent chlorophyll catabolite
pheide	pheophorbide
POD	peroxidase
PPH	pheophytinase
PPH2	PHEOPHORBIDASE 2
PsbO	PHOTOSYSTEM II SUBUNIT O
PSII	Photosystem II
QTL	quantitative trait loci
RCA	Rubisco activase
RCAL	RCA large isoform
RCAS	RCA small isoform
RCCR	Red chl catabolite reductase
ROS	Reactive oxygen species
Rubisco	Ribulose-1,5-bisphosphate carboxylase/oxygenase
RuBP	Ribulose-1,5-bisphosphate
SGR	STAY-GREEN
SID	senescence-induced degradation
SOD	superoxide dismutase
ZR	zeatin riboside

References

1.   Wahid, A.; Gelani, S.; Ashraf, M.; Foolad, M.R. Heat tolerance in plants: An overview. *Environ. Exp. Bot.* **2007**, *61*, 199–223. [CrossRef]

2.   Allakhverdiev, S.I.; Kreslavski, V.D.; Klimov, V.V.; Los, D.A.; Carpentier, R.; Mohanty, P. Heat stress: An overview of molecular responses in photosynthesis. *Photosynth. Res.* **2008**, *98*, 541–550. [CrossRef] [PubMed]

3.   Berry, J.; Bjorkman, O. Photosynthetic Response and Adaptation to Temperature in Higher-Plants. *Annu. Rev. Plant Physiol. Plant Mol. Biol.* **1980**, *31*, 491–543. [CrossRef]

4.   Yu, H.-D.; Yang, X.-F.; Chen, S.-T.; Wang, Y.-T.; Li, J.-K.; Shen, Q.; Liu, X.-L.; Guo, F.-Q. Downregulation of Chloroplast RPS1 Negatively Modulates Nuclear Heat-Responsive Expression of HsfA2 and Its Target Genes in *Arabidopsis. PLoS Genet.* **2012**, *8*, e1002669. [CrossRef] [PubMed]

5.   Chen, S.-T.; He, N.-Y.; Chen, J.-H.; Guo, F.-Q. Identification of core subunits of photosystem II as action sites of HSP21, which is activated by the GUN5-mediated retrograde pathway in *Arabidopsis. Plant J.* **2017**, *89*, 1106–1118. [CrossRef] [PubMed]

6.   Sun, A.-Z.; Guo, F.-Q. Chloroplast Retrograde Regulation of Heat Stress Responses in Plants. *Front. Plant Sci.* **2016**, *7*. [CrossRef] [PubMed]

7.   Havaux, M. Characterization of Thermal-Damage to the Photosynthetic Electron-Transport System in Potato Leaves. *Plant Sci.* **1993**, *94*, 19–33. [CrossRef]

8.   Sharkey, T.D. Effects of moderate heat stress on photosynthesis: Importance of thylakoid reactions, rubisco deactivation, reactive oxygen species, and thermotolerance provided by isoprene. *Plant Cell Environ.* **2005**, *28*, 269–277. [CrossRef]

9.   Yamada, M.; Hidaka, T.; Fukamachi, H. Heat tolerance in leaves of tropical fruit crops as measured by chlorophyll fluorescence. *Sci. Hortic.* **1996**, *67*, 39–48. [CrossRef]

10.   Wise, R.R.; Olson, A.J.; Schrader, S.M.; Sharkey, T.D. Electron transport is the functional limitation of photosynthesis in field-grown *Pima cotton* plants at high temperature. *Plant Cell Environ.* **2004**, *27*, 717–724. [CrossRef]

11.   Sharkey, T.D.; Schrader, S.M. High temperature stress. In *Physiology and Molecular Biology of Stress Tolerance in Plants*; Springer: New York, NY, USA, 2006; pp. 101–129.

12.   Havaux, M.; Tardy, F. Temperature-dependent adjustment of the thermal stability of photosystem II in vivo: Possible involvement of xanthophyll-cycle pigments. *Planta* **1996**, *198*, 324–333. [CrossRef]

13.   Klimov, V.V.; Baranov, S.V.; Allakhverdiev, S.I. Bicarbonate protects the donor side of photosystem II against photoinhibition and thermoinactivation. *FEBS Lett.* **1997**, *418*, 243–246. [CrossRef]

14.   Yamane, Y.; Kashino, Y.; Koike, H.; Satoh, K. Effects of high temperatures on the photosynthetic systems in spinach: Oxygen-evolving activities, fluorescence characteristics and the denaturation process. *Photosynth. Res.* **1998**, *57*, 51–59. [CrossRef]

15.   Salvucci, M.E.; Crafts-Brandner, S.J. Relationship between the heat tolerance of photosynthesis and the thermal stability of Rubisco activase in plants from contrasting thermal environments. *Plant Physiol.* **2004**, *134*, 1460–1470. [CrossRef] [PubMed]

16.   Gounaris, K.; Brain, A.R.R.; Quinn, P.J.; Williams, W.P. Structural Reorganization of Chloroplast Thylakoid Membranes in Response to Heat-Stress. *Biochim. Biophys. Acta* **1984**, *766*, 198–208. [CrossRef]

17.   Semenova, G.A. Structural reorganization of thylakoid systems in response to heat treatment. *Photosynthetica* **2004**, *42*, 521–527. [CrossRef]

18.   Yamamoto, Y.; Aminaka, R.; Yoshioka, M.; Khatoon, M.; Komayama, K.; Takenaka, D.; Yamashita, A.; Nijo, N.; Inagawa, K.; Morita, N.; et al. Quality control of photosystem II: Impact of light and heat stresses. *Photosynth. Res.* **2008**, *98*, 589–608. [CrossRef] [PubMed]

19.   Vani, B.; Saradhi, P.P.; Mohanty, P. Alteration in chloroplast structure and thylakoid membrane composition due to in vivo heat treatment of rice seedlings: Correlation with the functional changes. *J. Plant Physiol.* **2001**, *158*, 583–592. [CrossRef]

20.   Kmiecik, P.; Leonardelli, M.; Teige, M. Novel connections in plant organellar signalling link different stress responses and signalling pathways. *J. Exp. Bot.* **2016**, *67*, 3793–3807. [CrossRef] [PubMed]

21.   Apel, K.; Hirt, H. Reactive oxygen species: Metabolism, oxidative stress, and signal transduction. *Annu. Rev. Plant Biol.* **2004**, *55*, 373–399. [CrossRef] [PubMed]

22. Hortensteiner, S. Stay-green regulates chlorophyll and chlorophyll-binding protein degradation during senescence. *Trends Plant Sci.* **2009**, *14*, 155–162. [CrossRef] [PubMed]

23. Hortensteiner, S. Chlorophyll degradation during senescence. *Annu. Rev. Plant Biol.* **2006**, *57*, 55–77. [CrossRef] [PubMed]

24. Hortensteiner, S.; Feller, U. Nitrogen metabolism and remobilization during senescence. *J. Exp. Bot.* **2002**, *53*, 927–937. [CrossRef] [PubMed]

25. Ginsburg, S.; Schellenberg, M.; Matile, P. Cleavage of Chlorophyll-Porphyrin—Requirement for Reduced Ferredoxin and Oxygen. *Plant Physiol.* **1994**, *105*, 545–554. [CrossRef] [PubMed]

26. Tsuchiya, T.; Ohta, H.; Okawa, K.; Iwamatsu, A.; Shimada, H.; Masuda, T.; Takamiya, K. Cloning of chlorophyllase, the key enzyme in chlorophyll degradation: Finding of a lipase motif and the induction by methyl jasmonate. *Proc. Natl. Acad. Sci. USA* **1999**, *96*, 15362–15367. [CrossRef] [PubMed]

27. Schenk, N.; Schelbert, S.; Kanwischer, M.; Goldschmidt, E.E.; Doermann, P.; Hoertensteiner, S. The chlorophyllases AtCLH1 and AtCLH2 are not essential for senescence-related chlorophyll breakdown in *Arabidopsis thaliana*. *FEBS Lett.* **2007**, *581*, 5517–5525. [CrossRef] [PubMed]

28. Schelbert, S.; Aubry, S.; Burla, B.; Agne, B.; Kessler, F.; Krupinska, K.; Hoertensteiner, S. Pheophytin Pheophorbide Hydrolase (Pheophytinase) Is Involved in Chlorophyll Breakdown during Leaf Senescence in *Arabidopsis*. *Plant Cell* **2009**, *21*, 767–785. [CrossRef] [PubMed]

29. Rodoni, S.; Muhlecker, W.; Anderl, M.; Krautler, B.; Moser, D.; Thomas, H.; Matile, P.; Hortensteiner, S. Chlorophyll breakdown in senescent chloroplasts—Cleavage of pheophorbide a in two enzymic steps. *Plant Physiol.* **1997**, *115*, 669–676. [CrossRef] [PubMed]

30. Wuthrich, K.L.; Bovet, L.; Hunziker, P.E.; Donnison, I.S.; Hortensteiner, S. Molecular cloning, functional expression and characterisation of RCC reductase involved in chlorophyll catabolism. *Plant J.* **2000**, *21*, 189–198. [CrossRef] [PubMed]

31. Pruzinska, A.; Tanner, G.; Anders, I.; Roca, M.; Hortensteiner, S. Chlorophyll breakdown: Pheophorbide a oxygenase is a Rieske-type iron-sulfur protein, encoded by the accelerated cell death 1 gene. *Proc. Natl. Acad. Sci. USA* **2003**, *100*, 15259–15264. [CrossRef] [PubMed]

32. Pruzinska, A.; Anders, I.; Aubry, S.; Schenk, N.; Tapernoux-Luthi, E.; Muller, T.; Krautler, B.; Hortensteiner, S. In vivo participation of red chlorophyll catabolite reductase in chlorophyll breakdown. *Plant Cell* **2007**, *19*, 369–387. [CrossRef] [PubMed]

33. Hinder, B.; Schellenberg, M.; Rodon, S.; Ginsburg, S.; Vogt, E.; Martinoia, E.; Matile, P.; Hortensteiner, S. How plants dispose of chlorophyll catabolites—Directly energized uptake of tetrapyrrolic breakdown products into isolated vacuoles. *J. Biol. Chem.* **1996**, *271*, 27233–27236. [CrossRef] [PubMed]

34. Tommasini, R.; Vogt, E.; Fromenteau, M.; Hortensteiner, S.; Matile, P.; Amrhein, N.; Martinoia, E. An ABC-transporter of *Arabidopsis thaliana* has both glutathione-conjugate and chlorophyll catabolite transport activity. *Plant J.* **1998**, *13*, 773–780. [CrossRef] [PubMed]

35. Kusaba, M.; Ito, H.; Morita, R.; Iida, S.; Sato, Y.; Fujimoto, M.; Kawasaki, S.; Tanaka, R.; Hirochika, H.; Nishimura, M.; et al. Rice NON-YELLOW COLORING1 is involved in light-harvesting complex II and grana degradation during leaf senescence. *Plant Cell* **2007**, *19*, 1362–1375. [CrossRef] [PubMed]

36. Sato, Y.; Morita, R.; Katsuma, S.; Nishimura, M.; Tanaka, A.; Kusaba, M. Two short-chain dehydrogenase/reductases, NON-YELLOW COLORING 1 and NYC1-LIKE, are required for chlorophyll b and light-harvesting complex II degradation during senescence in rice. *Plant J.* **2009**, *57*, 120–131. [CrossRef] [PubMed]

37. Pruzinska, A.; Tanner, G.; Aubry, S.; Anders, I.; Moser, S.; Muller, T.; Ongania, K.H.; Krautler, B.; Youn, J.Y.; Liljegren, S.J.; et al. Chlorophyll breakdown in senescent *Arabidopsis* leaves. Characterization of chlorophyll catabolites and of chlorophyll catabolic enzymes involved in the degreening reaction. *Plant Physiol.* **2005**, *139*, 52–63. [CrossRef] [PubMed]

38. Tanaka, R.; Hirashima, M.; Satoh, S.; Tanaka, A. The Arabidopsis-accelerated cell death gene ACD1 is involved in oxygenation of pheophorbide a: Inhibition of the pheophorbide a oxygenase activity does not lead to the "Stay-Green" phenotype in *Arabidopsis*. *Plant Cell Physiol.* **2003**, *44*, 1266–1274. [CrossRef] [PubMed]

39. Thomas, H. Sid—A Mendelian Locus Controlling Thylakoid Membrane Disassembly in Senescing Leaves of Festuca-Pratensis. *Theor. Appl. Genet.* **1987**, *73*, 551–555. [CrossRef] [PubMed]

40. Armstead, I.; Donnison, I.; Aubry, S.; Harper, J.; Hoertensteiner, S.; James, C.; Mani, J.; Moffet, M.; Ougham, H.; Roberts, L.; et al. From crop to model to crop: Identifying the genetic basis of the staygreen mutation in the Lolium/Festuca forage and amenity grasses. *New Phytol.* **2006**, *172*, 592–597. [CrossRef] [PubMed]

41. Armstead, I.; Donnison, I.; Aubry, S.; Harper, J.; Hortensteiner, S.; James, C.; Mani, J.; Moffet, M.; Ougham, H.; Roberts, L.; et al. Cross-species identification of Mendel's/locus. *Science* **2007**, *315*, 73. [CrossRef] [PubMed]

42. Ren, G.; An, K.; Liao, Y.; Zhou, X.; Cao, Y.; Zhao, H.; Ge, X.; Kuai, B. Identification of a novel chloroplast protein AtNYE1 regulating chlorophyll degradation during leaf senescence in *Arabidopsis*. *Plant Physiol.* **2007**, *144*, 1429–1441. [CrossRef] [PubMed]

43. Jiang, H.; Li, M.; Liang, N.; Yan, H.; Wei, Y.; Xu, X.; Liu, J.; Xu, Z.; Chen, F.; Wu, G. Molecular cloning and function analysis of the stay green gene in rice. *Plant J.* **2007**, *52*, 197–209. [CrossRef] [PubMed]

44. Park, S.-Y.; Yu, J.-W.; Park, J.-S.; Li, J.; Yoo, S.-C.; Lee, N.-Y.; Lee, S.-K.; Jeong, S.-W.; Seo, H.S.; Koh, H.-J.; et al. The senescence-induced staygreen protein regulates chlorophyll degradation. *Plant Cell* **2007**, *19*, 1649–1664. [CrossRef] [PubMed]

45. Sato, Y.; Morita, R.; Nishimura, M.; Yamaguchi, H.; Kusaba, M. Mendel's green cotyledon gene encodes a positive regulator of the chlorophyll-degrading pathway. *Proc. Natl. Acad. Sci. USA* **2007**, *104*, 14169–14174. [CrossRef] [PubMed]

46. Aubry, S.; Mani, J.; Hortensteiner, S. Stay-green protein, defective in Mendel's green cotyledon mutant, acts independent and upstream of pheophorbide a oxygenase in the chlorophyll catabolic pathway. *Plant Mol. Biol.* **2008**, *67*, 243–256. [CrossRef] [PubMed]

47. Pogson, B.J.; Woo, N.S.; Foerster, B.; Small, I.D. Plastid signalling to the nucleus and beyond. *Trends Plant Sci.* **2008**, *13*, 602–609. [CrossRef] [PubMed]

48. Borovsky, Y.; Paran, I. Chlorophyll breakdown during pepper fruit ripening in the chlorophyll retainer mutation is impaired at the homolog of the senescence-inducible stay-green gene. *Theor. Appl. Genet.* **2008**, *117*, 235–240. [CrossRef] [PubMed]

49. Barry, C.S. The stay-green revolution: Recent progress in deciphering the mechanisms of chlorophyll degradation in higher plants. *Plant Sci.* **2009**, *176*, 325–333. [CrossRef]

50. Hoertensteiner, S. Update on the biochemistry of chlorophyll breakdown. *Plant Mol. Biol.* **2013**, *82*, 505–517. [CrossRef] [PubMed]

51. Lim, P.O.; Kim, H.J.; Nam, H.G. Leaf senescence. *Annu. Rev. Plant Biol.* **2007**, *58*, 115–136. [CrossRef] [PubMed]

52. Rossi, S.; Burgess, P.; Jespersen, D.; Huang, B. Heat-Induced Leaf Senescence Associated with Chlorophyll Metabolism in Bentgrass Lines Differing in Heat Tolerance. *Crop Sci.* **2017**, *57*, S169–S178. [CrossRef]

53. Todorov, D.T.; Karanov, E.N.; Smith, A.R.; Hall, M.A. Chlorophyllase activity and chlorophyll content in wild type and eti 5 mutant of *Arabidopsis thaliana* subjected to low and high temperatures. *Biol. Plant.* **2003**, *46*, 633–636. [CrossRef]

54. Djanaguiraman, M.; Prasad, P.V.V.; Boyle, D.L.; Schapaugh, W.T. High-Temperature Stress and Soybean Leaves: Leaf Anatomy and Photosynthesis. *Crop Sci.* **2011**, *51*, 2125–2131. [CrossRef]

55. Djanaguiraman, M.; Prasad, P.V.V.; Murugan, M.; Perumal, R.; Reddy, U.K. Physiological differences among sorghum (*Sorghum bicolor* L. Moench) genotypes under high temperature stress. *Environ. Exp. Bot.* **2014**, *100*, 43–54. [CrossRef]

56. Ristic, Z.; Momcilovic, I.; Fu, J.M.; Callegaric, E.; DeRidder, B.P. Chloroplast protein synthesis elongation factor, EF-Tu, reduces thermal aggregation of Rubisco activase. *J. Plant Physiol.* **2007**, *164*, 1564–1571. [CrossRef] [PubMed]

57. Akter, N.; Islam, M.R. Heat stress effects and management in wheat. A review. *Agron. Sustain. Dev.* **2017**, *37*, 37. [CrossRef]

58. Liu, X.Z.; Huang, B.R. Heat stress injury in relation to membrane lipid peroxidation in creeping bentgrass. *Crop Sci.* **2000**, *40*, 503–510. [CrossRef]

59. Jespersen, D.; Zhang, J.; Huang, B. Chlorophyll loss associated with heat-induced senescence in bentgrass. *Plant Sci.* **2016**, *249*, 1–12. [CrossRef] [PubMed]

60. Lin, Y.-P.; Wu, M.-C.; Charng, Y.-Y. Identification of a Chlorophyll Dephytylase Involved in Chlorophyll Turnover in *Arabidopsis*. *Plant Cell* **2016**, *28*, 2974–2990. [CrossRef] [PubMed]

61. Veerasamy, M.; He, Y.; Huang, B. Leaf senescence and protein metabolism in creeping bentgrass exposed to heat stress and treated with cytokinins. *J. Am. Soc. Hortic. Sci.* **2007**, *132*, 467–472.

62. Chen, Y.; Cothren, J.T.; Chen, D.-H.; Ibrahim, A.M.H.; Lombardini, L. Ethylene-inhibiting compound 1-MCP delays leaf senescence in cotton plants under abiotic stress conditions. *J. Integr. Agric.* **2015**, *14*, 1321–1331. [CrossRef]

63. Huberman, M.; Riov, J.; Goldschmidt, E.E.; Apelbaum, A.; Goren, R. The novel ethylene antagonist, 3-cyclopropyl-1-enyl-propanoic acid sodium salt (CPAS), increases grain yield in wheat by delaying leaf senescence. *Plant Growth Regul.* **2014**, *73*, 249–255. [CrossRef]

64. Abdelrahman, M.; El-Sayed, M.; Jogaiah, S.; Burritt, D.J.; Lam-Son Phan, T. The "STAY-GREEN" trait and phytohormone signaling networks in plants under heat stress. *Plant Cell Rep.* **2017**, *36*, 1009–1025. [CrossRef] [PubMed]

65. Suzuky Pinto, R.; Lopes, M.S.; Collins, N.C.; Reynolds, M.P. Modelling and genetic dissection of staygreen under heat stress. *Theor. Appl. Genet.* **2016**, *129*, 2055–2074. [CrossRef] [PubMed]

66. Teixeira, R.N.; Ligterink, W.; Franca-Neto, J.D.B.; Hilhorst, H.W.M.; da Silva, E.A.A. Gene expression profiling of the green seed problem in Soybean. *BMC Plant Biol.* **2016**, *16*, 37. [CrossRef] [PubMed]

67. Vijayalakshmi, K.; Fritz, A.K.; Paulsen, G.M.; Bai, G.; Pandravada, S.; Gill, B.S. Modeling and mapping QTL for senescence-related traits in winter wheat under high temperature. *Mol. Breed.* **2010**, *26*, 163–175. [CrossRef]

68. Chen, Y.-E.; Zhang, C.-M.; Su, Y.-Q.; Ma, J.; Zhang, Z.-W.; Yuan, M.; Zhang, H.-Y.; Yuan, S. Responses of photosystem II and antioxidative systems to high light and high temperature co-stress in wheat. *Environ. Exp. Bot.* **2017**, *135*, 45–55. [CrossRef]

69. Shirdelmoghanloo, H.; Cozzolino, D.; Lohraseb, I.; Collins, N.C. Truncation of grain filling in wheat (*Triticum aestivum*) triggered by brief heat stress during early grain filling: Association with senescence responses and reductions in stem reserves. *Funct. Plant Biol.* **2016**, *43*, 919–930. [CrossRef]

70. Shirdelmoghanloo, H.; Lohraseb, I.; Rabie, H.S.; Brien, C.; Parent, B.; Collins, N.C. Heat susceptibility of grain filling in wheat (*Triticum aestivum* L.) linked with rapid chlorophyll loss during a 3-day heat treatment. *Acta Physiol. Plant.* **2016**, *38*, 208. [CrossRef]

71. Asada, K. Production and scavenging of reactive oxygen species in chloroplasts and their functions. *Plant Physiol.* **2006**, *141*, 391–396. [CrossRef] [PubMed]

72. Suzuki, N.; Koussevitzky, S.; Mittler, R.; Miller, G. ROS and redox signalling in the response of plants to abiotic stress. *Plant Cell Environ.* **2012**, *35*, 259–270. [CrossRef] [PubMed]

73. Pospisil, P.; Prasad, A. Formation of singlet oxygen and protection against its oxidative damage in Photosystem II under abiotic stress. *J. Photochem. Photobiol. B Biol.* **2014**, *137*, 39–48. [CrossRef] [PubMed]

74. Mittler, R.; Vanderauwera, S.; Gollery, M.; Van Breusegem, F. Reactive oxygen gene network of plants. *Trends Plant Sci.* **2004**, *9*, 490–498. [CrossRef] [PubMed]

75. Foyer, C.H.; LopezDelgado, H.; Dat, J.F.; Scott, I.M. Hydrogen peroxide- and glutathione-associated mechanisms of acclimatory stress tolerance and signalling. *Physiol. Plant.* **1997**, *100*, 241–254. [CrossRef]

76. Dat, J.F.; Lopez-Delgado, H.; Foyer, C.H.; Scott, I.M. Parallel changes in H_2O_2 and catalase during thermotolerance induced by salicylic acid or heat acclimation in mustard seedlings. *Plant Physiol.* **1998**, *116*, 1351–1357. [CrossRef] [PubMed]

77. Vallelian-Bindschedler, L.; Schweizer, P.; Mosinger, E.; Metraux, J.P. Heat-induced resistance in barley to powdery mildew (*Blumeria graminis* f.sp. hordei) is associated with a burst of active oxygen species. *Physiol. Mol. Plant Pathol.* **1998**, *52*, 185–199. [CrossRef]

78. Foyer, C.H.; Noctor, G. Redox Signaling in Plants. *Antioxid. Redox Signal.* **2013**, *18*, 2087–2090. [CrossRef] [PubMed]

79. Vacca, R.A.; Valenti, D.; Bobba, A.; Merafina, R.S.; Passarella, S.; Marra, E. Cytochrome c is released in a reactive oxygen species-dependent manner and is degraded via caspase-like proteases in tobacco bright-yellow 2 cells en route to heat shock-induced cell death. *Plant Physiol.* **2006**, *141*, 208–219. [CrossRef] [PubMed]

80. Wang, P.; Duan, W.; Takabayashi, A.; Endo, T.; Shikanai, T.; Ye, J.Y.; Mi, H.L. Chloroplastic NAD(P)H dehydrogenase in tobacco leaves functions in alleviation of oxidative damage caused by temperature stress. *Plant Physiol.* **2006**, *141*, 465–474. [CrossRef] [PubMed]

81. Yamashita, A.; Nijo, N.; Pospisil, P.; Morita, N.; Takenaka, D.; Aminaka, R.; Yamamoto, Y.; Yamamoto, Y. Quality control of photosystem II—Reactive oxygen species are responsible for the damage to photosystem II under moderate heat stress. *J. Biol. Chem.* **2008**, *283*, 28380–28391. [CrossRef] [PubMed]

82. Lin, Y.-P.; Lee, T.-Y.; Tanaka, A.; Charng, Y.-Y. Analysis of an *Arabidopsis* heat-sensitive mutant reveals that chlorophyll synthase is involved in reutilization of chlorophyllide during chlorophyll turnover. *Plant J.* **2014**, *80*, 14–26. [CrossRef] [PubMed]

83. Edreva, A. Generation and scavenging of reactive oxygen species in chloroplasts: A submolecular approach. *Agric. Ecosyst. Environ.* **2005**, *106*, 119–133. [CrossRef]

84. Foyer, C.H.; Neukermans, J.; Queval, G.; Noctor, G.; Harbinson, J. Photosynthetic control of electron transport and the regulation of gene expression. *J. Exp. Bot.* **2012**, *63*, 1637–1661. [CrossRef] [PubMed]

85. Foyer, C.H.; Noctor, G. Oxidant and antioxidant signalling in plants: A re-evaluation of the concept of oxidative stress in a physiological context. *Plant Cell Environ.* **2005**, *28*, 1056–1071. [CrossRef]

86. Foyer, C.H.; Noctor, G. Redox homeostasis and antioxidant signaling: A metabolic interface between stress perception and physiological responses. *Plant Cell* **2005**, *17*, 1866–1875. [CrossRef] [PubMed]

87. Foyer, C.H.; Noctor, G. Redox Regulation in Photosynthetic Organisms: Signaling, Acclimation, and Practical Implications. *Antioxid. Redox Signal.* **2009**, *11*, 861–905. [CrossRef] [PubMed]

88. Foyer, C.H.; Noctor, G. Ascorbate and Glutathione: The Heart of the Redox Hub. *Plant Physiol.* **2011**, *155*, 2–18. [CrossRef] [PubMed]

89. Foyer, C.H.; Noctor, G. Managing the cellular redox hub in photosynthetic organisms. *Plant Cell Environ.* **2012**, *35*, 199–201. [CrossRef] [PubMed]

90. Gill, S.S.; Tuteja, N. Reactive oxygen species and antioxidant machinery in abiotic stress tolerance in crop plants. *Plant Physiol. Biochem.* **2010**, *48*, 909–930. [CrossRef] [PubMed]

91. Sainz, M.; Diaz, P.; Monza, J.; Borsani, O. Heat stress results in loss of chloroplast Cu/Zn superoxide dismutase and increased damage to Photosystem II in combined drought-heat stressed *Lotus japonicus*. *Physiol. Plant.* **2010**, *140*, 46–56. [CrossRef] [PubMed]

92. Martins, M.Q.; Rodrigues, W.P.; Fortunato, A.S.; Leitao, A.E.; Rodrigues, A.P.; Pais, I.P.; Martins, L.D.; Silva, M.J.; Reboredo, F.H.; Partelli, F.L.; et al. Protective Response Mechanisms to Heat Stress in Interaction with High [CO$_2$] Conditions in *Coffea* spp. *Front. Plant Sci.* **2016**, *7*. [CrossRef] [PubMed]

93. Mostofa, M.G.; Yoshida, N.; Fujita, M. Spermidine pretreatment enhances heat tolerance in rice seedlings through modulating antioxidative and glyoxalase systems. *Plant Growth Regul.* **2014**, *73*, 31–44. [CrossRef]

94. Wang, X.; Cai, J.; Liu, F.; Dai, T.; Cao, W.; Wollenweber, B.; Jiang, D. Multiple heat priming enhances thermo-tolerance to a later high temperature stress via improving subcellular antioxidant activities in wheat seedlings. *Plant Physiol. Biochem.* **2014**, *74*, 185–192. [CrossRef] [PubMed]

95. Du, H.; Zhou, P.; Huang, B. Antioxidant enzymatic activities and gene expression associated with heat tolerance in a cool-season perennial grass species. *Environ. Exp. Bot.* **2013**, *87*, 159–166. [CrossRef]

96. Kong, F.; Deng, Y.; Wang, G.; Wang, J.; Liang, X.; Meng, Q. LeCDJ1, a chloroplast DnaJ protein, facilitates heat tolerance in transgenic tomatoes. *J. Integr. Plant Biol.* **2014**, *56*, 63–74. [CrossRef] [PubMed]

97. Zou, M.; Yuan, L.; Zhu, S.; Liu, S.; Ge, J.; Wang, C. Effects of heat stress on photosynthetic characteristics and chloroplast ultrastructure of a heat-sensitive and heat-tolerant cultivar of wucai (*Brassica campestris* L.). *Acta Physiol. Plant.* **2017**, *39*. [CrossRef]

98. Ara, N.; Nakkanong, K.; Lv, W.; Yang, J.; Hu, Z.; Zhang, M. Antioxidant Enzymatic Activities and Gene Expression Associated with Heat Tolerance in the Stems and Roots of Two Cucurbit Species ("*Cucurbita maxima*" and "*Cucurbita moschata*") and Their Interspecific Inbred Line "Maxchata". *Int. J. Mol. Sci.* **2013**, *14*, 24008–24028. [CrossRef] [PubMed]

99. Miller, G.; Shulaev, V.; Mittler, R. Reactive oxygen signaling and abiotic stress. *Physiol. Plant.* **2008**, *133*, 481–489. [CrossRef] [PubMed]

100. Mittler, R.; Vanderauwera, S.; Suzuki, N.; Miller, G.; Tognetti, V.B.; Vandepoele, K.; Gollery, M.; Shulaev, V.; Van Breusegem, F. ROS signaling: The new wave? *Trends Plant Sci.* **2011**, *16*, 300–309. [CrossRef] [PubMed]

101. Xiao, Y.; Wang, J.; Dehesh, K. Review of stress specific organelles-to-nucleus metabolic signal molecules in plants. *Plant Sci.* **2013**, *212*, 102–107. [CrossRef] [PubMed]

102. Komayama, K.; Khatoon, M.; Takenaka, D.; Horie, J.; Yamashita, A.; Yoshioka, M.; Nakayama, Y.; Yoshida, M.; Ohira, S.; Morita, N.; et al. Quality control of photosystem II: Cleavage and aggregation of heat-damaged D1 protein in spinach thylakoids. *Biochim. Biophys. Acta Bioenerg.* **2007**, *1767*, 838–846. [CrossRef] [PubMed]

103. Yoshioka, M.; Uchida, S.; Mori, H.; Komayama, K.; Ohira, S.; Morita, N.; Nakanishi, T.; Yamamoto, Y. Quality control of photosystem II—Cleavage of reaction center D1 protein in spinach thylakoids by FtsH protease under moderate heat stress. *J. Biol. Chem.* **2006**, *281*, 21660–21669. [CrossRef] [PubMed]

104. Khatoon, M.; Inagawa, K.; Pospisil, P.; Yamashita, A.; Yoshioka, M.; Lundin, B.; Horie, J.; Morita, N.; Jajoo, A.; Yamamoto, Y.; et al. Quality Control of Photosystem II thylakoid unstacking is necessary to avoid further damage to the D1 protein and to facilitate D1 degradation under light stress in spinach thylakoids. *J. Biol. Chem.* **2009**, *284*, 25343–25352. [CrossRef] [PubMed]

105. Jarvi, S.; Suorsa, M.; Aro, E.M. Photosystem II repair in plant chloroplasts—Regulation, assisting proteins and shared components with photosystem II biogenesis. *Biochim. Biophys. Acta Bioenerg.* **2015**, *1847*, 900–909. [CrossRef] [PubMed]

106. Vass, I.; Cser, K. Janus-faced charge recombinations in photosystem II photoinhibition. *Trends Plant Sci.* **2009**, *14*, 200–205. [CrossRef] [PubMed]

107. Telfer, A.; Dhami, S.; Bishop, S.M.; Phillips, D.; Barber, J. Beta-Carotene Quenches Singlet Oxygen Formed by Isolated Photosystem-II Reaction Centers. *Biochemistry* **1994**, *33*, 14469–14474. [CrossRef] [PubMed]

108. Kruk, J.; Trebst, A. Plastoquinol as a singlet oxygen scavenger in photosystem II. *BBA-Bioenergetics* **2008**, *1777*, 154–162.

109. Weisz, D.A.; Gross, M.L.; Pakrasi, H.B. Reactive oxygen species leave a damage trail that reveals water channels in Photosystem II. *Sci. Adv.* **2017**, *3*. [CrossRef] [PubMed]

110. Pospisil, P.; Yamamoto, Y. Damage to photosystem II by lipid peroxidation products. *Biochim. Biophys. Acta Gen. Subj.* **2017**, *1861*, 457–466. [CrossRef] [PubMed]

111. Park, Y.I.; Chow, W.S.; Anderson, J.M. Light Inactivation of Functional Photosystem-Ii in Leaves of Peas Grown in Moderate Light Depends on Photon Exposure. *Planta* **1995**, *196*, 401–411. [CrossRef]

112. Tyystjarvi, E.; Aro, E.M. The rate constant of photoinhibition, measured in lincomycin-treated leaves, is directly proportional to light intensity. *Proc. Natl. Acad. Sci. USA* **1996**, *93*, 2213–2218. [CrossRef] [PubMed]

113. Nelson, C.J.; Alexova, R.; Jacoby, R.P.; Millar, A.H. Proteins with High Turnover Rate in Barley Leaves Estimated by Proteome Analysis Combined with in Planta Isotope Labeling. *Plant Physiol.* **2014**, *166*, 91–108. [CrossRef] [PubMed]

114. Bergantino, E.; Brunetta, A.; Touloupakis, E.; Segalla, A.; Szabo, I.; Giacometti, G.M. Role of the PSII-H subunit in photoprotection—Novel aspects of D1 turnover in Synechocystis 6803. *J. Biol. Chem.* **2003**, *278*, 41820–41829. [CrossRef] [PubMed]

115. Rokka, A.; Suorsa, M.; Saleem, A.; Battchikova, N.; Aro, E.M. Synthesis and assembly of thylakoid protein complexes: Multiple assembly steps of photosystem II. *Biochem. J.* **2005**, *388*, 159–168. [CrossRef] [PubMed]

116. Kato, Y.; Sakamoto, W. Protein Quality Control in Chloroplasts: A Current Model of D1 Protein Degradation in the Photosystem II Repair Cycle. *J. Biochem.* **2009**, *146*, 463–469. [CrossRef] [PubMed]

117. Barber, J.; Andersson, B. Too Much of a Good Thing—Light Can Be Bad for Photosynthesis. *Trends Biochem. Sci.* **1992**, *17*, 61–66. [CrossRef]

118. Aro, E.M.; Virgin, I.; Andersson, B. Photoinhibition of Photosystem-2—Inactivation, Protein Damage and Turnover. *Biochim. Biophys. Acta* **1993**, *1143*, 113–134. [CrossRef]

119. Adir, N.; Zer, H.; Shochat, S.; Ohad, I. Photoinhibition—A historical perspective. *Photosynth. Res.* **2003**, *76*, 343–370. [CrossRef] [PubMed]

120. Aro, E.M.; Suorsa, M.; Rokka, A.; Allahverdiyeva, Y.; Paakkarinen, V.; Saleem, A.; Battchikova, N.; Rintamaki, E. Dynamics of photosystem II: A proteomic approach to thylakoid protein complexes. *J. Exp. Bot.* **2005**, *56*, 347–356. [CrossRef] [PubMed]

121. Baena-Gonzalez, E.; Aro, E.M. Biogenesis, assembly and turnover of photosystem II units. *Philos. Trans. R. Soc. Lond. Ser. B Biol. Sci.* **2002**, *357*, 1451–1459. [CrossRef] [PubMed]

122. Cheregi, O.; Wagner, R.; Funk, C. Insights into the Cyanobacterial Deg/HtrA Proteases. *Front. Plant Sci.* **2016**, *7*. [CrossRef] [PubMed]

123. Huesgen, P.F.; Schuhmann, H.; Adamska, I. Deg/HtrA proteases as components of a network for photosystem II quality control in chloroplasts and cyanobacteria. *Res. Microbiol.* **2009**, *160*, 726–732. [CrossRef] [PubMed]

124. Schuhmann, H.; Huesgen, P.F.; Adamska, I. The family of Deg/HtrA proteases in plants. *BMC Plant Biol.* **2012**, *12*. [CrossRef] [PubMed]

125. Schuhmann, H.; Adamska, I. Deg proteases and their role in protein quality control and processing in different subcellular compartments of the plant cell. *Physiol. Plant.* **2012**, *145*, 224–234. [CrossRef] [PubMed]

126. Sun, X.; Wang, L.; Zhang, L. Involvement of DEG5 and DEG8 proteases in the turnover of the photosystem II reaction center D1 protein under heat stress in *Arabidopsis thaliana*. *Chin. Sci. Bull.* **2007**, *52*, 1742–1745. [CrossRef]

127. Kato, Y.; Sun, X.; Zhang, L.; Sakamoto, W. Cooperative D1 Degradation in the Photosystem II Repair Mediated by Chloroplastic Proteases in *Arabidopsis*. *Plant Physiol.* **2012**, *159*, 1428–1439. [CrossRef] [PubMed]

128. Chen, J.P.; Burke, J.J.; Velten, J.; Xin, Z.U. FtsH11 protease plays a critical role in *Arabidopsis thermotolerance*. *Plant J.* **2006**, *48*, 73–84. [CrossRef] [PubMed]

129. Yoshioka, M.; Yamamoto, Y. Quality control of Photosystem II: Where and how does the degradation of the D1 protein by FtsH proteases start under light stress?—Facts and hypotheses. *J. Photochem. Photobiol. B Biol.* **2011**, *104*, 229–235. [CrossRef] [PubMed]

130. Peltier, J.B.; Emanuelsson, O.; Kalume, D.E.; Ytterberg, J.; Friso, G.; Rudella, A.; Liberles, D.A.; Soderberg, L.; Roepstorff, P.; von Heijne, G.; et al. Central functions of the lumenal and peripheral thylakoid proteome of *Arabidopsis* determined by experimentation and genome-wide prediction. *Plant Cell* **2002**, *14*, 211–236. [CrossRef] [PubMed]

131. Schubert, M.; Petersson, U.A.; Haas, B.J.; Funk, C.; Schroder, W.P.; Kieselbach, T. Proteome map of the chloroplast lumen of *Arabidopsis thaliana*. *J. Biol. Chem.* **2002**, *277*, 8354–8365. [CrossRef] [PubMed]

132. Kapri-Pardes, E.; Naveh, L.; Adam, Z. The thylakoid lumen protease Deg1 is involved in the repair of photosystem II from photoinhibition in *Arabidopsis*. *Plant Cell* **2007**, *19*, 1039–1047. [CrossRef] [PubMed]

133. Sun, X.; Peng, L.; Guo, J.; Chi, W.; Ma, J.; Lu, C.; Zhang, L. Formation of DEG5 and DEG8 complexes and their involvement in the degradation of photodamaged photosystem II reaction center D1 protein in *Arabidopsis*. *Plant Cell* **2007**, *19*, 1347–1361. [CrossRef] [PubMed]

134. Kley, J.; Schmidt, B.; Boyanov, B.; Stolt-Bergner, P.C.; Kirk, R.; Ehrmann, M.; Knopf, R.R.; Naveh, L.; Adam, Z.; Clausen, T. Structural adaptation of the plant protease Deg1 to repair photosystem II during light exposure. *Nat. Struct. Mol. Biol.* **2011**, *18*, 728–731. [CrossRef] [PubMed]

135. Sun, X.; Ouyang, M.; Guo, J.; Ma, J.; Lu, C.; Adam, Z.; Zhang, L. The thylakoid protease Deg1 is involved in photosystem-II assembly in *Arabidopsis thaliana*. *Plant J.* **2010**, *62*, 240–249. [CrossRef] [PubMed]

136. Chassin, Y.; Kapri-Pardes, E.; Sinvany, G.; Arad, T.; Adam, Z. Expression and characterization of the thylakoid lumen protease DegP1 from *Arabidopsis*. *Plant Physiol.* **2002**, *130*, 857–864. [CrossRef] [PubMed]

137. Aro, E.M.; McCaffery, S.; Anderson, J.M. Photoinhibition and D1 Protein-Degradation in Peas Acclimated to Different Growth Irradiances. *Plant Physiol.* **1993**, *103*, 835–843. [CrossRef] [PubMed]

138. Tyedmers, J.; Mogk, A.; Bukau, B. Cellular strategies for controlling protein aggregation. *Nat. Rev. Mol. Cell Biol.* **2010**, *11*, 777–788. [CrossRef] [PubMed]

139. Vierling, E. The Roles of Heat-Shock Proteins in Plants. *Annu. Rev. Plant Physiol. Plant Mol. Biol.* **1991**, *42*, 579–620. [CrossRef]

140. Chen, Q.; Vierling, E. Analysis of Conserved Domains Identifies a Unique Structural Feature of a Chloroplast Heat-Shock Protein. *Mol. Gen. Genet.* **1991**, *226*, 425–431. [CrossRef] [PubMed]

141. Chen, Q.; Lauzon, L.M.; Derocher, A.E.; Vierling, E. Accumulation, Stability, and Localization of a Major Chloroplast Heat-Shock Protein. *J. Cell Biol.* **1990**, *110*, 1873–1883. [CrossRef] [PubMed]

142. Vierling, E.; Harris, L.M.; Chen, Q. The Major Low-Molecular-Weight Heat-Shock Protein in Chloroplasts Shows Antigenic Conservation among Diverse Higher-Plant Species. *Mol. Cell. Biol.* **1989**, *9*, 461–468. [CrossRef] [PubMed]

143. Heckathorn, S.A.; Downs, C.A.; Sharkey, T.D.; Coleman, J.S. The small, methionine-rich chloroplast heat-shock protein protects photosystem II electron transport during heat stress. *Plant Physiol.* **1998**, *116*, 439–444. [CrossRef] [PubMed]

144. Heckathorn, S.A.; Ryan, S.L.; Baylis, J.A.; Wang, D.F.; Hamilton, E.W.; Cundiff, L.; Luthe, D.S. In vivo evidence from an *Agrostis stolonifera* selection genotype that chloroplast small heat-shock proteins can protect photosystem II during heat stress. *Funct. Plant Biol.* **2002**, *29*, 933–944. [CrossRef]

145. Kim, K.-H.; Alam, I.; Kim, Y.-G.; Sharmin, S.A.; Lee, K.-W.; Lee, S.-H.; Lee, B.-H. Overexpression of a chloroplast-localized small heat shock protein OsHSP26 confers enhanced tolerance against oxidative and heat stresses in tall fescue. *Biotechnol. Lett.* **2012**, *34*, 371–377. [CrossRef] [PubMed]

146. Shakeel, S.; Ul Haq, N.; Heckathorn, S.A.; Hamilton, E.W.; Luthe, D.S. Ecotypic variation in chloroplast small heat-shock proteins and related thermotolerance in *Chenopodium album. Plant Physiol. Biochem.* **2011**, *49*, 898–908. [CrossRef] [PubMed]

147. Wang, D.F.; Luthe, D.S. Heat sensitivity in a bentgrass variant. Failure to accumulate a chloroplast heat shock protein isoform implicated in heat tolerance. *Plant Physiol.* **2003**, *133*, 319–327. [CrossRef] [PubMed]

148. Harndahl, U.; Hall, R.B.; Osteryoung, K.W.; Vierling, E.; Bornman, J.F.; Sundby, C. The chloroplast small heat shock protein undergoes oxidation-dependent conformational changes and may protect plants from oxidative stress. *Cell Stress Chaperones* **1999**, *4*, 129–138. [CrossRef]

149. Neta-Sharir, I.; Isaacson, T.; Lurie, S.; Weiss, D. Dual role for tomato heat shock protein 21: Protecting photosystem II from oxidative stress and promoting color changes during fruit maturation. *Plant Cell* **2005**, *17*, 1829–1838. [CrossRef] [PubMed]

150. Downs, C.A.; Coleman, J.S.; Heckathorn, S.A. The chloroplast 22-Ku heat-shock protein: A lumenal protein that associates with the oxygen evolving complex and protects photosystem II during heat stress. *J. Plant Physiol.* **1999**, *155*, 477–487. [CrossRef]

151. Bernfur, K.; Rutsdottir, G.; Emanuelsson, C. The chloroplast-localized small heat shock protein Hsp21 associates with the thylakoid membranes in heat-stressed plants. *Protein Sci. Publ. Protein Soc.* **2017**, *26*, 1773–1784. [CrossRef] [PubMed]

152. Spreitzer, R.J.; Salvucci, M.E. Rubisco: Structure, regulatory interactions, and possibilities for a better enzyme. *Annu. Rev. Plant Biol.* **2002**, *53*, 449–475. [CrossRef] [PubMed]

153. Perdomo, J.A.; Capo-Bauca, S.; Carmo-Silva, E.; Galmes, J. Rubisco and Rubisco Activase Play an Important Role in the Biochemical Limitations of Photosynthesis in Rice, Wheat, and Maize under High Temperature and Water Deficit. *Front. Plant Sci.* **2017**, *8*. [CrossRef] [PubMed]

154. Sage, R.F.; Way, D.A.; Kubien, D.S. Rubisco, Rubisco activase, and global climate change. *J. Exp. Bot.* **2008**, *59*, 1581–1595. [CrossRef] [PubMed]

155. Portis, A.R. Rubisco activase—Rubisco's catalytic chaperone. *Photosynth. Res.* **2003**, *75*, 11–27. [CrossRef] [PubMed]

156. Schrader, S.M.; Wise, R.R.; Wacholtz, W.F.; Ort, D.R.; Sharkey, T.D. Thylakoid membrane responses to moderately high leaf temperature in *Pima cotton. Plant Cell Environ.* **2004**, *27*, 725–735. [CrossRef]

157. Yamori, W.; Noguchi, K.; Kashino, Y.; Terashima, I. The role of electron transport in determining the temperature dependence of the photosynthetic rate in spinach leaves grown at contrasting temperatures. *Plant Cell Physiol.* **2008**, *49*, 583–591. [CrossRef] [PubMed]

158. Carmo-Silva, A.E.; Salvucci, M.E. The activity of Rubisco's molecular chaperone, Rubisco activase, in leaf extracts. *Photosynth. Res.* **2011**, *108*, 143–155. [CrossRef] [PubMed]

159. Salvucci, M.E.; Crafts-Brandner, S.J. Mechanism for deactivation of Rubisco under moderate heat stress. *Physiol. Plant.* **2004**, *122*, 513–519. [CrossRef]

160. Robinson, S.P.; Streusand, V.J.; Chatfield, J.M.; Portis, A.R. Purification and Assay of Rubisco activase from Leaves. *Plant Physiol.* **1988**, *88*, 1008–1014. [CrossRef] [PubMed]

161. Holbrook, G.P.; Galasinski, S.C.; Salvucci, M.E. Regulation of 2-Carboxyarabinitol 1-Phosphatase. *Plant Physiol.* **1991**, *97*, 894–899. [CrossRef] [PubMed]

162. Feller, U.; Crafts-Brandner, S.J.; Salvucci, M.E. Moderately high temperatures inhibit ribulose-1,5-bisphosphate carboxylase/oxygenase (Rubisco) activase-mediated activation of Rubisco. *Plant Physiol.* **1998**, *116*, 539–546. [CrossRef] [PubMed]

163. Salvucci, M.E.; Osteryoung, K.W.; Crafts-Brandner, S.J.; Vierling, E. Exceptional sensitivity of Rubisco activase to thermal denaturation in vitro and in vivo. *Plant Physiol.* **2001**, *127*, 1053–1064. [CrossRef] [PubMed]

164. Schrader, S.M.; Kleinbeck, K.R.; Sharkey, T.D. Rapid heating of intact leaves reveals initial effects of stromal oxidation on photosynthesis. *Plant Cell Environ.* **2007**, *30*, 671–678. [CrossRef] [PubMed]

165. Zhang, R.; Sharkey, T.D. Photosynthetic electron transport and proton flux under moderate heat stress. *Photosynth. Res.* **2009**, *100*, 29–43. [CrossRef] [PubMed]

166. Way, D.A.; Yamori, W. Thermal acclimation of photosynthesis: On the importance of adjusting our definitions and accounting for thermal acclimation of respiration. *Photosynth. Res.* **2014**, *119*, 89–100. [CrossRef] [PubMed]

167. Yamori, W.; Hikosaka, K.; Way, D.A. Temperature response of photosynthesis in C-3, C-4, and CAM plants: Temperature acclimation and temperature adaptation. *Photosynth. Res.* **2014**, *119*, 101–117. [CrossRef] [PubMed]

168. Salvucci, M.E.; Crafts-Brandner, S.J. Inhibition of photosynthesis by heat stress: The activation state of Rubisco as a limiting factor in photosynthesis. *Physiol. Plant.* **2004**, *120*, 179–186. [CrossRef] [PubMed]

169. DeRidder, B.P.; Salvucci, M.E. Modulation of Rubisco activase gene expression during heat stress in cotton (*Gossypium hirsutum* L.) involves post-transcriptional mechanisms. *Plant Sci.* **2007**, *172*, 246–254. [CrossRef]

170. Kumar, R.R.; Goswami, S.; Singh, K.; Dubey, K.; Singh, S.; Sharma, R.; Verma, N.; Kala, Y.K.; Rai, G.K.; Grover, M.; et al. Identification of Putative RuBisCo Activase (TaRca1)-The Catalytic Chaperone Regulating Carbon Assimilatory Pathway in Wheat (*Triticum aestivum*) under the Heat Stress. *Front. Plant Sci.* **2016**, *7*, 986. [CrossRef] [PubMed]

171. Kurek, I.; Chang, T.K.; Bertain, S.M.; Madrigal, A.; Liu, L.; Lassner, M.W.; Zhu, G. Enhanced thermostability of *Arabidopsis* Rubisco activase improves photosynthesis and growth rates under moderate heat stress. *Plant Cell* **2007**, *19*, 3230–3241. [CrossRef] [PubMed]

172. Kumar, A.; Li, C.; Portis, A.R., Jr. *Arabidopsis thaliana* expressing a thermostable chimeric Rubisco activase exhibits enhanced growth and higher rates of photosynthesis at moderately high temperatures. *Photosynth. Res.* **2009**, *100*, 143–153. [CrossRef] [PubMed]

173. Ristic, Z.; Momcilovic, I.; Bukovnik, U.; Prasad, P.V.V.; Fu, J.; DeRidder, B.P.; Elthon, T.E.; Mladenov, N. Rubisco activase and wheat productivity under heat-stress conditions. *J. Exp. Bot.* **2009**, *60*, 4003–4014. [CrossRef] [PubMed]

174. Shivhare, D.; Mueller-Cajar, O. In Vitro Characterization of Thermostable CAM Rubisco Activase Reveals a Rubisco Interacting Surface Loop. *Plant Physiol.* **2017**, *174*, 1505–1516. [CrossRef] [PubMed]

175. Salvucci, M.E. Association of Rubisco activase with chaperonin-60 beta: A possible mechanism for protecting photosynthesis during heat stress. *J. Exp. Bot.* **2008**, *59*, 1923–1933. [CrossRef] [PubMed]

176. Scales, J.C.; Parry, M.A.J.; Salvucci, M.E. A non-radioactive method for measuring Rubisco activase activity in the presence of variable ATP: ADP ratios, including modifications for measuring the activity and activation state of Rubisco. *Photosynth. Res.* **2014**, *119*, 355–365. [CrossRef] [PubMed]

177. Carmo-Silva, A.E.; Salvucci, M.E. The Regulatory Properties of Rubisco activase Differ among Species and Affect Photosynthetic Induction during Light Transitions. *Plant Physiol.* **2013**, *161*, 1645–1655. [CrossRef] [PubMed]

178. Wang, D.; Li, X.-F.; Zhou, Z.-J.; Feng, X.-P.; Yang, W.-J.; Jiang, D.-A. Two Rubisco activase isoforms may play different roles in photosynthetic heat acclimation in the rice plant. *Physiol. Plant.* **2010**, *139*, 55–67. [CrossRef] [PubMed]

179. Hozain, M.D.I.; Salvucci, M.E.; Fokar, M.; Holaday, A.S. The differential response of photosynthesis to high temperature for a boreal and temperate *Populus* species relates to differences in Rubisco activation and Rubisco activase properties. *Tree Physiol.* **2010**, *30*, 32–44. [CrossRef] [PubMed]

180. Feller, U. Drought stress and carbon assimilation in a warming climate: Reversible and irreversible impacts. *J. Plant Physiol.* **2016**, *203*, 69–79. [CrossRef] [PubMed]

14

The Complete Chloroplast Genome Sequence of Tree of Heaven (*Ailanthus altissima* (Mill.) (Sapindales: Simaroubaceae), an Important Pantropical Tree

Josphat K. Saina [1,2,3,4], **Zhi-Zhong Li** [2,3], **Andrew W. Gichira** [2,3,4] and **Yi-Ying Liao** [1,*]

[1]   Fairy Lake Botanical Garden, Shenzhen & Chinese Academy of Sciences, Shenzhen 518004, China;
      jksaina@wbgcas.cn
[2]   Key Laboratory of Aquatic Botany and Watershed Ecology, Wuhan Botanical Garden,
      Chinese Academy of Sciences, Wuhan 430074, China; wbg_georgelee@163.com (Z.-Z.L.);
      gichira@wbgcas.cn (A.W.G.)
[3]   University of Chinese Academy of Sciences, Beijing 100049, China
[4]   Sino-African Joint Research Center, Chinese Academy of Sciences, Wuhan 430074, China
[*]   Correspondence: liaoyiying666@163.com

Abstract: *Ailanthus altissima* (Mill.) Swingle (Simaroubaceae) is a deciduous tree widely distributed throughout temperate regions in China, hence suitable for genetic diversity and evolutionary studies. Previous studies in *A. altissima* have mainly focused on its biological activities, genetic diversity and genetic structure. However, until now there is no published report regarding genome of this plant species or Simaroubaceae family. Therefore, in this paper, we first characterized *A. altissima* complete chloroplast genome sequence. The tree of heaven chloroplast genome was found to be a circular molecule 160,815 base pairs (bp) in size and possess a quadripartite structure. The *A. altissima* chloroplast genome contains 113 unique genes of which 79 and 30 are protein coding and transfer RNA (tRNA) genes respectively and also 4 ribosomal RNA genes (rRNA) with overall GC content of 37.6%. Microsatellite marker detection identified A/T mononucleotides as majority SSRs in all the seven analyzed genomes. Repeat analyses of seven Sapindales revealed a total of 49 repeats in *A. altissima*, *Rhus chinensis*, *Dodonaea viscosa*, *Leitneria floridana*, while *Azadirachta indica*, *Boswellia sacra*, and *Citrus aurantiifolia* had a total of 48 repeats. The phylogenetic analysis using protein coding genes revealed that *A. altissima* is a sister to *Leitneria floridana* and also suggested that Simaroubaceae is a sister to Rutaceae family. The genome information reported here could be further applied for evolution and invasion, population genetics, and molecular studies in this plant species and family.

Keywords: *Ailanthus altissima*; chloroplast genome; microsatellites; Simaroubaceae; Sapindales

1. Introduction

Ailanthus altissima (Mill.) Swingle, a deciduous tree in the Simaroubaceae family, is widely distributed throughout temperate regions in China. It grows rapidly reaching heights of 15 m (49ft) in 25 years and can tolerate various levels of extreme environments (e.g., low temperatures, sterile soils, arid land). Besides, it reproduces through sexual (seeds disperse by wind) or asexual (sprouts) methods [1]. Two hundred years ago it was brought to Europe and North America. *A. altissima* being an early colonizer can survive high levels of natural or human disturbance [2]. Therefore, in recent years, it is commonly known as an exotic invasive tree developed into an invasive species expanding on all continents except Antarctica [1]. While previous studies in *A. altissima* have mainly focused on discovering the biological features of this plant to prevent its expansion, there is limited information to understand the impact of genetic diversity and evolution. Thus, genomic information of *A. altissima* is essential for further molecular studies, identification, and evolutionary studies.

Many studies have analyzed the genetic diversity of *A. altissima* using various markers, for example, chloroplast DNA [2,3], microsatellite primer [4,5]. These studies provided a detailed series of information about genetic structure and genetic diversity of *A. altissima* in native and invasive area. However, to understand the genetic diversity and population structure within *A. altissima* natural populations, more genetic resources are required.

It is well known that chloroplasts (cp) are key organelles in plants, with crucial functions in the photosynthesis and biosynthesis [6]. Current research shows that chloroplast genomes in angiosperms have highly conserved structure, gene content, organization, compared with either nuclear or mitochondrial genomes [7,8]. In general, cp genomes in angiosperms have circular structure consisting of two inverted repeat regions (IRa and IRb) that divides a large–single-copy (LSC) and a small-single-copy (SSC) regions [9]. Nevertheless, mutations, duplications, arrangements and gene loss have been observed, including the loss of the inverted repeat region in leguminous plants [7,10–12]. Some studies have applied plant plastomes to study population genetic analyses and basal phylogenetic relationships at different taxonomic levels [13], also to investigate the functional and structural evolution in plants [14–16]. At present, more cp genomes have been sequenced as a result of next-generation sequencing technologies advancement resulting in low sequencing costs.

More than 800 sequenced plastomes from various land plants have boosted our understanding of intracellular gene transfer, conservation, diversity, and genetic basis [17]. Although cp genomes have been sequenced in many trees such as *Castanea mollissima* [18]), *Liriodendron tulipifera* [19], *Eucalyptus globules* [20], and *Larix deciduas* [21], the plastome of *Leitneria floridana* (GenBank NC_030482) a member of Simaroubaceae has been sequenced but no analysis has been published at present despite the family containing many high economic value trees. Regardless of its potential use in crop or tree species improvement, studies on invasive species such as *A. altissima* which is also an important economic tree in the North China are too few. Chloroplast genome sequencing in invasive species could bring insights into evolutionary aspects in stress-tolerance related trait and genetic variation.

Simple sequence repeat (SSR) also called microsatellite markers are known to be more informative and versatile DNA-based markers used in plant genetic research [22]. These DNA markers are reliable molecular tools that can be used to examine plants genetic variation. SSR loci are evenly distributed and very abundant in angiosperms plastomes [23,24]. Chloroplast microsatellites are typically mononucleotide tandem repeats, and SSR in the fully sequenced genome could be used in plant species identification and diversity analysis. CpSSR in the fully sequenced plants plastomes such as; orchid genus *Chiloglottis* [25], *Cryptomeria japonica* [26], *Podocarpus lambertii* [27], *Actinidia chinensis* [28], have proven to be useful genetic tools in determining gene flow and population genetics of cp genomes. However, the lack of published plastome of *A. altissima* has limited the development of suitable SSR markers.

Here, we sequenced the complete chloroplast genome of *A. altissima*, and characterized its organization and structure. Furthermore, phylogenetic relationship using protein coding genes from selected species, consisting of 31 species from five families was uncovered for the Simaroubaceae family within the order Sapindales. Lastly, this resource will be used to develop SSR markers for analyzing genetic diversity and structure of several wild populations of *A. altissima*.

2. Results and Discussion

2.1. Ailanthus altissima Genome Size and Features

The *A. altissima* chloroplast genome has a quadripartite organizational structure with overall size of 160,815 base pairs (bp) including two copies of Inverted repeats (IRa and IRb) (27,051 bp each) separating the Large Single Copy (LSC) (88,382 bp) and Small Single Copy Region (SSC) (18,332 bp) (Figure 1). Notably, the genome content; gene order, orientation and organization of *A. altissima* is similar to the reference genome and other sequenced Sapindales plastomes [29,30] with genome size of about 160 kb. The overall guanine-cytosine (GC) content of the whole genome is 37.6%, while the average adenine-thymine A + T content is 62.36%. The relatively higher IR GC content and A + T bias in this chloroplast have been previously reported in genomes of relative species in order Sapindales [31]. The GC content of the LSC, SSC and IR regions are 35.7, 32.2 and 42.6% respectively. Moreover, the protein coding sequences had 38.3% GC content.

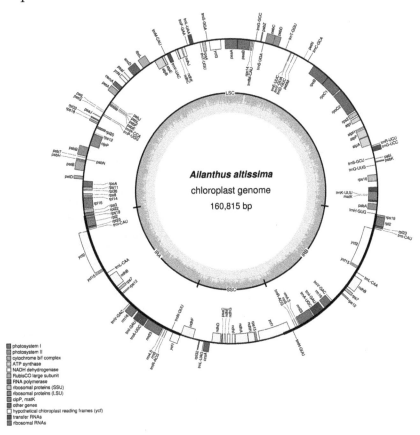

Figure 1. Circular gene map of *A. altissima* complete chloroplast genome. Genes drawn on the outside of the circle are transcribed clockwise, whereas those inside are transcribed clockwise. The light gray in the inner circle corresponds to AT content, while the darker gray corresponds to the GC content. Large single copy (LSC), Inverted repeats (IRa and IRb), and Small single copy (SSC) are indicated.

The tree of heaven (*A. altissima*) chloroplast genome harbored a total of 113 different genes, comprising 79 protein coding genes (PCGS), 30 transfer RNA (tRNA) genes, and four ribosomal RNA (rRNA) genes (Table 1). All the 77 PCGS started with the common initiation codon (ATG), but *rps19* and *ndhD* genes started with alternative codons GTG and ACG respectively, this unusual initiator codons have been observed to be common in other angiosperm chloroplast genomes [32–34]. Of the 79 protein coding sequences, 60 are located in the LSC, 11 in the SSC and eight genes were duplicated the IR region, while 22 tRNA genes were found in LSC, seven replicated in the IR region and one located in the SSC region.

Similar to some closely related plant species in order Sapindales, the chloroplast genome of *A. altissima* has maintained intron content [35]. Among the 113 unique genes, the *rps16*, *rpoC1*, *petB*, *rpl2*, *ycf15*, *ndhB*, *ndhA*, *atpF*, six tRNA genes (*trnG-GCC**, *trnA-UGC**, *trnL-UAA**, *trnI-GAU**, *trnK-UUU**, *trnV-UAC**) possess one intron, and *ycf3* and *clpP* genes harbored two introns. The *rps12* trans-splicing gene has two 3′ end exons repeated in the IRs and the 5′ end exon located in the LSC region, which is similar to that in *C. platymamma* [30], *C. aurantiifolia* [29], *Dipteronia* species [35]. The *ycf1* gene crossed the IR/SSC junction forming a pseudogene *ycf1* on the corresponding IR region. The *rps19* gene in *A. altissima* was completely duplicated in the inverted repeat (IR) region, which most other chloroplast genomes have presented [29,30].

Table 1. List of genes found in *Ailanthus altissima* Chloroplast genome.

Functional Category	Group of Genes	Gene Name	Number
Self-replication	rRNA genes	*rrn16*(×2), *rrn23*(×2), *rrn4.5*(×2), *rrn5*(×2),	4
	tRNA genes	*trnA-UGC**(×2), *trnC-GCA*, *trnD-GUC*, *trnE-UUC*, *trnF-GAA*, *trnG-UCC*, *trnH-GUG*, *trnI-CAU*(×2), *trnI-GAU**(×2), *trnK-UUU**, *trnL-CAA*(×2), *trnL-UAA**, *trnL-UAG*, *trnG-GCC**, *trnM-CAU*, *trnN-GUU*(×2), *trnP-GGG*, *trnP-UGG*, *trnQ-UUG*, *trnR-ACG*(×2), *trnR-UCU*, *trnS-GCU*, *trnS-GGA*, *trnS-UGA*, *trnT-GGU*, *trnT-UGU*, *trnV-GAC*(×2), *trnV-UAC**, *trnW-CCA*, *trnY-GUA*	30
	Ribosomal small subunit	*rps2*, *rps3*, *rps4*, *rps7*(×2), *rps8*, *rps11*, *rps12*, *rps14*, *rps15*, *rps16**, *rps18*, *rps19*	12
	Ribosomal large subunit	*rpl2**(×2), *rpl14*, *rpl16*, *rpl20*, *rpl22*, *rpl23*(×2), *rpl32*, *rpl33*, *rpl36*	9
	DNA-dependent RNA polymerase	*rpoA*, *rpoB*, *rpoC1**, *rpoC2*	4
Photosynthesis	Large subunit of rubisco	*rbcL*	1
	Photosystem I	*psaA*, *psaB*, *psaC*, *psaI*, *psaJ*, *ycf3***	6
	Photosystem II	*psbA*, *psbB*, *psbC*, *psbD*, *psbE*, *psbF*, *psbH*, *psbI*, *psbJ*, *psbK*, *psbL*, *psbM*, *psbN*, *psbT*, *psbZ*	15
	NADH dehydrogenase	*ndhA**, *ndhB**(×2), *ndhC*, *ndhD*, *ndhE*, *ndhF*, *ndhG*, *ndhH*, *ndhI*, *ndhJ*, *ndhK*	11
	Cytochrome b/f complex	*petA*, *petB**, *petD*, *petG*, *petL*, *petN*	6
	ATP synthase	*atpA*, *atpB*, *atpE*, *atpF**, *atpH*, *atpI*	6
Other	Maturase	*matK*	1
	Subunit of acetyl-CoA carboxylase	*accD*	1
	Envelope membrane protein	*cemA*	1
	Protease	*clpP***	1
	c-type cytochrome synthesis	*ccsA*	1
Functions unknown	Conserved open reading frames (*ycf*)	*ycf1*, *ycf2*(×2), *ycf4*, *ycf15*(×2)	4
Total			113

Note: * Gene with one intron, ** Genes with two introns. (×2) Genes with two copies.

2.2. IR Expansion and Contraction and Genome Rearrangement

The angiosperms chloroplast genomes are highly conserved, but slightly vary as a result of either expansion or contraction of the single-copy (SC) and IR boundary regions [36]. The expansion and contraction of the IR causes size variations and rearrangements in the SSC/IRa/IRb/LSC junctions [37]. Therefore, in this study, exact IR boundary positions and their adjacent genes of seven representative species from different families in order Sapindales were compared (Figure 2). The functional *ycf1* gene crossed the IRa/SSC boundary creating *ycf1* pseudogene fragment at the IRb region in all the genomes. Besides, *ycf1* pseudogene overlapped with the *ndhF* gene in the SSC and IRa junctions in four genomes with a stretch of 9 to 85 bp, but *ndhF* gene is located in SSC region in *L. floridana*, *R. chinensis* and *A. altissima*.

The *rpl22* gene crossed the LSC/IRb junction in all the chloroplast genomes except in *R. chinensis*. Furthermore, this gene was partially duplicated forming a pseudogene fragment at the corresponding IRA/LSC junction in *L. floridana* and *B. sacra*, but completely duplicated in *D. viscosa*. In all the seven chloroplast genomes, the *trnH-GUG* gene was located in the LSC regions, however this gene overlapped with *rpl22* gene in *D. viscosa*. The results reported here are congruent with the recent studies which showed that the *trnH-GUG* gene was situated in the LSC region in some species from order Sapindales, while the SSC/IRa border extends into the protein coding gene *ycf1* with subsequent formation of a *ycf1* pseudogene [29,30]. Despite the seven chloroplast genomes of Sapindales having well-conserved genomic structure in terms of gene order and number, length variation of the whole chloroplast genome sequences and LSC, SSC and IR regions was detected among these genomes. This sequence variation might have been as a result of boundary expansion and contraction between the single copy and IR boundary regions among plant lineages as suggested by Wang and Messing 2011 [38].

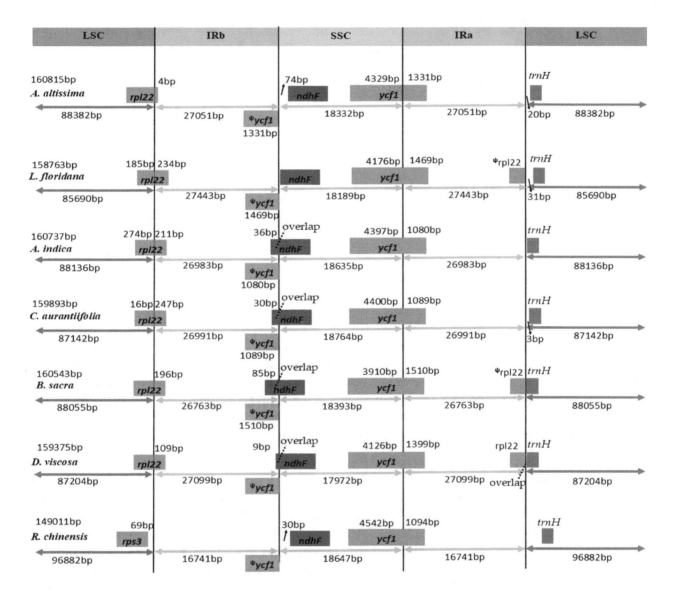

Figure 2. Comparison of IR, LSC and SSC junction positions among seven Chloroplast genomes. The features drawn are not to scale. The symbol $^{\varphi}$ means pseudogene created by IRb/SSC border extension into *ycf1* genes. Colored boxes for genes represent the gene position.

The mauve alignment for seven species revealed that all the genomes formed collinear blocks (LCBs). In particular, all the seven species; *A. altissima, Leitneria floridana, Azadirachta indica, Citrus aurantiifolia, Boswellia Sacra, Spondias bahiensis* and *Dodonaea viscosa* reveal a syntenic structure, however block two was inverted (from *rpl20* to *rbcL* genes) compared to the reference genome (*Aquilaria sinensis*). The collinear blocks of the genes including ribosomal RNA, tRNA, and protein coding genes revealed that all the seven genomes were relatively conserved with no gene rearrangement (Figure 3). Some other studies have revealed homology in genome organization and no gene rearrangements, thus our findings support their conclusions [31,39,40].

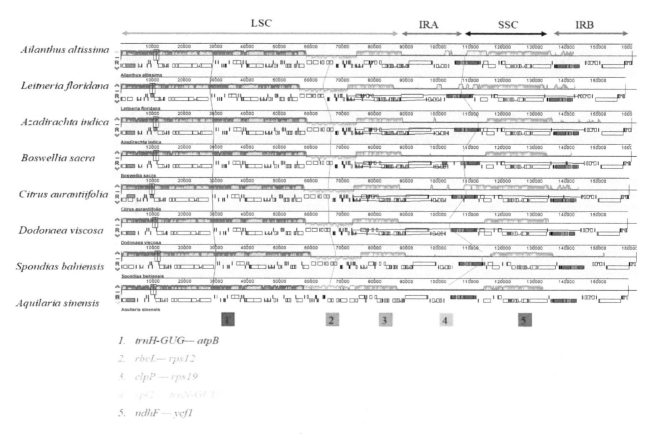

1. *trnH-GUG— atpB*
2. *rbcL— rps12*
3. *clpP — rps19*
4. *rpl2   trnN-GUU*
5. *ndhF — ycf1*

Figure 3. Gene arrangement map of seven chloroplast genomes representing families from Sapindales, and one reference species (*Aquilaria sinensis*) aligned using Mauve software Local collinear blocks within each alignment are represented in as blocks of similar color connected with lines. Annotations of rRNA, protein coding and tRNA genes are shown in red, white and green boxes respectively.

2.3. Codon Usage and Putative RNA Editing Sites in Chloroplast Genes of A. altissima

In this study, we analyzed codon usage frequency and the relative synonymous codon usage (RSCU) in the *A. altissima* plastome. All the protein coding genes presented a total of 68,952 bp and 22,964 codons in *A. altissima* chloroplast genome. Of 22,964 codons, leucine (Leu) being the most abundant amino acid had a frequency of 10.56%, then isoleucine (Ile) with 8.54%, while cysteine (Cys) was rare with a proportion of 1.12% (Tables S1 and S2, Figure 4). Our study species genome is like other previously reported genomes which showed that leucine and isoleucine are more common [41–45]. Furthermore, comparable to other angiosperm chloroplast genomes, our results followed the trend of codon preference towards A/T ending which was observed in plastomes of two *Aristolochia* species [46], *Scutellaria baicalensis* [47], *Decaisnea insignis* [34], *Papaver rhoeas* and *Papaver orientale* [48] *Cinnamomum camphora* [49], and *Forsythia suspensa* [41]. All the twenty-eight A/U—ending codons had RSCU values of more than one (RSCU > 1), whereas the C/G—ending codons had RSCU values of less than one. Two amino acids, Methionine (Met) and tryptophan (Trp) showed no codon bias. The results for number of codons (Nc) of each protein coding gene ranged from 38.94 (*rps14* gene) to 58.37 (*clpP* gene).

The potential RNA editing sites in tree of heaven chloroplast genome was done using PREP program which revealed that most conversions at the codon positions change from serine (S) to leucine (L) (Table 2). In addition, 15 (27.78%), 39 (72.22%), and 0 editing locations were used in the first, second and third codons respectively. One RNA editing site converted the amino acid from apolar to polar (proline (P) to serine (S)). Overall, the PREP program identified a total of 54 editing sites in 21 protein coding genes, with *ndhB* and *ndhD* genes predicted to have the highest number of editing sites (9). Followed by *ndhA*, *matK*, *rpoC2*, and *rpoB* with four editing sites, whereas *ndhF* had three sites. Interestingly, fifty three of fifty four RNA editing conversions in the *A. altissima* chloroplast genome resulted into hydrophobic products comprising; isoleucine, leucine, tryptophan, tyrosine valine, methionine, and phenylalanine. In general our results are congruent with the preceding reports which also found that most RNA editing sites led to amino acid change from polar to apolar, resulting in increase in protein hydrophobicity [41,46,50].

Figure 4. Amino acid frequencies in *A. altissima* chloroplast genome protein coding sequences.

Table 2. Predicted RNA editing site in the *A. altissima* chloroplast genome.

Gene	Nucleotide Position	Amino Acid Position	Codon Conversion	Amino Acid Conversion	Score
accD	818	273	TCG ≥ TTG	S ≥ L	0.80
atpF	92	31	CCA ≥ CTA	P ≥ L	0.86
	353	118	TCA ≥ TTA	S ≥ L	1.00
atpB	403	135	CCA ≥ TCA	P ≥ S	0.86
rps14	80	27	TCA ≥ TTA	S ≥ L	1.00
	149	50	TCA ≥ TTA	S ≥ L	1.00
ccsA	145	49	CTT ≥ TTT	L ≥ F	1.00
clpP	556	186	CAT ≥ TAT	H ≥ Y	1.00
MatK	319	107	CTT ≥ TTT	L ≥ F	0.86
	457	153	CAC ≥ TAC	H ≥ Y	1.00
	643	215	CAT ≥ TAT	H ≥ Y	1.00
	1246	416	CAC ≥ TAC	H ≥ Y	1.00
ndhA	107	36	CCT ≥ CTT	P ≥ L	1.00
	341	114	TCA ≥ TTA	S ≥ L	1.00
	566	189	TCA ≥ TTA	S ≥ L	1.00
	1073	358	TCC ≥ TTC	S ≥ F	1.00
ndhB	149	50	TCA ≥ TTA	S ≥ L	1.00
	467	156	CCA ≥ CTA	P ≥ L	1.00
	586	196	CAT ≥ TAT	H ≥ Y	1.00
	611	204	TCA ≥ TTA	S ≥ L	0.80
	746	249	TCT ≥ TTT	S ≥ F	1.00
	830	277	TCA ≥ TTA	S ≥ L	1.00
	836	279	TCA ≥ TTA	S ≥ L	1.00
	1255	419	CAT ≥ TAT	H ≥ Y	1.00
	1481	494	CCA ≥ CTA	P ≥ L	1.00

Table 2. *Cont.*

Gene	Nucleotide Position	Amino Acid Position	Codon Conversion	Amino Acid Conversion	Score
ndhD	2	1	ACG ≥ ATG	T ≥ M	1.00
	313	105	CGG ≥ TGG	R ≥ W	0.80
	383	128	TCA ≥ TTA	S ≥ L	1.00
	674	225	TCA ≥ TTA	S ≥ L	1.00
	878	293	TCA ≥ TTA	S ≥ L	1.00
	887	296	CCT ≥ CTT	P ≥ L	1.00
	1076	359	GCT ≥ GTT	A ≥ V	1.00
	1298	433	TCA ≥ TTA	S ≥ L	0.80
	1310	437	TCA ≥ TTA	S ≥ L	0.80
ndhF	290	97	TCA ≥ TTA	S ≥ L	1.00
	586	196	CTT ≥ TTT	L ≥ F	0.80
	1919	640	GCT ≥ GTT	A ≥ V	0.80
ndhG	166	56	CAT ≥ TAT	H ≥ Y	0.80
	320	107	ACA ≥ ATA	T ≥ I	0.80
petL	119	40	CCT ≥ CTT	P ≥ L	0.86
psbF	77	26	TCT ≥ TTT	S ≥ F	1.00
rpl20	308	103	TCA ≥ TTA	S ≥ L	0.86
rpoA	830	277	TCA ≥ TTA	S ≥ L	1.00
rpoB	338	113	TCT ≥ TTT	S ≥ F	1.00
	551	184	TCA ≥ TTA	S ≥ L	1.00
	566	189	TCG ≥ TTG	S ≥ L	1.00
	2426	809	TCA ≥ TTA	S ≥ L	0.86
rpoC1	41	14	TCA ≥ TTA	S ≥ L	1.00
rpoC2	1681	561	CAT ≥ TAT	H ≥ Y	0.86
	2030	677	ACT ≥ ATT	T ≥ I	1.00
	2314	772	CGG ≥ TGG	R ≥ W	1.00
	4183	1395	CTT ≥ TTT	L ≥ F	0.80
rps2	248	83	TCA ≥ TTA	S ≥ L	1.00
rps16	209	70	TCA ≥ TTA	S ≥ L	0.83

The cytidines marked are putatively edited to uredines.

Comparisons of RNA editing sites with other six species from other families revealed that *R. chinensis* and *D. viscosa* have high RNA editing sites (61 each distributed in 20 and 17 genes respectively) followed by *B. sacra* (57 in 20 genes), *A. altissima* (54 in 21 genes), *A. indica* (53 in 21 genes), *C. aurantiifolia* (52 in 21 genes), and *L. floridana* 48 in 20 genes. As shown in Table S3, these results are consistent with several studies in that all the RNA editing sites predicted among the seven species are cytidine (C) to uridine (U) conversions [41,50–52]. Majority of RNA editing occurred at the second positions of the codons with a frequency from 62.30% (38/61) in *D. viscosa* to 81.28% (39/48) in *L. floridana*, which concurs with previous plastid genome studies in other land plants [53,54]. All the species shared 19 editing sites distributed in twelve genes (Table 3), whereas the two species from Simaroubaceae family (*L. floridana* and *A. altissima*) shared 33 editing sites in 16 genes this implies that the RNA editing sites in these two species are highly conserved (Table S4). Like previous studies [41,51,55], the *ndhB* gene in most of species analyzed here have the highest number of editing sites. Notably, a RNA editing event was detected at the initiator codon (ACG) resulting in ATG translational start codon in the *ndhD* gene.

Table 3. List of RNA editing sites shared by the seven plastomes predicted by PREP program.

Gene	A.A Position	Citrus aurantiifolia Codon (A.A) Conversion	Rhus chinensis	Dodonaea viscosa	Boswellia Sacra	Leitneria floridana	Azadirachta indica	Ailanthus altissima
atpF	31	CCA (P) ≥ CTA (L)	CCA (P) ≥ CTA (L)	CCA (P) ≥ CTA (L)	CCA (P) ≥ CTA (L)	CCA (P) ≥ CTA (L)	CCA (P) ≥ CTA (L)	CCA (P) ≥ CTA (L)
clpP	187	CAT (H) ≥ TAT (Y)	CAT (H) ≥ TAT (Y)	CAT (H) ≥ TAT (Y)	CAT (H) ≥ TAT (Y)	CAT (H) ≥ TAT (Y)	CAT (H) ≥ TAT (Y)	CAT (H) ≥ TAT (Y)
MatK			CAT (H) ≥ TAT (Y)	CAT (H) ≥ TAT (Y)	CAT (H) ≥ TAT (Y)	CAT (H) ≥ TAT (Y)	CAT (H) ≥ TAT (Y)	CAT (H) ≥ TAT (Y)
ndhA	358	TCA (S) ≥ TTA (L)	TCA (S) ≥ TTA (L)	TCA (S) ≥ TTA (L)	TCA (S) ≥ TTA (L)	TCA (S) ≥ TTA (L)	TCA (S) ≥ TTA (L)	TCA (S) ≥ TTA (L)
ndhB	50	TCA (S) ≥ TTA (L)	TCA (S) ≥ TTA (L)	TCA (S) ≥ TTA (L)	TCA (S) ≥ TTA (L)	TCA (S) ≥ TTA (L)	TCA (S) ≥ TTA (L)	TCA (S) ≥ TTA (L)
ndhB	156	CCA (P) ≥ CTA (L)	CCA (P) ≥ CTA (L)	CCA (P) ≥ CTA (L)	CCA (P) ≥ CTA (L)	CCA (P) ≥ CTA (L)	CCA (P) ≥ CTA (L)	CCA (P) ≥ CTA (L)
ndhB	196	CAT (H) ≥ TAT (Y)	CAT (H) ≥ TAT (Y)	CAT (H) ≥ TAT (Y)	CAT (H) ≥ TAT (Y)	CAT (H) ≥ TAT (Y)	CAT (H) ≥ TAT (Y)	CAT (H) ≥ TAT (Y)
ndhB	249	TCT (S) ≥ TTT (F)	TCT (S) ≥ TTT (F)	TCT (S) ≥ TTT (F)	TCT (S) ≥ TTT (F)	TCT (S) ≥ TTT (F)	TCT (S) ≥ TTT (F)	TCT (S) ≥ TTT (F)
ndhB	419	CAT (H) ≥ TAT (Y)	CAT (H) ≥ TAT (Y)	CAT (H) ≥ TAT (Y)	CAT (H) ≥ TAT (Y)	CAT (H) ≥ TAT (Y)	CAT (H) ≥ TAT (Y)	CAT (H) ≥ TAT (Y)
ndhD	1	ACG (T) ≥ ATG (M)	ACG (T) ≥ ATG (M)	ACG (T) ≥ ATG (M)	ACG (T) ≥ ATG (M)	ACG (T) ≥ ATG (M)	ACG (T) ≥ ATG (M)	ACG (T) ≥ ATG (M)
ndhD	128	TCA (S) ≥ TTA (L)	TCA (S) ≥ TTA (L)	TCA (S) ≥ TTA (L)	TCA (S) ≥ TTA (L)	TCA (S) ≥ TTA (L)	TCA (S) ≥ TTA (L)	TCA (S) ≥ TTA (L)
ndhG	107	ACA (T) ≥ ATA (I)	ACA (T) ≥ ATA (I)	ACA (T) ≥ ATA (I)	ACA (T) ≥ ATA (I)	ACA (T) ≥ ATA (I)	ACA (T) ≥ ATA (I)	ACA (T) ≥ ATA (I)
rpoA	278	TCA (S) ≥ TTA (L)	TCA (S) ≥ TTA (L)	TCA (S) ≥ TTA (L)	TCA (S) ≥ TTA (L)	TCA (S) ≥ TTA (L)	TCA (S) ≥ TTA (L)	TCA (S) ≥ TTA (L)
rpoB	113	TCT (S) ≥ TTT (F)	TCT (S) ≥ TTT (F)	TCT (S) ≥ TTT (F)	TCT (S) ≥ TTT (F)	TCT (S) ≥ TTT (F)	TCT (S) ≥ TTT (F)	TCT (S) ≥ TTT (F)
rpoB	184	TCA (S) ≥ TTA (L)	TCA (S) ≥ TTA (L)	TCA (S) ≥ TTA (L)	TCA (S) ≥ TTA (L)	TCA (S) ≥ TTA (L)	TCA (S) ≥ TTA (L)	TCA (S) ≥ TTA (L)
rpoB	809	TCG (S) ≥ TTG (L)	TCG (S) ≥ TTG (L)	TCG (S) ≥ TTG (L)	TCG (S) ≥ TTG (L)	TCG (S) ≥ TTG (L)	TCG (S) ≥ TTG (L)	TCG (S) ≥ TTG (L)
rpoC1	14	TCA (S) ≥ TTA (L)	TCA (S) ≥ TTA (L)	TCA (S) ≥ TTA (L)	TCA (S) ≥ TTA (L)	TCA (S) ≥ TTA (L)	TCA (S) ≥ TTA (L)	TCA (S) ≥ TTA (L)
rpoC2	563	CAT (H) ≥ TAT (Y)	CAT (H) ≥ TAT (Y)	CAT (H) ≥ TAT (Y)	CAT (H) ≥ TAT (Y)	CAT (H) ≥ TAT (Y)	CAT (H) ≥ TAT (Y)	CAT (H) ≥ TAT (Y)
rps14	27	TCA (S) ≥ TTA (L)	TCA (S) ≥ TTA (L)	TCA (S) ≥ TTA (L)	TCA (S) ≥ TTA (L)	TCA (S) ≥ TTA (L)	TCA (S) ≥ TTA (L)	TCA (S) ≥ TTA (L)

2.4. Repeat Sequence Analysis

Microsatellites are usually 1–6 bp tandem repeat DNA sequences and are distributed throughout the genome. The presence of microsatellites was detected in the chloroplast genome of *A. altissima* (Figure 5). A total of 219 simple sequence repeats (SSRs) loci were detected, of which mononucleotide repeats occurred with high frequency constituting 190 (86.76%) of all the SSRs. Majority of mononucleotides composed of poly A (polyadenine) (39.27%) and poly T (polythymine) (47.49%) repeats, whereas poly G (polyguanine)or polyC (polycytosine) repeats were rather rare (2.74%). Among the dinucleotide repeat motifs AT/AT were more abundant, while AG/CT were less frequent. One trinucleotide motif (AAT/ATT), five tetra-(AAAG/CTTT, AAAT/ATTT, AACT/AGTT, AATC/ATTG and AGAT/ATCT) and two pentanucleotide repeats (AAAAG/CTTTT and AATAG/ATTCT) were identified. Hexanucleotide repeats were not detected in the *A. altissima* chloroplast genome.

As shown in Figure 5, the SSR analysis for seven species showed that *Leitneria floridana* had the highest number of SSRs (256) while *Dodonaea viscosa* and *Rhus chinensis* had the lowest (186). In all the seven species, mononucleotide repeats were more abundant with A/T repeats being the most common repeats. This result is consistent with earlier studies in [31,35,46] which revealed that many angiosperm chloroplast genomes are rich in poly A and poly T. Moreover, in the seven analyzed species, hexanucleotide repeats were not detected, whereas *Azadirachta indica*, *Dodonaea viscosa* and *Leitneria floridana* had no pentanucleotide repeats.

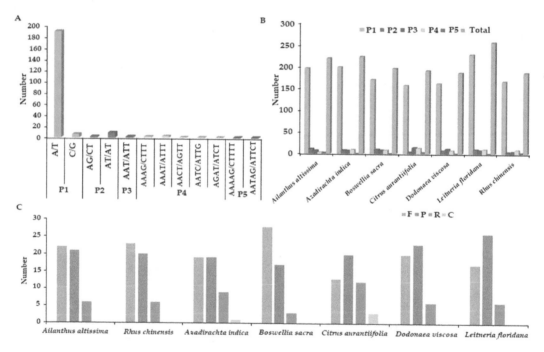

Figure 5. Simple sequence repeat (SSRs) type, distribution and presence in *A. altissima* and other representative species from Sapindales. (**A**) Number of detected SSR motifs in different repeat types in *A. altissima* Chloroplast genome. (**B**) Number of identified repeat sequences in seven chloroplast genomes. (**C**) Number of different SSR types in seven representative species. F, indicate (forward), P (palindromic), R (reverse), and C (complement), while P1, P2, P3, P4, P5 indicates Mono-, di-, tri-, tetra-, and penta-nucleotides respectively. F: forward; P: palindromic, R: reverse; C: complement.

The REPuter program revealed that *A. altissima* chloroplast genome contains 21 palindromic (p), 22 forward (f) and six reverse (r) repeats, however the complement repeats were not detected (Table 4). We notice that all the identified tandem repeats in *A. altissima* were more than 20 bp, while thirteen had length of more than 30 bp. Repeat analyses of seven Sapindales revealed a total of 48 or 49 repeats for each species, with all species containing forward, palindromic and reverse repeats

(Figure 5). Compliment repeats were not identified in other species except for *Azadirachta indica* and *Citrus aurantiifolia* which had one and three repeats respectively. *Citrus aurantiifolia* had the highest number of reverse repeats but also lowest number of forward repeats. Most of the repeat lengths were less than 50 bp, however *Boswellia sacra* chloroplast had seven forward repeats with length of between 65 to 251 bp. Besides, we found out that almost all the repeat sequences were located in either IR or LSC region.

Table 4. Distribution and localization of repetitive sequences F, forward: P, palindromic, R; reverse in *A. altissima* chloroplast genome.

Number	Size	Position 1	Type	Position 2	Location 1 (2)	Region
1	48	95,957	F	95,975	ycf2	IRa
2	48	153,174	F	153,192	ycf2	IRb
3	37	103,326	F	125,821	rps12/trnV-GAC(ndhA*)	IRa/SSC
4	30	95,957	F	95,993	ycf2	IRa
5	30	153,174	F	153,210	ycf2	IRb
6	29	50,944	F	50,972	trnL-UAA*	LSC
7	29	58,040	F	58,078	rbcL	LSC
8	28	115,434	F	115,460	ycf1	SSC
9	26	39,399	F	39,625	psbZ/trnG-UCC	LSC
10	25	71,153	F	71,178	trnP-GGG/psaJ	LSC
11	23	47,036	F	103,323	ycf3**(rps12/trnV-GAC)	LSC/IRa
12	23	112,456	F	112,488	rrn4.5/rrn5	IRa
13	23	136,686	F	136,718	rrn5/rrn4.5	IRb
14	22	11,749	F	11,771	trnR-UCU/atpA	LSC
15	21	248	F	270	trnH-GUG/psbA	LSC
16	21	9541	F	38,293	trnS-GCU (trnS-UGA)	LSC
17	21	41,956	F	44,180	psaB(psaA)	LSC
18	21	49,678	F	49,699	trnL-UAA*	LSC
19	20	1945	F	1965	trnK-UUU	LSC
20	20	15,166	F	92,503	atpH/atpI(ycf2)	LSC
21	20	47,039	F	125,821	ycf3**(rps15)	LSC/IRa
22	20	88,907	F	160,270	rpl2	IRa/IRb
25	48	31,790	P	31,790	petN/psbM	LSC
26	48	95,957	P	153,174	ycf2	IRa/IRb
27	48	95,975	P	153,192	ycf2	IRa/IRb
28	37	125,821	P	145,834	ndhA*(trnV-GAC/rps12)	SSC/IRb
29	36	30,970	P	30,970	petN/psbM	LSC
30	30	72,117	P	72,117	rpl33/rps18	LSC
31	30	95,957	P	153,174	ycf2	IRa/IRb
32	30	95,993	P	153,210	ycf2	IRa/IRb
33	27	542	P	571	trnH-GUG/psbA	LSC
34	25	11,403	P	11,430	trnS-GCU/trnR-UCU	LSC
35	24	4867	P	4897	trnK-UUU/rps16	LSC
36	24	9535	P	48,164	trnS-GCU(psaA/ycf3)	LSC
37	23	47,036	P	145,851	ycf3**(trnV-GAC/rps12)	LSC/IRb
38	23	51,804	P	119,066	trnF-GAA/ndhJ(rpl32/trnL-UAG)	LSC/SSC
39	23	112,456	P	136,686	rrn4.5/rrn5	IRa/IRb
40	23	112,488	P	136,718	rrn4.5/rrn5	IRa/IRb
41	22	39,195	P	39,195	psbZ/trnG-UCC	LSC
42	20	15,166	P	156,674	atpH(ycf2)	LSC/IRb
43	20	38,361	P	48,100	trnS-UGA(trnS-GGA)	LSC
44	20	88,907	P	88,907	rpl2	IRa
45	20	107,097	P	107,130	rrn16/trnI-GAU	IRa
46	23	39,184	R	39,184	psbZ/trnG-UCC	LSC
47	21	9751	R	9751	trnS-GCU/trnR-UCU	LSC
48	21	51,281	R	51,281	trnL-UAA/trnF-GAA	LSC
49	21	85,055	R	85,055	rps8/rpl14	LSC
50	20	53,712	R	53,712	ndhC	LSC
51	20	9385	R	13,356	psbI(atpA/atpF)	LSC

F: forward; P: palindrome; R; reverse* intron or ** introns.

2.5. Phylogenetic Tree

The phylogenetic position of *A. altissima* within Sapindales was carried out using 75 protein coding sequences shared by thirty-one taxa from Sapindales (Table S5). Three remaining species were from Thymelaeaceae family (*Aquilaria sinensis*) and Malvaceae (*Theobroma cacao* and *Abelmoschus esculentus*) from order Malvales selected as outgroups (Figure 6). The maximum likelihood (ML) analysis produced a phylogenetic tree which fully supported *A. altissima* to be closely related with *Leitneria floridana* with 100% bootstrap value. The ML resolved 26 nodes with high branch support (over 98% bootstrap values), however six nodes were moderately supported perhaps as a result of less samples use (59 to 95%). Concerning relationships among families within Sapindales order, family Simaroubaceae early diverged and formed a sister clade/relationship with a 95% bootstrap support to Rutaceae family. Interestingly, the placement of families within Sapindales in our phylogenetic tree supports the one reported by previous studies [30,56,57] based on some chloroplast and nuclear markers. The families Anacardiaceae and Burseraceae formed a sister clade/ group, this clade further branched forming a sister clade with families Sapindaceae, Meliaceae, Simaroubaceae and Rutaceae analyzed in our study. Therefore, it is crucial to use more species for better understanding of Simaroubaceae phylogeny and evolution. This study provides a basis for future phylogenetic of Simaroubaceae species.

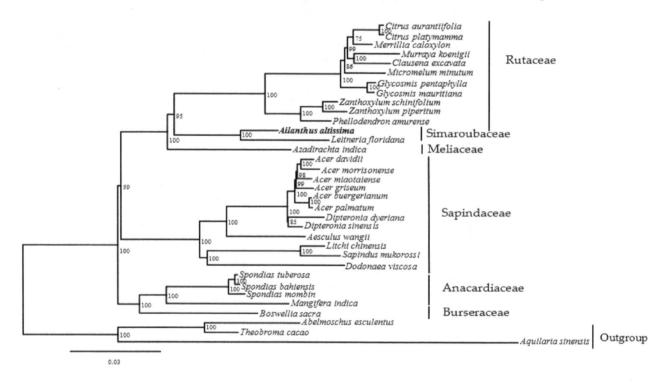

Figure 6. Phylogenetic tree of 31 Sapindales species with three outgroup Malvales species inferred from ML (Maximum likelihood) based on common protein coding genes. The position of *A. altissima* is shown in bold, while bootstrap support values are shown at each node.

3. Materials and Methods

3.1. Plant Materials and DNA Extraction

Fresh leaves of *Ailanthus altissima* were collected in Wuhan Botanical Garden, Chinese Academy of Sciences in China. Total genomic DNA isolation was carried out using MagicMag Genomic DNA Micro Kit (Sangon Biotech Co., Shanghai, China) based on the manufactures protocol. The quality and integrity of DNA were checked and inspected using spectrophotometry and agarose gel electrophoresis respectively. The voucher specimen (HIB-LZZ-CC003) has been deposited at the Wuhan Botanical Garden herbarium (HIB) Wuhan, China.

3.2. The Tree of Heaven Plastome Sequence Assembly and Annotation

Library preparation was constructed using the Illumina Hiseq 2500 platform at NOVOgene Company (Beijing, China) with an average insert size of approximately 350 bp. The high-quality data (5 Gb) were filtered from raw sequence data (5.2 Gb) using the PRINSEQ lite v0.20.4 (San Diego State University, San Diego, CA, USA) [58] (phredQ ≥ 20, Length ≥ 50), followed by de novo assembling using NOVOPlasty [59] with default sets (K-mer = 31). The seeds and reference plastome used was from the closely related species *Leitneria floridana* (NC_030482) with high coverage of chloroplast reads ~1500×. Lastly, one contig of *Ailanthus altissima* was generated and mapped with reference plastome using GENEIOUS 8.1 (Biomatters Ltd., Auckland, New Zealand) [60]. Finally, online web-based server local blast was used to verify the inverted repeat (IR) single copy (SC) junctions.

Preliminary gene annotation of assembled genome was done using the program DOGMA (Dual Organellar GenoMe Annotator, University of Texas at Austin, Austin, TX, USA) [61], and BLAST (http://blast.ncbi.nlm.nih.gov/). The positions of start and stop codons together with position of introns were confirmed by comparing with homologous genes of other chloroplast genomes available at the GenBank database. Moreover, tRNA genes were verified with tRNAscan-SE server (http://lowelab.ucsc.edu/tRNAscan-SE/) [62]. The chloroplast genome physical circular map was drawn using program OGDRAW (Organellar Genome DRAW) [63] Max planck Institute of Molecular Plant Physiology, Potsdam, Germany) accompanied by manual corrections. The chloroplast genome sequence of *A. altissima* was deposited in the GenBank database, accession (MG799542).

3.3. Genome Comparison and Gene Rearrangement

The border region between Inverted repeat (IR) and large single copy (LSC), also between inverted repeats and small single copy (SSC) junction were compared among seven representative species from Sapindales order. Additionally, alignments of seven chloroplast with one reference genome to determined gene rearrangements was carried out using Mauve v.4.0 [64].

3.4. Repeat Analysis in A. altissima Chloroplast Genome

Microsatellites were identified in the tree of heaven chloroplast genome and other selected representative genomes belonging to order Sapindales using an online software MIcroSAtellite (MISA) [65].The minimum number of repetitions were set to eight repeat units for mononucleotide SSR motifs, five repeat units for dinucleotide SSRs, four for trinucleotide SSRs and three repeat units for tetra-, penta-, and hexanucleotide motifs. The REPuter (https://bibiserv.cebitec.uni-bielefeld.de/reputer) program [66] (University of Bielefeld, Bielefeld, Germany) with default parameters was used to identify the location and sizes of forward, palindromic, complement and reverse repeat sequences in *A. altissima* chloroplast genome.

3.5. Codon Usage and RNA Editing Sites

CodonW1.4.2 (http://downloads.fyxm.net/CodonW-76666.html) was used to analyze codon usage. Subsequently, possible RNA editing sites in *A. altissima* protein coding genes were predicted using the program predictive RNA editor for plants (PREP) suite [67] with the cutoff value set to 0.8. PREP server uses 35 genes as reference for potential RNA editing sites prediction by comparing the predicted protein genes to homologous proteins from other plants.

3.6. Phylogenetic Analysis

Seventy five protein coding sequences present in 31 species from order Sapindales and three species from Thymelaeaceae (*Aquilaria sinensis*), Malvaceae (*Theobroma cacao* and *Abelmoschus esculentus*) as outgroups were used for the phylogenetic reconstruction. These species chloroplast genomes were downloaded from GenBank (Table S5). The protein coding sequences alignment was done using GENEIOUS v8.0.2 (Biomatters Ltd., Auckland, New Zealand) [60]. Maximum likelihood (ML) analysis was carried out using RAxMLversion 8.0.20 (Scientific Computing Group, Heidelberg Institute for Theoretical Studies, Institute of Theoretical Informatics, Karlsruhe Institute of Technology, Karlsruhe, Germany) [68] with 1000 replicates for bootstrap test. Lastly, the jModelTest v2.1.7 [69] was used to select the best substitution model (GTR + I + G).

4. Conclusions

In this study, we present the plastome of tree of heaven from family Simaroubaceae which contains about 22 genera and over 100 species. *A. altissima* chloroplast genome possess circular and quadripartite structure which is well conserved similar to other plants chloroplast genomes. Nonetheless, the plastome showed slight variations at the four boundary junctions due to expansion and contraction in SC and IR borders. About 219 SSR loci and 49 repeats sequences were identified in *A. altissima* genome, this provides genetic information for designing DNA molecular markers for analyzing gene pool dynamics and genetic diversity of *A. altissima* natural populations aiming dispersal mechanism of this invasive tree. The phylogenetic analysis performed using 75 protein coding genes of 34 species available at the GenBank database, comprising 3 outgroup species from Malvales and 31 species representing families from order Sapindales. The two species from family Simaroubaceae formed a cluster and were group together with other families to form a single clade (Sapindales). In addition, the RNA editing analysis in *A. altissima* genome identified a total of 54 possible editing sites in 21 chloroplast genes with C-to-U transitions being the most. The availability of this chloroplast genome provides a tool to advance the study of evolution and invasion in *A. altissima* in order to address present evolutionary, ecological and genetic questions regarding this species.

Acknowledgments: This work was supported by the National Natural Scientific Foundation of China (Grant No. 31500457) and CAS-TWAS President's Fellowship for International PhD Students.

Author Contributions: Yi-Ying Liao conceived and designed the experiment; Josphat K. Saina, Zhi-Zhong Li and Andrew W. Gichira assembled plastome sequence and revised the manuscript. Josphat K. Saina performed the experiments, analyzed data and wrote the manuscript. Yi-Ying Liao and Zhi-Zhong Li collected the plant material. All the authors read and approved submission of the final manuscript.

Abbreviations

SC　　Single copy
LSC　　Large single copy
SSC　　Small single copy
IR　　Inverted repeat

References

1. Kowarik, I.; Säumel, I. Biological flora of central Europe: *Ailanthus altissima* (Mill.) swingle. *Perspect. Plant Ecol. Evol. Syst.* **2007**, *8*, 207–237. [CrossRef]
2. Liao, Y.Y.; Guo, Y.H.; Chen, J.M.; Wang, Q.F. Phylogeography of the widespread plant *Ailanthus altissima* (Simaroubaceae) in China indicated by three chloroplast DNA regions. *J. Syst. Evol.* **2014**, *52*, 175–185. [CrossRef]
3. Kurokochi, H.; Saito, Y.; Ide, Y. Genetic structure of the introduced heaven tree (*Ailanthus altissima*) in Japan: Evidence for two distinct origins with limited admixture. *Botany* **2014**, *93*, 133–139. [CrossRef]
4. Dallas, J.F.; Leitch, M.J.; Hulme, P.E. Microsatellites for tree of heaven (*Ailanthus altissima*). *Mol. Ecol. Resour.* **2005**, *5*, 340–342. [CrossRef]
5. Aldrich, P.R.; Briguglio, J.S.; Kapadia, S.N.; Morker, M.U.; Rawal, A.; Kalra, P.; Huebner, C.D.; Greer, G.K. Genetic structure of the invasive tree *Ailanthus altissima* in eastern United States cities. *J. Bot.* **2010**, *2010*, 795735.
6. Neuhaus, H.; Emes, M. Nonphotosynthetic metabolism in plastids. *Annu. Rev. Plant Biol.* **2000**, *51*, 111–140. [CrossRef] [PubMed]

7.   Henry, R.J. *Plant Diversity and Evolution: Genotypic and Phenotypic Variation in Higher Plants*; CABI Publishing: Cambridge, MA, USA, 2005.

8.   Raubeson, L.A.; Jansen, R.K. Chloroplast genomes of plants. In *Plant Diversity and Evolution: Genotypic and Phenotypic Variation in Higher Plants*; CABI Publishing: Cambridge, MA, USA, 2005; pp. 45–68.

9.   Wicke, S.; Schneeweiss, G.M.; Müller, K.F.; Quandt, D. The evolution of the plastid chromosome in land plants: Gene content, gene order, gene function. *Plant Mol. Biol.* **2011**, *76*, 273–297. [CrossRef] [PubMed]

10. Yue, F.; Cui, L.; Moret, B.M.; Tang, J. Gene rearrangement analysis and ancestral order inference from chloroplast genomes with inverted repeat. *BMC Genom.* **2008**, *9*, S25. [CrossRef] [PubMed]

11. Jansen, R.K.; Wojciechowski, M.F.; Sanniyasi, E.; Lee, S.-B.; Daniell, H. Complete plastid genome sequence of the chickpea (*Cicer arietinum*) and the phylogenetic distribution of *rps12* and *clpP* intron losses among legumes (Leguminosae). *Mol. Phylogen. Evol.* **2008**, *48*, 1204–1217. [CrossRef] [PubMed]

12. Lee, H.-L.; Jansen, R.K.; Chumley, T.W.; Kim, K.-J. Gene relocations within chloroplast genomes of *Jasminum* and *Menodora* (Oleaceae) are due to multiple, overlapping inversions. *Mol. Biol. Evol.* **2007**, *24*, 1161–1180. [CrossRef] [PubMed]

13. Parks, M.; Cronn, R.; Liston, A. Increasing phylogenetic resolution at low taxonomic levels using massively parallel sequencing of chloroplast genomes. *BMC Biol.* **2009**, *7*, 84. [CrossRef] [PubMed]

14. Yi, X.; Gao, L.; Wang, B.; Su, Y.-J.; Wang, T. The complete chloroplast genome sequence of *Cephalotaxus oliveri* (Cephalotaxaceae): Evolutionary comparison of *Cephalotaxus* chloroplast DNAs and insights into the loss of inverted repeat copies in gymnosperms. *Genome Biol. Evol.* **2013**, *5*, 688–698. [CrossRef] [PubMed]

15. Moore, M.J.; Bell, C.D.; Soltis, P.S.; Soltis, D.E. Using plastid genome-scale data to resolve enigmatic relationships among basal angiosperms. *Proc. Natl. Acad. Sci. USA* **2007**, *104*, 19363–19368. [CrossRef] [PubMed]

16. Wu, C.-S.; Wang, Y.-N.; Hsu, C.-Y.; Lin, C.-P.; Chaw, S.-M. Loss of different inverted repeat copies from the chloroplast genomes of Pinaceae and Cupressophytes and influence of heterotachy on the evaluation of gymnosperm phylogeny. *Genome Biol. Evol.* **2011**, *3*, 1284–1295. [CrossRef] [PubMed]

17. Daniell, H.; Lin, C.-S.; Yu, M.; Chang, W.-J. Chloroplast genomes: Diversity, evolution, and applications in genetic engineering. *Genome Biol.* **2016**, *17*, 134. [CrossRef] [PubMed]

18. Jansen, R.K.; Saski, C.; Lee, S.-B.; Hansen, A.K.; Daniell, H. Complete plastid genome sequences of three rosids (*Castanea*, *Prunus*, *Theobroma*): Evidence for at least two independent transfers of *rpl22* to the nucleus. *Mol. Biol. Evol.* **2010**, *28*, 835–847. [CrossRef] [PubMed]

19. Cai, Z.; Penaflor, C.; Kuehl, J.V.; Leebens-Mack, J.; Carlson, J.E.; Boore, J.L.; Jansen, R.K. Complete plastid genome sequences of drimys, *Liriodendron*, and *Piper*: Implications for the phylogenetic relationships of magnoliids. *BMC Evol. Biol.* **2006**, *6*, 77. [CrossRef] [PubMed]

20. Steane, D.A. Complete nucleotide sequence of the chloroplast genome from the tasmanian blue gum, *Eucalyptus globulus* (myrtaceae). *DNA Res.* **2005**, *12*, 215–220. [CrossRef] [PubMed]

21. Wu, C.-S.; Lin, C.-P.; Hsu, C.-Y.; Wang, R.-J.; Chaw, S.-M. Comparative chloroplast genomes of Pinaceae: Insights into the mechanism of diversified genomic organizations. *Genome Biol. Evol.* **2011**, *3*, 309–319. [CrossRef] [PubMed]

22. Zalapa, J.E.; Cuevas, H.; Zhu, H.; Steffan, S.; Senalik, D.; Zeldin, E.; McCown, B.; Harbut, R.; Simon, P. Using next-generation sequencing approaches to isolate simple sequence repeat (SSR) loci in the plant sciences. *Am. J. Bot.* **2012**, *99*, 193–208. [CrossRef] [PubMed]

23. Buschiazzo, E.; Gemmell, N.J. The rise, fall and renaissance of microsatellites in eukaryotic genomes. *Bioessays* **2006**, *28*, 1040–1050. [CrossRef] [PubMed]

24. Kelkar, Y.D.; Tyekucheva, S.; Chiaromonte, F.; Makova, K.D. The genome-wide determinants of human and chimpanzee microsatellite evolution. *Genome Res.* **2008**, *18*, 30–38. [CrossRef] [PubMed]

25. Ebert, D.; Peakall, R. A new set of universal de novo sequencing primers for extensive coverage of noncoding chloroplast DNA: New opportunities for phylogenetic studies and CPSSR discovery. *Mol. Ecol. Resour.* **2009**, *9*, 777–783. [CrossRef] [PubMed]

26. Hirao, T.; Watanabe, A.; Miyamoto, N.; Takata, K. Development and characterization of chloroplast microsatellite markers for *Cryptomeria japonica* D. Don. *Mol. Ecol. Resour.* **2009**, *9*, 122–124. [CrossRef] [PubMed]

27. Do Nascimento Vieira, L.; Faoro, H.; Rogalski, M.; de Freitas Fraga, H.P.; Cardoso, R.L.A.; de Souza, E.M.; de Oliveira Pedrosa, F.; Nodari, R.O.; Guerra, M.P. The complete chloroplast genome sequence of *Podocarpus lambertii*: Genome structure, evolutionary aspects, gene content and SSR detection. *PLoS ONE* **2014**, *9*, e90618.

28. Yao, X.; Tang, P.; Li, Z.; Li, D.; Liu, Y.; Huang, H. The first complete chloroplast genome sequences in Actinidiaceae: Genome structure and comparative analysis. *PLoS ONE* **2015**, *10*, e0129347. [CrossRef] [PubMed]

29. Su, H.-J.; Hogenhout, S.A.; Al-Sadi, A.M.; Kuo, C.-H. Complete chloroplast genome sequence of Omani lime (*Citrus aurantiifolia*) and comparative analysis within the rosids. *PLoS ONE* **2014**, *9*, e113049. [CrossRef] [PubMed]

30. Lee, M.; Park, J.; Lee, H.; Sohn, S.-H.; Lee, J. Complete chloroplast genomic sequence of *Citrus platymamma* determined by combined analysis of sanger and NGS data. *Hortic. Environ. Biotechnol.* **2015**, *56*, 704–711. [CrossRef]

31. Saina, J.K.; Gichira, A.W.; Li, Z.-Z.; Hu, G.-W.; Wang, Q.-F.; Liao, K. The complete chloroplast genome sequence of *Dodonaea viscosa*: Comparative and phylogenetic analyses. *Genetica* **2018**, *146*, 101–113. [CrossRef] [PubMed]

32. Raman, G.; Park, S. The complete chloroplast genome sequence of *Ampelopsis*: Gene organization, comparative analysis, and phylogenetic relationships to other angiosperms. *Front. Plant Sci.* **2016**, *7*, 341. [CrossRef] [PubMed]

33. Park, I.; Kim, W.J.; Yeo, S.-M.; Choi, G.; Kang, Y.-M.; Piao, R.; Moon, B.C. The complete chloroplast genome sequences of *Fritillaria ussuriensis* maxim. In addition, *Fritillaria cirrhosa* D. Don, and comparative analysis with other *Fritillaria* species. *Molecules* **2017**, *22*, 982. [CrossRef] [PubMed]

34. Li, B.; Lin, F.; Huang, P.; Guo, W.; Zheng, Y. Complete chloroplast genome sequence of *Decaisnea insignis*: Genome organization, genomic resources and comparative analysis. *Sci. Rep.* **2017**, *7*, 10073. [CrossRef] [PubMed]

35. Zhou, T.; Chen, C.; Wei, Y.; Chang, Y.; Bai, G.; Li, Z.; Kanwal, N.; Zhao, G. Comparative transcriptome and chloroplast genome analyses of two related *Dipteronia* species. *Front. Plant Sci.* **2016**, *7*, 1512. [CrossRef] [PubMed]

36. Kim, K.-J.; Lee, H.-L. Complete chloroplast genome sequences from Korean ginseng (*Panax schinseng* nees) and comparative analysis of sequence evolution among 17 vascular plants. *DNA Res.* **2004**, *11*, 247–261. [CrossRef] [PubMed]

37. Wang, R.-J.; Cheng, C.-L.; Chang, C.-C.; Wu, C.-L.; Su, T.-M.; Chaw, S.-M. Dynamics and evolution of the inverted repeat-large single copy junctions in the chloroplast genomes of monocots. *BMC Evol. Biol.* **2008**, *8*, 36. [CrossRef] [PubMed]

38. Wang, W.; Messing, J. High-throughput sequencing of three Lemnoideae (duckweeds) chloroplast genomes from total DNA. *PLoS ONE* **2011**, *6*, e24670. [CrossRef] [PubMed]

39. Khan, A.L.; Al-Harrasi, A.; Asaf, S.; Park, C.E.; Park, G.-S.; Khan, A.R.; Lee, I.-J.; Al-Rawahi, A.; Shin, J.-H. The first chloroplast genome sequence of *Boswellia sacra*, a resin-producing plant in Oman. *PLoS ONE* **2017**, *12*, e0169794. [CrossRef] [PubMed]

40. Yang, J.; Yue, M.; Niu, C.; Ma, X.-F.; Li, Z.-H. Comparative analysis of the complete chloroplast genome of four endangered herbals of *Notopterygium*. *Genes* **2017**, *8*, 124. [CrossRef] [PubMed]

41. Wang, W.; Yu, H.; Wang, J.; Lei, W.; Gao, J.; Qiu, X.; Wang, J. The complete chloroplast genome sequences of the medicinal plant *Forsythia suspensa* (oleaceae). *Int. J. Mol. Sci.* **2017**, *18*, 2288. [CrossRef] [PubMed]

42. Yang, Y.; Zhu, J.; Feng, L.; Zhou, T.; Bai, G.; Yang, J.; Zhao, G. Plastid genome comparative and phylogenetic analyses of the key genera in Fagaceae: Highlighting the effect of codon composition bias in phylogenetic inference. *Front. Plant Sci.* **2018**, *9*, 82. [CrossRef] [PubMed]

43. Guo, S.; Guo, L.; Zhao, W.; Xu, J.; Li, Y.; Zhang, X.; Shen, X.; Wu, M.; Hou, X. Complete chloroplast genome sequence and phylogenetic analysis of *Paeonia ostii*. *Molecules* **2018**, *23*, 246. [CrossRef] [PubMed]

44. Shen, X.; Wu, M.; Liao, B.; Liu, Z.; Bai, R.; Xiao, S.; Li, X.; Zhang, B.; Xu, J.; Chen, S. Complete chloroplast genome sequence and phylogenetic analysis of the medicinal plant *Artemisia annua*. *Molecules* **2017**, *22*, 1330. [CrossRef] [PubMed]

45. Li, Z.-Z.; Saina, J.K.; Gichira, A.W.; Kyalo, C.M.; Wang, Q.-F.; Chen, J.-M. Comparative genomics of the Balsaminaceae sister genera *Hydrocera triflora* and *Impatiens pinfanensis*. *Int. J. Mol. Sci.* **2018**, *19*, 319. [CrossRef] [PubMed]

46. Zhou, J.; Chen, X.; Cui, Y.; Sun, W.; Li, Y.; Wang, Y.; Song, J.; Yao, H. Molecular structure and phylogenetic analyses of complete chloroplast genomes of two *Aristolochia* medicinal species. *Int. J. Mol. Sci.* **2017**, *18*, 1839. [CrossRef] [PubMed]

47. Jiang, D.; Zhao, Z.; Zhang, T.; Zhong, W.; Liu, C.; Yuan, Q.; Huang, L. The chloroplast genome sequence of *Scutellaria baicalensis* provides insight into intraspecific and interspecific chloroplast genome diversity in *Scutellaria*. *Genes* **2017**, *8*, 227. [CrossRef] [PubMed]

48. Zhou, J.; Cui, Y.; Chen, X.; Li, Y.; Xu, Z.; Duan, B.; Li, Y.; Song, J.; Yao, H. Complete chloroplast genomes of *Papaver rhoeas* and *Papaver orientale*: Molecular structures, comparative analysis, and phylogenetic analysis. *Molecules* **2018**, *23*, 437. [CrossRef] [PubMed]

49. Chen, C.; Zheng, Y.; Liu, S.; Zhong, Y.; Wu, Y.; Li, J.; Xu, L.-A.; Xu, M. The complete chloroplast genome of *Cinnamomum camphora* and its comparison with related *Lauraceae* species. *PeerJ* **2017**, *5*, e3820. [CrossRef] [PubMed]

50. De Santana Lopes, A.; Pacheco, T.G.; Nimz, T.; do Nascimento Vieira, L.; Guerra, M.P.; Nodari, R.O.; de Souza, E.M.; de Oliveira Pedrosa, F.; Rogalski, M. The complete plastome of macaw palm [*Acrocomia aculeata* (jacq.) lodd. Ex mart.] and extensive molecular analyses of the evolution of plastid genes in arecaceae. *Planta* **2018**, 1–20. [CrossRef] [PubMed]

51. Kumbhar, F.; Nie, X.; Xing, G.; Zhao, X.; Lin, Y.; Wang, S.; Weining, S. Identification and characterisation of rna editing sites in chloroplast transcripts of einkorn wheat (*Triticum monococcum*). *Ann. Appl. Biol.* **2018**, *172*, 197–207. [CrossRef]

52. Huang, Y.-Y.; Cho, S.-T.; Haryono, M.; Kuo, C.-H. Complete chloroplast genome sequence of common Bermuda grass (*Cynodon dactylon* (L.) pers.) and comparative analysis within the family poaceae. *PLoS ONE* **2017**, *12*, e0179055.

53. Park, M.; Park, H.; Lee, H.; Lee, B.-H.; Lee, J. The complete plastome sequence of an antarctic bryophyte *Sanionia uncinata* (Hedw.) loeske. *Int. J. Mol. Sci.* **2018**, *19*, 709. [CrossRef] [PubMed]

54. Chen, H.; Deng, L.; Jiang, Y.; Lu, P.; Yu, J. RNA editing sites exist in protein-coding genes in the chloroplast genome of *Cycas taitungensis*. *J. Integr. Plant Biol.* **2011**, *53*, 961–970. [CrossRef] [PubMed]

55. De Santana Lopes, A.; Pacheco, T.G.; dos Santos, K.G.; do Nascimento Vieira, L.; Guerra, M.P.; Nodari, R.O.; de Souza, E.M.; de Oliveira Pedrosa, F.; Rogalski, M. The *Linum usitatissimum* L. Plastome reveals atypical structural evolution, new editing sites, and the phylogenetic position of Linaceae within malpighiales. *Plant Cell Rep.* **2018**, *37*, 307–328. [CrossRef] [PubMed]

56. Clayton, J.W.; Fernando, E.S.; Soltis, P.S.; Soltis, D.E. Molecular phylogeny of the tree-of-heaven family (Simaroubaceae) based on chloroplast and nuclear markers. *Int. J. Plant Sci.* **2007**, *168*, 1325–1339. [CrossRef]

57. Lee, Y.S.; Kim, I.; Kim, J.-K.; Park, J.Y.; Joh, H.J.; Park, H.-S.; Lee, H.O.; Lee, S.-C.; Hur, Y.-J.; Yang, T.-J. The complete chloroplast genome sequence of *Rhus chinensis* mill (Anacardiaceae). *Mitochondrial DNA Part B* **2016**, *1*, 696–697. [CrossRef]

58. Schmieder, R.; Edwards, R. Quality control and preprocessing of metagenomic datasets. *Bioinformatics* **2011**, *27*, 863–864. [CrossRef] [PubMed]

59. Dierckxsens, N.; Mardulyn, P.; Smits, G. NOVOPlasty: De novo assembly of organelle genomes from whole genome data. *Nucleic Acids Res.* **2016**, *45*, e18.

60. Kearse, M.; Moir, R.; Wilson, A.; Stones-Havas, S.; Cheung, M.; Sturrock, S.; Buxton, S.; Cooper, A.; Markowitz, S.; Duran, C. Geneious basic: An integrated and extendable desktop software platform for the organization and analysis of sequence data. *Bioinformatics* **2012**, *28*, 1647–1649. [CrossRef] [PubMed]

61. Wyman, S.K.; Jansen, R.K.; Boore, J.L. Automatic annotation of organellar genomes with DOGMA. *Bioinformatics* **2004**, *20*, 3252–3255. [CrossRef] [PubMed]

62. Schattner, P.; Brooks, A.N.; Lowe, T.M. The tRNAscan-SE, snoscan and snoGPS web servers for the detection of tRNAs and snoRNAs. *Nucleic Acids Res.* **2005**, *33*, W686–W689. [CrossRef] [PubMed]

63. Lohse, M.; Drechsel, O.; Bock, R. OrganellarGenomeDRAW (OGDRAW): A tool for the easy generation of high-quality custom graphical maps of plastid and mitochondrial genomes. *Curr. Genet.* **2007**, *52*, 267–274. [CrossRef] [PubMed]

64. Darling, A.C.; Mau, B.; Blattner, F.R.; Perna, N.T. Mauve: Multiple alignment of conserved genomic sequence with rearrangements. *Genome Res.* **2004**, *14*, 1394–1403. [CrossRef] [PubMed]

65. Thiel, T.; Michalek, W.; Varshney, R.; Graner, A. Exploiting EST databases for the development and characterization of gene-derived SSR-markers in barley (*Hordeum vulgare* L.). *Theor. Appl. Genet.* **2003**, *106*, 411–422. [CrossRef] [PubMed]

66. Kurtz, S.; Choudhuri, J.V.; Ohlebusch, E.; Schleiermacher, C.; Stoye, J.; Giegerich, R. Reputer: The manifold applications of repeat analysis on a genomic scale. *Nucleic Acids Res.* **2001**, *29*, 4633–4642. [CrossRef] [PubMed]

67. Mower, J.P. The prep suite: Predictive RNA editors for plant mitochondrial genes, chloroplast genes and user-defined alignments. *Nucleic Acids Res.* **2009**, *37*, W253–W259. [CrossRef] [PubMed]

68. Stamatakis, A. Raxml version 8: A tool for phylogenetic analysis and post-analysis of large phylogenies. *Bioinformatics* **2014**, *30*, 1312–1313. [CrossRef] [PubMed]

69. Darriba, D.; Taboada, G.L.; Doallo, R.; Posada, D. jModelTest 2: More models, new heuristics and parallel computing. *Nat. Methods* **2012**, *9*, 772. [CrossRef] [PubMed]

15

Comparative Chloroplast Genome Analyses of Species in *Gentiana* section *Cruciata* (Gentianaceae) and the Development of Authentication Markers

Tao Zhou [1], Jian Wang [1], Yun Jia [2], Wenli Li [1], Fusheng Xu [1] and Xumei Wang [1,*]

[1] School of Pharmacy, Xi'an Jiaotong University, Xi'an 710061, China; zhoutao196@mail.xjtu.edu.cn (T.Z.); wangjian6318@126.com (J.W.); lwl3659003@stu.xjtu.edu.cn (W.L.); xfs19940903@stu.xjtu.edu.cn (F.X.)

[2] Key Laboratory of Resource Biology and Biotechnology in Western China (Ministry of Education), School of Life Sciences, Northwest University, Xi'an 710069, China; jy878683@163.com

*   Correspondence: wangxumei@mail.xjtu.edu.cn

Abstract: *Gentiana* section *Cruciata* is widely distributed across Eurasia at high altitudes, and some species in this section are used as traditional Chinese medicine. Accurate identification of these species is important for their utilization and conservation. Due to similar morphological and chemical characteristics, correct discrimination of these species still remains problematic. Here, we sequenced three complete chloroplast (cp) genomes (*G. dahurica*, *G. siphonantha* and *G. officinalis*). We further compared them with the previously published plastomes from sect. *Cruciata* and developed highly polymorphic molecular markers for species authentication. The eight cp genomes shared the highly conserved structure and contained 112 unique genes arranged in the same order, including 78 protein-coding genes, 30 tRNAs, and 4 rRNAs. We analyzed the repeats and nucleotide substitutions in these plastomes and detected several highly variable regions. We found that four genes (*accD*, *clpP*, *matK* and *ycf1*) were subject to positive selection, and sixteen InDel-variable loci with high discriminatory powers were selected as candidate barcodes. Our phylogenetic analyses based on plastomes further confirmed the monophyly of sect. *Cruciata* and primarily elucidated the phylogeny of Gentianales. This study indicated that cp genomes can provide more integrated information for better elucidating the phylogenetic pattern and improving discriminatory power during species authentication.

Keywords: *Gentiana* section *Cruciata*; chloroplast genome; molecular markers; species authentication

1. Introduction

Gentiana is the largest genus in the family Gentianaceae and widely distributed throughout the northern Hemisphere [1]. Approximately 362 species are recognized in genus *Gentiana* which have been divided into 15 sections [2]. Section *Cruciata* contains 21 species which are mainly distributed in eastern Eurasia [3]. Most species of this section are restricted to alpine regions, although some of them could be found at altitudes below 1000 m at higher latitudes [1]. Four species (*G. macrophylla*, *G. crassicaulis*, *G. straminea*, and *G. dahurica*) in sect. *Cruciata* are used as the original plants of traditional Chinese medicine named Qin-jiao [4]. The roots of these plants contain abundant secoiridoid active compounds which could be used for the treatment of diabetes, apoplexy, paralysis, and rheumatism [5–8].

Recently, the wild resources of some *Gentiana* species are dramatically declined due to overexploitation and some of them have been listed in the National Key Protected Wild Herbs in China [5,7]. However, the demand of natural sources for these plants remains high due to the high pharmacological and economical values. Therefore, many economically motivated adulterants of Qin-jiao products with similar morphological characters have been developed to substitute the genuine medicinal materials. Generally, the authentication of herbs was based on the morphological and histological inspection. But these methods may not be suitable for authenticating some species in sect. *Cruciata* due to the following reasons. Firstly, most species of sect. *Cruciata* shared the similar morphological characters especially in terms of leaf shape. Secondly, some species in this section are usually located in the sympatric distributions, thus intermediate morphology could be detected due to interspecific hybridization [9,10]. Thirdly, pharmacognostical studies showed that some species such as *G. siphonantha* and *G. straminea* usually shared similar chemical profiles [11]. Some other factors, such as growth conditions, developmental stage, and internal metabolism may affect the secondary metabolite accumulation in Qin-jiao and limit the application of such chemical analyses for authenticating the species of sect. *Cruciata*. In addition, chemical methods for identifying the medicinal plants are also expensive and not suitable for high-throughput analysis [12]. Therefore, reliable and cost-efficient methods are needed to authenticate the medical plants of sect. *Cruciata*.

Chloroplast (cp) genome of angiosperm is characterized by a typical quadripartite structure that contains a pair of inverted repeat (IR) regions separated by a large single-copy (LSC) and a small single-copy (SSC) region [13], and it is highly conserved compared to nuclear and mitochondrial genomes. Although chloroplast genomes are highly conserved, some hotspot regions with single nucleotide polymorphisms and insertion/deletions could be found and these regions may provide enough information for species identification [14,15]. Due to low recombination, uniparental inheritance, and low nucleotide substitution rates, many cp genetic markers have been used for plant phylogenetic, phylogeographic, and population genetic analyses [16]. It has been proven that some chloroplast sequences such as *trnH-psbA*, *rbcL*, and *matK* were commonly used as DNA barcodes for plants discrimination [17]. But in some cases, above commonly used DNA barcodes were not suitable to distinguish closely related plants due to limited variation loci [16,18]. Recently, it has been proposed that the complete cp genome could be used as a plant barcode, and various research have demonstrated that complete cp genome can greatly increase resolution for resolving difficult phylogenetic relationships at lower taxonomic levels [16,19–22]. In addition, using the cp genome as a genetic marker for identifying the plant will avoid the problems such as gene deletion and low Polymerase Chain Reaction (PCR) efficiency [23].

Most species in section *Cruciata* were recently diverged and originated from a common radiation in the Qinghai-Tibet Plateau (QTP) before the Pleistocene [1,10], therefore these species were usually closely related and showed parallel evolutionary relationships [1]. Previous research showed that commonly used DNA barcodes in some cases may not be suitable to identify the medicinal plant of this section [24,25]. Therefore, more specific barcodes with enough variation are needed to discriminate closely related species belong to sect. *Cruciata*. Nowadays, with the improvement of sequencing and assembly technologies, it is comparatively simple to obtain comprehensive chloroplast sequences for identifying *Gentiana* species. By utilizing the variable information provided from cp genomes, we can not only obtain more specific barcodes for species authentication in sect. *Cruciata*, but also shed light on the complex evolutionary relationships of the species in this section.

In the present study, we obtained the chloroplast genome sequences of *G. dahurica*, *G. siphonantha* and *G. officinalis* by using de novo assembly of whole-genome sequencing (WGS) data derived from high throughput sequencing technology. We also comparatively analyzed the chloroplast genomes of eight species in sect. *Cruciata* and developed credible cp genome derived InDel markers to authenticate these species. These markers are not only valuable tools for further evolutionary and population genetic studies on *Gentiana*, but also could be used as standardized barcodes for authenticating the original plants of Qin-jiao.

2. Results

2.1. Complete Chloroplast Genome Features of Sect. Cruciata

The chloroplast genomes of G. dahurica, G. siphonantha, and G. officinalis were sequenced with approximately 5.2, 5.8, and 5.6 Gb of paired-end reads, respectively. The raw reads with a sequence length of 125 bp were trimmed to generate the clean reads for the next assembly. After quality filtering, 10,114,902, 11,405,694, and 11,288,676 clean reads were recovered for G. dahurica, G. siphonantha and G. officinalis, respectively. Combined with the de novo and reference guided assembly, the cp genomes were obtained. The four junction regions between the IRs and SSC/LSC regions were confirmed by PCR amplification and Sanger sequencing. We mapped the obtained sequences to the new assembled genomes and no mismatch or InDel was observed.

We compared the basic genome features of three newly sequenced cp genomes with five previously published cp genomes [26–28] and found that all the chloroplast genomes possessed the typical quadripartite structure with the length range from 148,765 to 149,916 bp (Table 1, Figure 1). The whole cp genome contained a pair of inverted repeat regions (IRs: 24,955–25,337 bp) which were separated by a small single copy region (SSC: 17,070–17,095 bp) and a large single copy region (LSC: 81,119–82,911 bp) (Table 1). Although genomic structure and size were highly conserved in eight cp genomes, the IR/SC boundary regions still varied slightly (Figure 2). All the eight chloroplast genomes contained 112 unique genes arranged in the same order, including 78 protein-coding genes, 30 tRNA genes, and 4 rRNA genes. Two genes (rps16, infA) were inferred to be pseudogenes (Figure S1). The overall guanine and cytosine (GC) content in each chloroplast genome is identically 37.7% (Table 1).

Table 1. Summary of complete chloroplast genomes for eight Gentiana species.

Name of Taxon	G. dahurica	G. siphonantha	G. officinalis	G. straminea
Genome length	148,803	148,908	148,879	148,991
LSC length	81,154	81,121	81,119	81,240
SSC length	17,093	17,113	17,088	17,085
IR length	25,278	25,337	25,336	25,333
Total gene number	112	112	112	112
No. of protein coding genes	78	78	78	78
No. of tRNA genes	30	30	30	30
No. of rRNA genes	4	4	4	4
GC content in genome (%)	37.7	37.7	37.7	37.7
Name of Taxon	**G. crassicaulis**	**G. robusta**	**G. tibetica**	**G. macrophylla**
Genome length	148,776	148,911	148,765	149,916
LSC length	81,164	81,164	81,163	82,911
SSC length	17,071	17,085	17,070	17,095
IR length	25,271	25,333	25,266	24,955
Total gene number	112	112	112	112
No. of protein coding genes	78	78	78	78
No. of tRNA genes	30	30	30	30
No. of rRNA genes	4	4	4	4
GC content in genome (%)	37.7	37.7	37.7	37.7

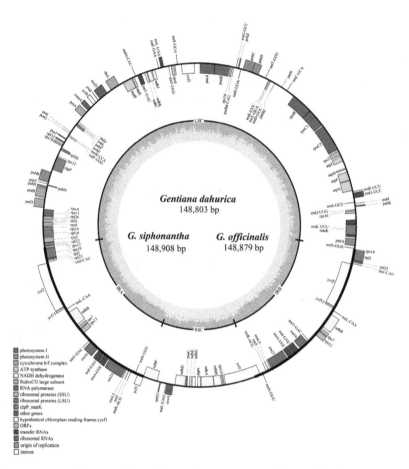

Figure 1. Merged gene map of the complete chloroplast genomes of three *Gentiana* species. Genes belonging to different functional groups are classified by different colors. The genes drawn outside of the circle are transcribed counterclockwise, while those inside are clockwise. Dashed area in the inner circle represent GC content of chloroplast genome.

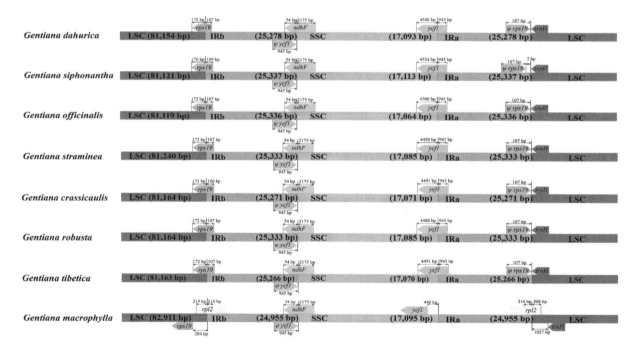

Figure 2. Comparison of chloroplast genome borders of LSC, SSC, and IRs among eight species in *Gentiana* sect. *Cruciata*. Ψ indicates a pseudogene.

2.2. Comparative Analyses of the Chloroplast Genomes of Species of Sect. Cruciata

Repeat analyses of three newly sequenced cp genomes showed 13/13/13 (*G. siphonantha*/ *G. officinalis*/*G. dahurica*) palindromic repeats, 12/11/11 dispersed repeats, and 7/6/6 tandem repeats (Figure 3A,B) with the repeat length range from 15 to 38 bp (Tables S1 and S2). The numbers and distribution of all repeat types were similar and conserved in these three cp genomes. Overall, 32/30/30 repeats were detected in three cp genomes. Similarly, 37, 34, 34, and 37 repeats were found in previously reported *G. crassicaulis*, *G. robusta*, *G. straminea*, and *G. tibetica* cp genomes (Figure 3A,B). Unexpectedly, 61 repeats, including 28 dispersed repeats, 18 palindromic repeats and 15 tandem repeats, were found in the cp genome of *G. macrophylla*. We found most of repeats in eight cp genomes were located in the intergenic or intron regions, and only a few repeats were distributed in protein-coding regions (*ycf1*, *ycf2*, and *psaA*) (Tables S1 and S2). Simple sequence repeats (SSRs) consisting of 1–6 bp repeat unit are distributed throughout the genome. In our study, perfect SSRs in eight *Gentiana* cp genomes were detected. The results showed that Mono-nucleotide repeats were most abundant type, followed by Tetra-nucleotides, Di-nucleotides and Tri-nucleotides. The penta- and hexa-nucleotides were very rare across the cp genomes (Figure 3C,D). Most SSRs are located in intergenic regions, but some were found in *rpoC2*, *rpoC1*, *atpB*, *ndhF*, and *ycf1* coding genes (Table S3). To investigate the evolutionary characteristics of cpDNA genes in eight *Gentiana* cp genomes and estimate selection pressures, nonsynonymous (dN), synonymous substitution rates (dS), and the ratio of dN/dS were calculated for 78 protein-coding genes (Table S4). We obtained 771 pairwise comparison results of dN/dS values and the remaining could not be calculated due to dS = 0. Only four genes (*accD*, *clpP*, *matK*, and *ycf1*) had dN/dS values ≥1 indicating that they had undergone positive selection.

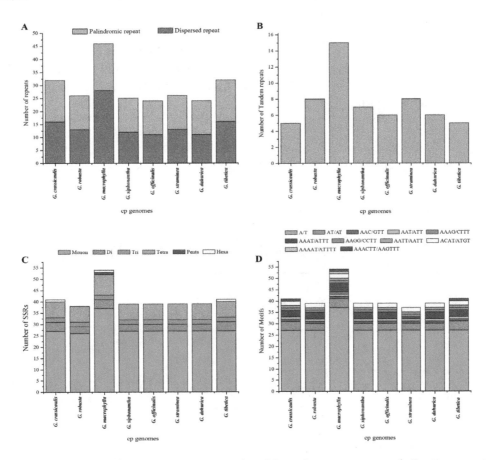

Figure 3. Analysis of different repeats in eight chloroplast genomes of *Gentiana* sect. *Cruciata*. (**A**) Histogram showing the number of palindromic repeats and dispersed repeats; (**B**) histogram showing the number of tandem repeats; (**C**) number of different simple sequence repeat (SSR) types detected in eight chloroplast genomes; (**D**) total numbers of different SSR motifs in eight chloroplast genomes.

To understand the level of sequence divergence, comparative analysis among eight *Gentiana* cp genomes was performed using mVISTA with the annotation of *G. crassicaulis* as a reference. The cp genomes within sect. *Cruciata* showed high sequence similarities with identities of only a few regions below 90%, indicating a high conservatism of these chloroplast genomes (Figure 4). The single-copy regions and intergenic regions were more divergent than the IR regions and genic regions (Figure 5). According to the comparative analyses, some hotspot regions for genome divergence that could be utilized as potential genetic markers to elucidate the phylogenies and to discriminate the species in sect. *Cruciata*. These regions were *psbA-trnH*, *trnK-rps16*, *rps16-trnQ*, *trnS-trnG*, *trnE-trnT*, *psbM-trnD*, *trnT-psbD*, *trnS-psbZ*, *ndhC-trnV*, *atpB-rbcL*, *rbcL-accD*, *accD-psbI*, *rpl33-rps18*, *trnR-trnA*, and *trnV-rps7* (Figure 4).

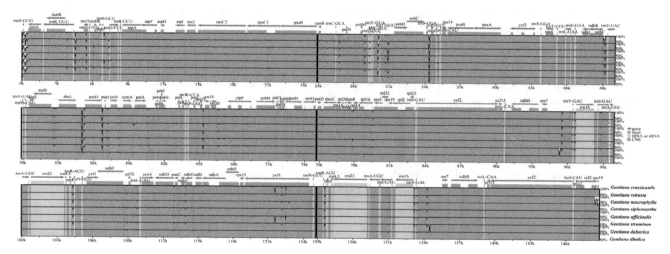

Figure 4. mVISTA percent identity plot comparing the eight chloroplast genomes of *Gentiana* sect. *Cruciata* with *G. crassicaulis* as a reference. The *y*-axis represents the percent identity within 50–100%. Genome regions are color-coded as protein coding (purple), rRNA, or tRNA coding genes (blue), and noncoding sequences (pink).

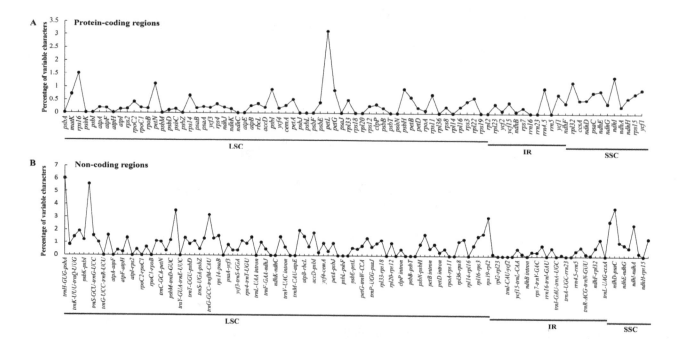

Figure 5. Percentage of variable characters in eight aligned chloroplast genomes of *Gentiana* sect. *Cruciata*. (**A**) Coding region; (**B**) Noncoding region.

2.3. Development of InDel Markers to Discriminate Species of Sect. Cruciata

Based on the alignment of complete cp genome sequences, the 16 most InDel-variable loci were selected as candidate DNA markers for authentication (Table S5). After PCR amplification, these 16 markers could successfully amplify the expected polymorphic band sizes (Figure 6). Some of these 16 markers had unique amplicon sizes specific to different *Gentiana* species (Figure 6). Especially five markers (QJcpm9, QJcpm12, QJcpm14, QJcpm15, and QJcpm16) were specific to *G. crassicaulis*, which all derived from long InDels in the intergenic regions including *rps16-trnQ*, *psbM-trnD*, *trnS-psbZ*, *accD-psbI*, and *trnK-rps16*. The marker QJcpm1 was specific to *G. robusta* and *G. crassicaulis* and was derived from a 54 and 64 bp InDel in the *ndhC-trnV* region. The QJcpm2 marker derived from 14 bp tandem repeat (TR) in *cemA-petA* region was specific to *G. siphonantha* and *G. crassicaulis*. QJcpm3 marker, which was specific to *G. officinalis* and *G. crassicaulis*, was derived from 72, 14 bp InDels, and 7 bp TR in *rbcL-accD* region. Three markers (QJcpm4, QJcpm10, and QJcpm11) were specific to *G. straminea*, *G. robusta*, and *G. crassicaulis*. QJcpm4 marker was derived from 12 bp InDels and 6 bp TR in the *rpl33-rps18* region; QJcpm10 marker was derived from 9 bp TR and 33 bp InDel in the *trnT-psbD*; QJcpm11 marker was derived from 18 bp InDel in *rrn5-trnA* region. The QJcpm5 marker, which was derived from 14, 4, and 7 bp TR in *atpB-rbcL*, was specific to *G. macrophylla*, *G. robusta*, and *G. crassicaulis*. Three markers QJcpm6, QJcpm8, and QJcpm13 were derived from a 42 bp InDel in *ycf1*, 9 bp InDel in *rps8-rpl14* region, and 24 bp TR in the *trnS-trnG* region, respectively, and were specific to *G. straminea* and *G. robusta*. The marker QJcpm7, which was specific *G. dahurica* and *G. siphonantha*, was also derived from 24 InDel in *ycf1* CDS region. Our validation results indicated all these markers can be used to identify species in sect. *Cruciata*.

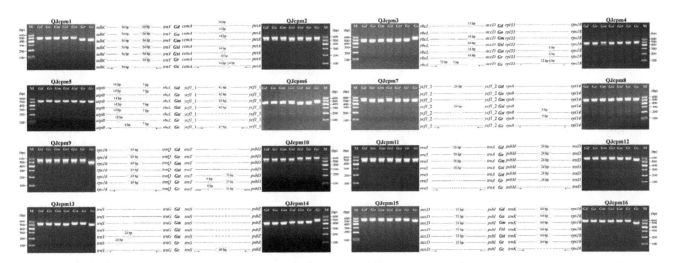

Figure 6. Validation of 16 molecular markers derived from InDel regions of eight chloroplast genomes of *Gentiana* sect. *Cruciata*. Inserted sequences and tandem repeats are designated by diamonds and triangle, respectively. Solid and dotted lines indicate conserved and deleted sequences, respectively. Left and right black arrows indicate forward and reverse primers, respectively. Abbreviated species names were shown on schematic diagrams: *Gd, G. dahurica; Go, G. officinalis; Gm, G. macrophylla; Gsi, G. siphonantha; Gst, G. straminea; Gr, G. robusta; Gc, G. crassicaulis;* M, D600 DNA ladder.

2.4. Phylogenetic Relationships of Species Belong to Sect. Cruciata

Here, 27 cp genomes were retrieved to infer the interspecific relationships of eight species in sect. *Cruciata* as well as to clarify the phylogenetic relationships of some Gentianales species (Table S6). Phylogenetic analyses were performed using Maximum parsimony (MP), Maximum likelihood (ML) and Bayesian inference (BI) methods, and *Arabidopsis thaliana* was set as outgroup. Three different datasets including complete cp genomes, 70 shared protein-coding genes (PCGs) and the most conserved regions (TMCRs) of cp genomes were used to construct the phylogenetic trees. The results showed the same phylogenetic signals for these three datasets and the phylogenetic trees inferred from MP/ML/BI methods also shared identical topologies (Figure 7, Figures S2 and S3). In these phylogenetic trees, we found all the species of sect. *Cruciata* formed a monophyletic clade a with high bootstrap and BI support values and clustered with another two Gentianaceae species (*G. lawrencei* and *Swertia mussotii*) in the same clade [29,30]. Of these species, *G. macrophylla*, *G. officinalis*, and *G. siphonantha* showed paraphyletic relationships with each other and formed a monophyletic clade with *G. dahurica*. *G. tibetica* and *G. crassicaulis* formed a monophyletic clade and located in the basal position of these eight species in sect. *Cruciata*. Interestingly, *G. robusta* and *G. straminea* with similar morphological characteristics were clustered in a monophyletic clade with a high resolution value. In addition, our phylogenetic results supported the monophyly of two families, including Apocynaceae and Rubiaceae, in the order Gentianales. Unexpectedly, *Gynochthodes nanlingensis* (*Morinda nanlingensis*) belongs to Rubiaceae was embed in the Apocynaceae species.

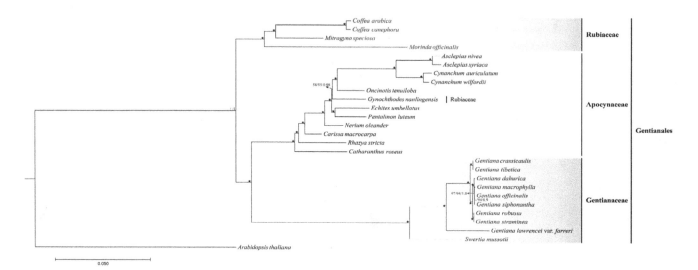

Figure 7. Phylogenetic relationships of species belong to *Gentiana* sect. *Cruciata* inferred from MP/ML/BI analysis based on complete chloroplast genome sequences. The numbers associated with each node are bootstrap support and posterior probability values, and the symbol ★ in the phylogenetic tree indicated that the support value of branch is 100/100/1.0.

3. Discussion

Three cp genomes of sect. *Cruciata* were sequenced using Illumina Hiseq platform, which provided more resources for evolutionary and genetics studies of *Gentiana*. The cp genomic information presented in this study will also contribute to the conservation and management of wild resources of sect. *Cruciata*. Although a recent research reported that 11 ndh genes had been lost in the cp genomes of *Gentiana* sect. *Kudoa* [31], eight cp genomes of sect. *Cruciata* analyzed in present study are rather conserved in gene structures, contents and arrangement, and no significant structural rearrangements, such as inversions or gene relocations, were detected. Of these eight species, *G. macrophylla* has the largest cp genome size and other species showed minor differences in genome size. The length variations of these cp genomes may result from the length of intergenic regions, similar result has been reported for *Paris* (Melanthiaceae) cp genomes [18].

All the eight cp genomes of sect. *Cruciata* had the same protein-coding genes, tRNA and rRNA genes. We found that exon2 of *rps16* gene was lost in three newly sequenced cp genomes, and *rps16* in other cp genomes also showed same structure. Therefore, *rps16* pseudogene may commonly exist in the genus *Gentiana* [26]. And *infA* gene, which contains internal stop codons, was also inferred as pseudogene in these species. This pseudogene had been reported in many species [32–35]. Except for cp genome of *G. macrophylla*, the remained cp genomes showed minor variations in the junctions between the SSC and IRs regions. As most species of sect. *Cruciata* derived from a common radiation and usually showed closely interspecific relationships, we thus speculated that highly conserved nature of cp genomes resulted in the similar gene distributions at SC/IR boundaries.

Repeat structure plays an important role in genomic rearrangement, recombination, and sequence divergence in plastomes [36–38]. In the present study, cp genome of *G. macrophylla* has the largest number of repeats, while the number of repeats was similar in other cp genomes. Most of the repeated regions in different species showed similar characteristics and most repeats were located in intergenic regions or in *ycf1/pasA*. Repeats in these genes are commonly observed in other angiosperm lineages [22,32,39]. Cp microsatellites (cpSSR) usually showed high polymorphism within the same species and which are potentially useful markers for population genetics [40]. Here, 326 SSRs varying in number and type between eight major *Gentiana* species, and the most abundant repeat type was found to be stretches of mononucleotides (A/T). Similar to the distribution status of dispersed and tandem repeats, most cpSSRs were observed in noncoding regions, and only small proportion were found in coding regions. CpSSRs located in noncoding regions of the cp genome are generally short mononucleotide tandem repeats and commonly showed intraspecific variation in repeat number [15]. Therefore, cpSSRs derived from eight *Gentiana* species in this study are expected to be useful for the genetic diversity studies in *Gentiana*. As the wild resources of some *Gentiana* species were dramatically declined due to overexploitation, we thought these species need to transplant or cultivate in order to preserve their germplasm resources. We believe the obtained SSRs among these chloroplast genomes will also be useful for the domestication and breeding of *Gentiana* species.

Sequence divergence of the coding genes was observed between different species. Our analyses indicated that all of cp genes showed a low sequence divergence (dS < 0.1) and most cp genes were under purifying selection (dN/dS < 1); similar results were reported for other cp genomes [32,41,42]. Only four genes (*accD*, *clpP*, *matK*, and *ycf1*) were under positive selection. Previous research reported that *accD* and *clpP* genes had a high evolution rate in *Fagopyrum* species [43,44], we thus presumed that these genes may have a high evolution rate in *Gentiana* species. One other gene (*matK*) was highly divergent in Caryophyllaceae, and comparative cp genomes analyses of Myrtaceae also indicated *matK* was under positive pressure [45,46]. The *ycf1* gene with unknown functions showed a biased higher value for dN/dS ratio compared to dS value indicating that this gene evolved at a faster rate. It has also been shown to be subject to positive selection in many angiosperms [20,22,32,44,45].

DNA barcodes are defined as the short DNA sequences with a sufficiently high mutation rate to discriminate a species within a given taxonomic group and are confirmed as reliable tools for the identification of plant species [16,47]. Previously, *rbcL*, *trnH-psbA*, and *matK* were considered as "core" plant barcodes for species identification, but they often have limited resolutions at species level [18]. Previous research showed that three commonly used barcodes in some cases may not be suitable to authenticate the medicinal plant in section *Cruciata* [24,25]. Therefore, seeking for more effective DNA barcodes with high evolutionary rates is very important for the molecular identification of species in *Gentiana* sect. *Cruciata*. The complete cp genome has a conserved sequence from 110k to 160k bp, which far exceeds the length of commonly used molecular markers and provides more variation to distinguish closely related species [12,16]. Therefore, some mutation hotspot regions, including *trnK-rps16*, *rps16-trnQ*, *trnS-trnG*, *trnE-trnT*, *trnT-psbD*, *trnS-psbZ*, *ndhC-trnV*, *rbcL-accD*, *accD-psbI*, *trnR-trnA*, *trnV-rps7*, and *ycf1*, detected from the cp genomes can provide more specific DNA barcodes for the authentication of medicinal materials of sect. *Cruciata* and also provide sufficient genetic markers for resolving the phylogeny of Gentianaceae.

We developed the specific markers for species authentication of sect. *Cruciata* based on the hotspot regions derived from cp genomes. Most of these markers were derived from the intergenic regions of cp genomes and showed high interspecific polymorphism. Previous molecular identification of *Panax*, *Zanthoxylum*, and *Eclipta* species also indicated that chloroplast-derived genetic markers had high discriminatory powers [12,14,48]. Therefore, specific markers developed from the comparative cp genomes were superior than the commonly used markers for identifying the closely related species. Especially for medicinal plants, these specific genetic markers are more effective in the authentication of their source plants. We found two InDels (42 and 24 bp) in the *ycf1* gene, which can be used to distinguish species in sect. *Cruciata*. *Ycf1*, which encodes a protein of approximately 1800 amino acids with unknown function, is the second largest gene in the cp genome. Because the sequence of *ycf1* is too long and too variable for designing universal primers, it has received little attention for DNA barcodes at low taxonomy [18,49]. But two markers derived from *ycf1* gene showed high PCR efficiency and polymorphism in species of sect. *Cruciata*, and could be used as specific barcodes for the authentication of *Gentiana* species. Although our study provided 16 genetic markers which had enough interspecies polymorphism for species identification, some of the markers were usually specific to two species. We thus suggest a combination of several markers should be considered for credible authentication between different species in genus *Gentiana*.

We inferred the phylogenetic relationships of sect. *Cruciata* using complete cp genomes. Three different methods (MP/ML/BI) were used to rebuilt the phylogenetic trees based on different datasets (cp genomes, 70 shared PCGs, and TMCRs), and the derived phylogenetic trees shared identical topology. All the species of sect. *Cruciata* formed a monophyletic clade with high bootstrap and BI support values. This result is comparable with the previous phylogenetic research based on four cpDNA fragments [1]. Four species, including *G. dahurica*, *G. macrophylla*, *G. siphonantha*, and *G. officinalis*, were clustered in the same clade with high support values. Although the flower color of *G. officinalis* was different from other three species, it shared similar morphological and chemical characters with *G. macrophylla* [50]. We found that *G. straminea* was closely related to *G. robusta*. *G. robusta* may have originated from introgression between *G. straminea* and another relative species, and these two species are usually closer to each other [26,51]. Two species, *G. tibetica* and *G. crassicaulis* were clustered in the same clade and located in the basal position in the clade of sect. *Cruciata*. However, a previous phylogenetic result indicated that *G. tibetica* was closely related to *G. straminea* and *G. robusta* [1]. As *G. tibetica* and *G. crassicaulis* distributed sympatrically in Tibet and intermediate types were produced by introgression between these two species [52], we thus inferred these two species should be closely related. In addition, based on the phylogenetic results, we found that the family Gentianaceae was closer to family Apocynaceae than to family Rubiaceae in order Gentianales. Previous phylogenetic studies of order Gentianales resulted in similar findings, but with relatively low support values [53,54]. Although our result confirmed the monophyly of section *Cruciata* and primarily elucidated the phylogeny of Gentianales based on available cp genomes, more complete cp genome sequences are needed to resolve the comprehensive phylogenies of this section, especially since limited taxon sampling may produce discrepancies in tree topologies [15,55].

4. Materials and Methods

4.1. Plant Materials and DNA Isolation

Samples of *G. dahurica*, *G. siphonantha* and *G. officinalis* were collected from Tianzhu (102.54° E, 37.01° N), Sunan (98.05° E, 39.55° N) and Yuzhong (104.05° E, 35.78° N) Counties in Gansu Province, China. Young leaves of three species were collected and immediately dried with silica gel for further DNA isolation. Total genomic DNA was isolated from each sample using the modified Cetyl Trimethyl Ammonium Bromide (CTAB) method [56]. The quantity and quality of extracted genomic DNA was determined by gel electrophoresis and NanoDrop 2000 Spectrophotometer (Thermo Scientific, Carlsbad, CA, USA).

4.2. Chloroplast Genome Sequencing, Assembly and Annotation

The DNA Library with insert size of 200 bp was prepared according to the description by Zhou et al. [32], and sequenced using Illumina Hiseq™ 2500 platform (Illumina Inc., San Diego, CA, USA) with the average read length of 125 bp. The obtained raw reads were filtered with the NGS QC Toolkit_v2.3.3 (National Institute of Plant Genome Research, New Delhi, India) [57]. Adapter sequences and low-quality reads with Q-value ≤ 20 were removed. Filtered paired-end reads were firstly mapped to the chloroplast genome of *Gentiana straminea* (KJ657732) by using the Bowtie 2-2.2.6 (University of Maryland,

College Park, MD, USA.) with default parameter [58]. And then the matched paired-end reads were de novo assembled using SPAdes-3.6.0 (St. Petersburg Academic University, St. Petersburg, Russia) [59]. After de novo assembly, the resultant scaffolds were further assembled using a baiting and iteration method based on Perl script MITObim_1.9.pl (University of Oslo, Oslo, Norway) [60]. Finally, all obtained reads were mapped to the spliced cp genome sequence using Geneious 10.1 (Biomatters Ltd., Auckland, New Zealand) in order to avoid assembly errors. The four junction regions between the IRs and SSC/LSC were confirmed by PCR amplification and Sanger sequencing (Primers and sequencing results are listed in Table S7). The cp genome genes were annotated with the online program Organellar Genome Annotator (DOGMA) [61], and the primary annotated results were manually verified according to the annotation information from other closely related species. The circular plastid genome maps were drawn using the online program OrganellarGenome DRAW (Max planck Institute of Molecular Plant Physiology, Potsdam, Germany) [62] and three newly sequenced cp genome were deposited in GenBank (MH261259–MH261261).

4.3. Repeat Structure, Genome Comparison and Sequence Divergence

Dispersed and palindromic repeats within the cp genomes were identified using REPuter (University of Bielefeld, Bielefeld, Germany) with a minimum repeat size of 30 bp and a sequence identity > 90% [63]. Tandem repeat sequences were searched using the Tandem Repeats Finder program (Mount Sinai School of Medicine, New York, NY, USA) with the following parameters: 2 for alignment parameters match, 7 for mismatch and InDel, respectively [64]. Simple sequence repeats (SSRs) were predicted using MISA perl script (Institute of Plant Genetics and Crop Plant Research, Gatersleben, Germany) with the parameters of ten for mono, five for di-, four for tri-, and three for tetra-, penta, and hexa-nucleotide motifs [65]. The nonsynonymous (dN), synonymous (dS), and dN/dS values of each protein coding gene were calculated using PAML packages 4.0 (University College London, London, UK) with Yang and Nielsen (YN) algorithm to detect whether selective pressure exists for plastid genes [66]. The cp genome gene distribution of eight *Gentiana* species was compared and visualized using mVISTA software with the annotation of *G. crassicaulis* as a reference [67]. To examine mutation hotspot regions of the cp genomes of eight *Gentiana* species, the percentages of variable characters for each coding and noncoding regions were analyzed using the method described by Zhang et al. [68].

4.4. Development and Validation of the InDel Molecular Marker

In order to validate interspecies polymorphisms within the chloroplast genomes and develop DNA genetic markers for identifying species belong to sect. *Cruciata*, specific primers were designed using Primer 3 based on the mutational hotspot regions found in these *Gentiana* chloroplast genomes [69]. PCR amplifications were performed in a reaction volume of 25 μL with 12.5 μL 2× Taq PCR Master Mix, 0.4 μM of each primer, 2 μL template DNA and 10.1 μL ddH$_2$O. All amplifications were carried out in SimpliAmp™ Thermal Cycler (Applied Biosystems, Carlsbad, CA, USA) as follow: denaturation at 94 °C for 5 min, followed by 30 cycles of 94 °C for 50 s, at specific annealing temperature (Tm) for 40 s, 72 °C for 90 s and 72 °C for 7 min as final extension. PCR products were visualized on 2% agarose gels after staining with ethidium bromide and then the DNA fragments were sequenced by Sangon Biotech (Shanghai, China) (Sequencing results are listed in Table S8).

4.5. Phylogenetic Analysis

The complete chloroplast genomes of 26 Gentianales species were recovered to clarify the phylogenetic relationships of sect. *Cruciata* and the cp genome of *Arabidopsis thaliana* was set as outgroup. In order to obtain a reliable result, phylogenetic analyses were implemented based on different cp genome datasets. On the one hand, whole cp genome sequences and 70 common cp protein-coding genes (PCGs) were separately used to infer the phylogenetic relationships of these species. On the other hand, multi-gene alignment matrix, which contained the most conserved regions (TMCRs) of cp genome was generated using HomBlocks (Ocean University of China, Qingdao, China) [70], was used to understand the phylogenetic relationships at cp genome level. Alignments were constructed using

MAFFT v7.308 (Osaka University, Suita, Japan) with default parameters and the best-fit nucleotide substitution model (General Time Reversible + Invariant + Gamma, GTR + I + G) was determined with Modeltest 3.7 (Brigham Young University, Provo, UT, USA) [71,72]. Maximum parsimony (MP) analyses of the resulting alignments from different datasets were performed using PAUP 4.0b10 (Smithsonian Institution, Washington, DC, USA) [73]. Maximum likelihood (ML) analyses were performed using RAxML 8.1.24 (Heidelberg Institute for Theoretical Studies, Heidelberg, Germany) with GTR + I + G nucleotide substitution model [74]. The reliability of each tree node was tested by bootstrap analysis with 1000 replicates. Bayesian analyses were also conducted with MrBayes v3.2.6 (Swedish Museum of Natural History, Stockholm, Sweden) [75] under the same substitution model (GTR + I + G). The Markov chain Monte Carlo (MCMC) algorithm was run for one million generations, with one tree sampled every 100 generations. The first 25% of trees were discarded as burn-in to construct majority-rule consensus tree and estimate posterior probabilities (PP) for each node.

Author Contributions: T.Z. and X.W. conceived and designed the work; T.Z. and J.W. collected samples; T.Z., J.W., Y.J., W.L., and F.X. performed the experiments and analyzed the data; T.Z. wrote the manuscript; X.W. revised the manuscript. All authors gave final approval of the paper.

Acknowledgments: This work was financially co-supported by the National Natural Science Foundation of China (31770364) and Scientific Research Supporting Project for New Teacher of Xi'an Jiaotong University (YX1K105).

References

1.   Zhang, X.L.; Wang, Y.J.; Ge, X.J.; Yuan, Y.M.; Yang, H.L.; Liu, J.Q. Molecular phylogeny and biogeography of *Gentiana* sect. *Cruciata* (Gentianaceae) based on four chloroplast DNA datasets. *Taxon* **2009**, *58*, 862–870.

2.   Ho, T.N.; Liu, S.W. *A Worldwide Monograph of Gentiana*; Science Press: Beijing, China, 2001.

3.   Ho, T.N.; Pringle, S.J. *"Gentianaceae," Flora of China*; Science Press: Beijing, China, 1995; Volume 16, pp. 1–140.

4.   State Pharmacopoeia Commission of the PRC. *Pharmacopoeia of P.R. China, Part 1*; Chemical Industry Publishing House: Beijing, China, 2015; pp. 270–271.

5.   Hua, W.; Zheng, P.; He, Y.; Cui, L.; Kong, W.; Wang, Z. An insight into the genes involved in secoiridoid biosynthesis in *Gentiana macrophylla* by RNA-seq. *Mol. Biol. Rep.* **2014**, *41*, 4817–4825. [CrossRef] [PubMed]

6.   Chang-Liao, W.-L.; Chien, C.-F.; Lin, L.-C.; Tsai, T.-H. Isolation of gentiopicroside from *Gentianae* Radix and its pharmacokinetics on liver ischemia/reperfusion rats. *J. Ethnopharmacol.* **2012**, *141*, 668–673. [CrossRef] [PubMed]

7.   Yin, H.; Zhao, Q.; Sun, F.-M.; An, T. Gentiopicrin-producing endophytic fungus isolated from *Gentiana macrophylla*. *Phytomedicine* **2009**, *16*, 793–797. [CrossRef] [PubMed]

8.   Yu, F.; Yu, F.; Li, R.; Wang, R. Inhibitory effects of the *Gentiana macrophylla* (Gentianaceae) extract on rheumatoid arthritis of rats. *J. Ethnopharmacol.* **2004**, *95*, 77–81. [CrossRef] [PubMed]

9.   Li, X.; Wang, L.; Yang, H.; Liu, J. Confirmation of natural hybrids between *Gentiana straminea* and *G. siphonantha* (Gentianaceae) based on molecular evidence. *Front. Biol. China* **2008**, *3*, 470–476. [CrossRef]

10. Hu, Q.; Peng, H.; Bi, H.; Lu, Z.; Wan, D.; Wang, Q.; Mao, K. Genetic homogenization of the nuclear ITS loci across two morphologically distinct gentians in their overlapping distributions in the Qinghai-Tibet Plateau. *Sci. Rep.* **2016**, *6*, 34244. [CrossRef] [PubMed]

11. Zhao, Z.; Su, J.; Wang, Z. Pharmacognostical studies on root of *Gentiana siphonantha*. *Chin. Tradit. Herbal Drugs* **2006**, *37*, 1875–1878.

12. Nguyen, V.B.; Park, H.-S.; Lee, S.-C.; Lee, J.; Park, J.Y.; Yang, T.-J. Authentication markers for five major *Panax* species developed via comparative analysis of complete chloroplast genome sequences. *J. Agric. Food Chem.* **2017**, *65*, 6298–6306. [CrossRef] [PubMed]

13. Bendich, A.J. Circular chloroplast chromosomes: The grand illusion. *Plant Cell* **2004**, *16*, 1661–1666. [CrossRef] [PubMed]

14. Lee, H.J.; Koo, H.J.; Lee, J.; Lee, S.-C.; Lee, D.Y.; Giang, V.N.L.; Kim, M.; Shim, H.; Park, J.Y.; Yoo, K.-O.; et al. Authentication of *Zanthoxylum* species based on integrated analysis of complete chloroplast genome sequences and metabolite profiles. *J. Agric. Food Chem.* **2017**, *65*, 10350–10359. [CrossRef] [PubMed]

15. Eguiluz, M.; Rodrigues, N.F.; Guzman, F.; Yuyama, P.; Margis, R. The chloroplast genome sequence from *Eugenia uniflora*, a Myrtaceae from Neotropics. *Plant Syst. Evol.* **2017**, *303*, 1199–1212. [CrossRef]

16. Li, X.; Yang, Y.; Henry, R.J.; Rossetto, M.; Wang, Y.; Chen, S. Plant DNA barcoding: From gene to genome. *Biol. Rev.* **2015**, *90*, 157–166. [CrossRef] [PubMed]

17. Hollingsworth, P.M.; Forrest, L.L.; Spouge, J.L.; Hajibabaei, M.; Ratnasingham, S.; van der Bank, M.; Chase, M.W.; Cowan, R.S.; Erickson, D.L.; Fazekas, A.J.; et al. A DNA barcode for land plants. *Proc. Natl. Acad. Sci. USA* **2009**, *106*, 12794–12797.

18. Song, Y.; Wang, S.; Ding, Y.; Xu, J.; Li, M.F.; Zhu, S.; Chen, N. Chloroplast genomic resource of *Paris* for species discrimination. *Sci. Rep.* **2017**, *7*, 3427. [CrossRef] [PubMed]

19. Ma, P.-F.; Zhang, Y.-X.; Zeng, C.-X.; Guo, Z.-H.; Li, D.-Z. Chloroplast phylogenomic analyses resolve deep-level relationships of an intractable bamboo tribe Arundinarieae (Poaceae). *Systematic Biol.* **2014**, *63*, 933–950. [CrossRef] [PubMed]

20. Carbonell-Caballero, J.; Alonso, R.; Ibañez, V.; Terol, J.; Talon, M.; Dopazo, J. A phylogenetic analysis of 34 chloroplast genomes elucidates the relationships between wild and domestic species within the genus *Citrus*. *Mol. Biol. Evol.* **2015**, *32*, 2015–2035. [CrossRef] [PubMed]

21. Dong, W.; Xu, C.; Li, W.; Xie, X.; Lu, Y.; Liu, Y.; Jin, X.; Suo, Z. Phylogenetic resolution in *Juglans* based on complete chloroplast genomes and nuclear DNA sequences. *Front. Plant Sci.* **2017**, *8*, 1148. [CrossRef] [PubMed]

22. Yang, Y.; Zhou, T.; Duan, D.; Yang, J.; Feng, L.; Zhao, G. Comparative analysis of the complete chloroplast genomes of five *Quercus* species. *Front. Plant Sci.* **2016**, *7*, 959. [CrossRef] [PubMed]

23. Huang, C.-Y.; Gruenheit, N.; Ahmadinejad, N.; Timmis, J.; Martin, W. Mutational decay and age of chloroplast and mitochondrial genomes transferred recently to angiosperm nuclear chromosomes. *Plant Physiol.* **2005**, *138*, 1723–1733. [CrossRef] [PubMed]

24. Liu, J.; Yan, H.-F.; Ge, X.-J. The use of DNA barcoding on recently diverged species in the genus *Gentiana* (Gentianaceae) in China. *PLoS ONE* **2016**, *11*, e0153008. [CrossRef] [PubMed]

25. Zhang, D.; Gao, Q.; Li, F.; Li, Y. DNA molecular identification of botanical origin in Chinese herb Qingjiao. *J. Anhui Agric. Sci.* **2011**, *39*, 14609–14612.

26. Ni, L.; Zhao, Z.; Xu, H.; Chen, S.; Dorje, G. Chloroplast genome structures in *Gentiana* (Gentianaceae), based on three medicinal alpine plants used in Tibetan herbal medicine. *Curr. Genet.* **2017**, *63*, 241–252. [CrossRef] [PubMed]

27. Ni, L.; Zhao, Z.; Xu, H.; Chen, S.; Dorje, G. The complete chloroplast genome of *Gentiana straminea* (Gentianaceae), an endemic species to the Sino-Himalayan subregion. *Gene* **2016**, *577*, 281–288. [CrossRef] [PubMed]

28. Wang, X.; Yang, N.; Su, J.; Zhang, H.; Cao, X. The complete chloroplast genome of *Gentiana macrophylla*. *Mitochondrial DNA B* **2017**, *2*, 395–396. [CrossRef]

29. Xiang, B.; Li, X.; Qian, J.; Wang, L.; Ma, L.; Tian, X.; Wang, Y. The complete chloroplast genome sequence of the medicinal plant *Swertia mussotii* using the PacBio RS II platform. *Molecules* **2016**, *21*, 1029. [CrossRef] [PubMed]

30. Fu, P.-C.; Zhang, Y.-Z.; Geng, H.-M.; Chen, S.-L. The complete chloroplast genome sequence of *Gentiana lawrencei* var. *farreri* (Gentianaceae) and comparative analysis with its congeneric species. *PeerJ* **2016**, *4*, e2540. [CrossRef] [PubMed]

31. Sun, S.-S.; Fu, P.-C.; Zhou, X.-J.; Cheng, Y.-W.; Zhang, F.-Q.; Chen, S.-L.; Gao, Q.-B. The complete plastome sequences of seven species in *Gentiana* sect. *Kudoa* (Gentianaceae): Insights into plastid gene loss and molecular evolution. *Front. Plant Sci.* **2018**, *9*, 493. [CrossRef] [PubMed]

32. Zhou, T.; Chen, C.; Wei, Y.; Chang, Y.; Bai, G.; Li, Z.; Kanwal, N.; Zhao, G. Comparative transcriptome and chloroplast genome analyses of two related *Dipteronia* Species. *Front. Plant Sci.* **2016**, *7*, 1512. [CrossRef] [PubMed]

33. Yang, J.-B.; Li, D.-Z.; Li, H.-T. Highly effective sequencing whole chloroplast genomes of angiosperms by nine novel universal primer pairs. *Mol. Ecol. Resour.* **2014**, *14*, 1024–1031. [CrossRef] [PubMed]

34. Hu, Y.; Woeste, K.E.; Zhao, P. Completion of the chloroplast genomes of five Chinese *Juglans* and their contribution to chloroplast phylogeny. *Front. Plant Sci.* **2016**, *7*, 1955. [CrossRef] [PubMed]

35. Sun, Y.; Moore, M.J.; Zhang, S.; Soltis, P.S.; Soltis, D.E.; Zhao, T.; Meng, A.; Li, X.; Li, J.; Wang, H. Phylogenomic and structural analyses of 18 complete plastomes across nearly all families of early-diverging eudicots, including an angiosperm-wide analysis of IR gene content evolution. *Mol. Phylogenet. Evol.* **2016**, *96*, 93–101. [CrossRef] [PubMed]

36. Weng, M.-L.; Blazier, J.C.; Govindu, M.; Jansen, R.K. Reconstruction of the ancestral plastid genome in Geraniaceae reveals a correlation between genome rearrangements, repeats and nucleotide substitution rates. *Mol. Biol. Evol.* **2013**, *31*, 645–659. [CrossRef] [PubMed]

37. Lu, L.; Li, X.; Hao, Z.; Yang, L.; Zhang, J.; Peng, Y.; Xu, H.; Lu, Y.; Zhang, J.; Shi, J.; et al. Phylogenetic studies and comparative chloroplast genome analyses elucidate the basal position of halophyte *Nitraria sibirica* (Nitrariaceae) in the Sapindales. *Mitochondrial DNA A* **2017**, 1–11. [CrossRef] [PubMed]

38.   Asano, T.; Tsudzuki, T.; Takahashi, S.; Shimada, H.; Kadowaki, K. Complete nucleotide sequence of the sugarcane (*Saccharum officinarum*) chloroplast genome: A comparative analysis of four monocot chloroplast genomes. *DNA Res.* **2004**, *11*, 93–99. [CrossRef] [PubMed]

39.   Curci, P.L.; De Paola, D.; Danzi, D.; Vendramin, G.G.; Sonnante, G. Complete chloroplast genome of the multifunctional crop globe artichoke and comparison with other asteraceae. *PLoS ONE* **2015**, *10*, e0120589. [CrossRef] [PubMed]

40.   Provan, J.; Powell, W.; Hollingsworth, P.M. Chloroplast microsatellites: New tools for studies in plant ecology and evolution. *Trends Ecol. Evol.* **2001**, *16*, 142–147. [CrossRef]

41.   Rousseau-Gueutin, M.; Bellot, S.; Martin, G.E.; Boutte, J.; Chelaifa, H.; Lima, O.; Michon-Coudouel, S.; Naquin, D.; Salmon, A.; Ainouche, K. The chloroplast genome of the hexaploid *Spartina maritima* (Poaceae, Chloridoideae): Comparative analyses and molecular dating. *Mol. Phylogenet. Evol.* **2015**, *93*, 5–16. [CrossRef] [PubMed]

42.   Xu, J.-H.; Liu, Q.; Hu, W.; Wang, T.; Xue, Q.; Messing, J. Dynamics of chloroplast genomes in green plants. *Genomics* **2015**, *106*, 221–231. [CrossRef] [PubMed]

43.   Yamane, K.; Yasui, Y.; Ohnishi, O. Intraspecific cpDNA variations of diploid and tetraploid perennial buckwheat, *Fagopyrum cymosum* (Polygonaceae). *Am. J. Bot.* **2003**, *90*, 339–346. [CrossRef] [PubMed]

44.   Cho, K.-S.; Yun, B.-K.; Yoon, Y.-H.; Hong, S.-Y.; Mekapogu, M.; Kim, K.-H.; Yang, T.-J. Complete chloroplast genome sequence of tartary buckwheat (*Fagopyrum tataricum*) and comparative analysis with common buckwheat (*F. esculentum*). *PLoS ONE* **2015**, *10*, e0125332. [CrossRef] [PubMed]

45.   Machado, L.D.O.; Vieira, L.D.N.; Stefenon, V.M.; Oliveira Pedrosa, F.D.; Souza, E.M.D.; Guerra, M.P.; Nodari, R.O. Phylogenomic relationship of feijoa (*Acca sellowiana* (O.Berg) Burret) with other Myrtaceae based on complete chloroplast genome sequences. *Genetica* **2017**, *145*, 163–174. [CrossRef] [PubMed]

46.   Cuenoud, P.; Savolainen, V.; Chatrou, L.W.; Powell, M.; Grayer, R.J.; Chase, M.W. Molecular phylogenetics of Caryophyllales based on nuclear 18S rDNA and plastid *rbcL*, *atpB*, and *matK* DNA sequences. *Am. J. Bot.* **2002**, *89*, 132–144. [CrossRef] [PubMed]

47.   Techen, N.; Parveen, I.; Pan, Z.; Khan, I.A. DNA barcoding of medicinal plant material for identification. *Curr. Opin. Biotech.* **2014**, *25*, 103–110. [CrossRef] [PubMed]

48.   Kim, I.; Young Park, J.; Sun Lee, Y.; Lee, H.; Park, H.-S.; Jayakodi, M.; Waminal, N.; Hwa Kang, J.; Joo Lee, T.; Sung, S.; et al. Discrimination and authentication of *Eclipta prostrata* and *E. alba* based on the complete chloroplast genomes. *Plant Breed. Biotech.* **2017**, *5*, 334–343. [CrossRef]

49.   Dong, W.; Xu, C.; Li, C.; Sun, J.; Zuo, Y.; Shi, S.; Cheng, T.; Guo, J.; Zhou, S. *ycf1*, the most promising plastid DNA barcode of land plants. *Sci. Rep.* **2015**, *5*, 8348. [CrossRef] [PubMed]

50.   Liu, L.; Wu, D.; Zhang, X. Pharmacognostical studies on root of *Gentiana officinalis*. *J. Chin. Med. Mater.* **2008**, *31*, 1635–1638.

51.   Xiong, B.; Zhao, Z.; Ni, L.; Gaawe, D.; Mi, M. DNA-based identification of *Gentiana robusta* and related species. *Chin. Med. Mater.* **2015**, *40*, 4680–4685.

52.   Zhang, X.; Ge, X.; Liu, J.; Yuan, Y. Morphological, karyological and molecular delimitation of two gentians: *Gentiana crassicaulis* versus *G. tibetica* (Gentianaceae). *Acta Phytotaxon. Sin.* **2006**, *44*, 627–640. [CrossRef]

53.   Maria, B.; Bengt, O.; Birgitta, B. Phylogenetic relationships within the Gentianales based on *ndhF* and *rbcL* sequences, with particular reference to the Loganiaceae. *Am. J. Bot.* **2000**, *87*, 1029–1043.

54.   Yang, L.L.; Li, H.L.; Wei, L.; Yang, T.; Kuang, D.Y.; Li, M.H.; Liao, Y.Y.; Chen, Z.D.; Wu, H.; Zhang, S.Z. A supermatrix approach provides a comprehensive genus-level phylogeny for Gentianales. *J. Syst. Evol.* **2016**, *54*, 400–415. [CrossRef]

55.   Leebens-Mack, J.; Raubeson, L.A.; Cui, L.; Kuehl, J.V.; Fourcade, M.H.; Chumley, T.W.; Boore, J.L.; Jansen, R.K.; Depamphilis, C.W. Identifying the basal angiosperm node in chloroplast genome phylogenies: Sampling one's way out of the Felsenstein zone. *Mol. Biol. Evol.* **2005**, *22*, 1948–1963. [CrossRef] [PubMed]

56.   Doyle, J.J. A rapid DNA isolation procedure for small quantities of fresh leaf tissue. *Phytochem. Bull.* **1987**, *19*, 11–15.

57.   Patel, R.K.; Jain, M. NGS QC Toolkit: A toolkit for quality control of next generation sequencing data. *PLoS ONE* **2012**, *7*, e30619. [CrossRef] [PubMed]

58.   Langmead, B.; Salzberg, S.L. Fast gapped-read alignment with Bowtie 2. *Nat. Methods* **2012**, *9*, 357–359. [CrossRef] [PubMed]

59.   Bankevich, A.; Nurk, S.; Antipov, D.; Gurevich, A.A.; Dvorkin, M.; Kulikov, A.S.; Lesin, V.M.; Nikolenko, S.I.;

Pham, S.; Prjibelski, A.D. SPAdes: A new genome assembly algorithm and its applications to single-cell sequencing. *J. Comput. Biol.* **2012**, *19*, 455–477. [CrossRef] [PubMed]

60. Hahn, C.; Bachmann, L.; Chevreux, B. Reconstructing mitochondrial genomes directly from genomic next-generation sequencing reads—A baiting and iterative mapping approach. *Nucleic Acids Res.* **2013**, *41*, e129. [CrossRef] [PubMed]

61. Wyman, S.K.; Jansen, R.K.; Boore, J.L. Automatic annotation of organellar genomes with DOGMA. *Bioinformatics* **2004**, *20*, 3252–3255. [CrossRef] [PubMed]

62. Lohse, M.; Drechsel, O.; Kahlau, S.; Bock, R. OrganellarGenomeDRAW—A suite of tools for generating physical maps of plastid and mitochondrial genomes and visualizing expression data sets. *Nucleic Acids Res.* **2013**, *41*, W575–W581. [CrossRef] [PubMed]

63. Kurtz, S.; Choudhuri, J.V.; Ohlebusch, E.; Schleiermacher, C.; Stoye, J.; Giegerich, R. REPuter: The manifold applications of repeat analysis on a genomic scale. *Nucleic Acids Res.* **2001**, *29*, 4633–4642. [CrossRef] [PubMed]

64. Benson, G. Tandem repeats finder: A program to analyze DNA sequences. *Nucleic Acids Res.* **1999**, *27*, 573. [CrossRef] [PubMed]

65. Thiel, T.; Michalek, W.; Varshney, R.; Graner, A. Exploiting EST databases for the development and characterization of gene-derived SSR-markers in barley (*Hordeum vulgare* L.). *Theor. Appl. Genet.* **2003**, *106*, 411–422. [CrossRef] [PubMed]

66. Yang, Z. PAML 4: Phylogenetic analysis by maximum likelihood. *Mol. Biol. Evol.* **2007**, *24*, 1586–1591. [CrossRef] [PubMed]

67. Frazer, K.A.; Pachter, L.; Poliakov, A.; Rubin, E.M.; Dubchak, I. VISTA: Computational tools for comparative genomics. *Nucleic Acids Res.* **2004**, *32* (Suppl. 2), W273–W279. [CrossRef] [PubMed]

68. Zhang, Y.-J.; Ma, P.-F.; Li, D.-Z. High-throughput sequencing of six bamboo chloroplast genomes: Phylogenetic implications for temperate woody bamboos (Poaceae: Bambusoideae). *PLoS ONE* **2011**, *6*, e20596. [CrossRef] [PubMed]

69. Koressaar, T.; Remm, M. Enhancements and modifications of primer design program Primer3. *Bioinformatics* **2007**, *23*, 1289–1291. [CrossRef] [PubMed]

70. Bi, G.; Mao, Y.; Xing, Q.; Cao, M. HomBlocks: A multiple-alignment construction pipeline for organelle phylogenomics based on locally collinear block searching. *Genomics* **2018**, *110*, 18–22. [CrossRef] [PubMed]

71. Katoh, K.; Standley, D.M. MAFFT multiple sequence alignment software version 7: Improvements in performance and usability. *Mol. Biol. Evol.* **2013**, *30*, 772–780. [CrossRef] [PubMed]

72. Posada, D.; Crandall, K.A. Modeltest: Testing the model of DNA substitution. *Bioinformatics* **1998**, *14*, 817–818. [CrossRef] [PubMed]

73. Swofford, D.L. *Commands Used in the PAUP Block in PAUP 4.0: Phylogenetic Analysis Using Parsimony 132–135*; Smithsonian Institution: Washington, DC, USA, 1998.

74. Stamatakis, A. RAxML version 8: A tool for phylogenetic analysis and post-analysis of large phylogenies. *Bioinformatics* **2014**, *30*, 1312–1313. [CrossRef] [PubMed]

75. Ronquist, F.; Teslenko, M.; van der Mark, P.; Ayres, D.L.; Darling, A.; Höhna, S.; Larget, B.; Liu, L.; Suchard, M.A.; Huelsenbeck, J.P. MrBayes 3.2: Efficient Bayesian phylogenetic inference and model choice across a large model space. *Syst. Biol.* **2012**, *61*, 539–542. [CrossRef] [PubMed]

Permissions

The contributors of this book come from diverse backgrounds, making this book a truly international effort. This book will bring forth new frontiers with its revolutionizing research information and detailed analysis of the nascent developments around the world.

We would like to thank all the contributing authors for lending their expertise to make the book truly unique. They have played a crucial role in the development of this book. Without their invaluable contributions this book wouldn't have been possible. They have made vital efforts to compile up to date information on the varied aspects of this subject to make this book a valuable addition to the collection of many professionals and students.

This book was conceptualized with the vision of imparting up-to-date information and advanced data in this field. To ensure the same, a matchless editorial board was set up. Every individual on the board went through rigorous rounds of assessment to prove their worth. After which they invested a large part of their time researching and compiling the most relevant data for our readers.

The editorial board has been involved in producing this book since its inception. They have spent rigorous hours researching and exploring the diverse topics which have resulted in the successful publishing of this book. They have passed on their knowledge of decades through this book. To expedite this challenging task, the publisher supported the team at every step. A small team of assistant editors was also appointed to further simplify the editing procedure and attain best results for the readers.

Apart from the editorial board, the designing team has also invested a significant amount of their time in understanding the subject and creating the most relevant covers. They scrutinized every image to scout for the most suitable representation of the subject and create an appropriate cover for the book.

The publishing team has been an ardent support to the editorial, designing and production team. Their endless efforts to recruit the best for this project, has resulted in the accomplishment of this book. They are a veteran in the field of academics and their pool of knowledge is as vast as their experience in printing. Their expertise and guidance has proved useful at every step. Their uncompromising quality standards have made this book an exceptional effort. Their encouragement from time to time has been an inspiration for everyone.

The publisher and the editorial board hope that this book will prove to be a valuable piece of knowledge for researchers, students, practitioners and scholars across the globe.

List of Contributors

Mary Ann C. Bautista
South China Botanical Garden, Chinese Academy of Sciences, Guangzhou 510650, China
Fairy Lake Botanical Garden, Chinese Academy of Sciences, Shenzhen 518004, China
Graduate School, University of Chinese Academy of Sciences, Beijing 100049, China

Yunfei Deng
South China Botanical Garden, Chinese Academy of Sciences, Guangzhou 510650, China
Graduate School, University of Chinese Academy of Sciences, Beijing 100049, China

Yan Zheng
Fairy Lake Botanical Garden, Chinese Academy of Sciences, Shenzhen 518004, China

Tao Chen
Fairy Lake Botanical Garden, Chinese Academy of Sciences, Shenzhen 518004, China
Graduate School, University of Chinese Academy of Sciences, Beijing 100049, China

Zhangli Hu
School of Life Sciences and Oceanology, Shenzhen University, Shenzhen 518060, China

Li Gu and Ting Su
College of Life Sciences, Guizhou University, Guiyang 550025, China
The Key Laboratory of Plant Resources Conservation and Germplasm Innovation in Mountainous Region Ministry of Education, Guizhou University, Guiyang 550025, China
Institute of Agro-Bioengineering, Guizhou University, Guiyang 550025, China

Guo-Xiong Hu
College of Life Sciences, Guizhou University, Guiyang 550025, China
The Key Laboratory of Plant Resources Conservation and Germplasm Innovation in Mountainous Region Ministry of Education, Guizhou University, Guiyang 550025, China

Ming-Tai An
College of Forestry, Guizhou University, Guiyang 550025, China

Andrew W. Gichira and Cornelius M. Kyalo
Key Laboratory of Aquatic Botany and Watershed Ecology, Wuhan Botanical Garden, Chinese Academy of Sciences, Wuhan 430074, China
University of Chinese Academy of Sciences, Beijing 100049, China
Sino-African Joint Research Center, Chinese Academy of Sciences, Wuhan 430074, China

Zhi-Zhong Li
Key Laboratory of Aquatic Botany and Watershed Ecology, Wuhan Botanical Garden, Chinese Academy of Sciences, Wuhan 430074, China
University of Chinese Academy of Sciences, Beijing 100049, China

Qing-Feng Wang and Jin-Ming Chen
Key Laboratory of Aquatic Botany and Watershed Ecology, Wuhan Botanical Garden, Chinese Academy of Sciences, Wuhan 430074, China
Sino-African Joint Research Center, Chinese Academy of Sciences, Wuhan 430074, China

Kangquan Yin, Yue Zhang, Yuejuan Li and Fang K. Du
College of Forestry, Beijing Forestry University, Beijing 100083, China

Yamato Yoshida
Department of Science, College of Science, Ibaraki University, Ibaraki 310-8512, Japan

Jianguo Zhou, Xinlian Chen, Yingxian Cui, Yu Wang, Jingyuan Song and Hui Yao
Key Lab of Chinese Medicine Resources Conservation, State Administration of Traditional Chinese Medicine of the People's Republic of China, Institute of Medicinal Plant Development, Chinese Academy of Medical Sciences & Peking Union Medical College, Beijing 100193, China

Wei Sun
Institute of Chinese Materia Medica, China Academy of Chinese Medicinal Sciences, Beijing 100700, China

Yonghua Li
Department of Pharmacy, Guangxi Traditional Chinese Medicine University, Nanning 530200, China

Wan-Lin Dong, Ruo-Nan Wang, Na-Yao Zhang, Wei-Bing Fan, Min-Feng Fang and Zhong-Hu Li
Key Laboratory of Resource Biology and Biotechnology in Western China, Ministry of Education, College of Life Sciences, Northwest University, Xi'an 710069, China

Julia Legen and Christian Schmitz-Linneweber
Institut of Biology, Department of the Life Sciences, Humboldt-Universität Berlin, Philippstraße 11–13, Grüne Amöbe, 10115 Berlin, Germany

Qixiang Lu, Wenqing Ye, Ruisen Lu, Wuqin Xu and Yingxiong Qiu
Key Laboratory of Conservation Biology for Endangered Wildlife of the Ministry of Education, and College of Life Sciences, Zhejiang University, Hangzhou 310058, China

Noriyuki Suetsugu
Graduate School of Biostudies, Kyoto University, Kyoto 606-8502, Japan

Masamitsu Wada
School of Science and Engineering, Tokyo Metropolitan University, Tokyo 192-0397, Japan

Akira Kawabe, Hiroaki Nukii and Hazuka Y. Furihata
Faculty of Life Sciences, Kyoto Sangyo University, Kyoto, Kyoto 603-8555, Japan

Ting Ren, Yanci Yang and Zhan-Lin Liu
Key Laboratory of Resource Biology and Biotechnology in Western China (Ministry of Education), College of Life Sciences, Northwest University, Xi'an 710069, China

Tao Zhou
School of Pharmacy, Xi'an Jiaotong University, Xi'an 710061, China

Qing-Long Wang, Juan-Hua Chen and Ning-Yu He
The National Key Laboratory of Plant Molecular Genetics, Institute of Plant Physiology & Ecology, Chinese Academy of Sciences, 300 Fenglin Road, Shanghai 200032, China

Fang-Qing Guo
The National Key Laboratory of Plant Molecular Genetics, Institute of Plant Physiology & Ecology, Chinese Academy of Sciences, 300 Fenglin Road, Shanghai 200032, China
CAS Center for Excellence in Molecular Plant Sciences, Institute of Plant Physiology & Ecology, Chinese Academy of Sciences, 300 Fenglin Road, Shanghai 200032, China

Yi-Ying Liao
Fairy Lake Botanical Garden, Shenzhen & Chinese Academy of Sciences, Shenzhen 518004, China

Josphat K. Saina
Fairy Lake Botanical Garden, Shenzhen & Chinese Academy of Sciences, Shenzhen 518004, China
Key Laboratory of Aquatic Botany and Watershed Ecology, Wuhan Botanical Garden, Chinese Academy of Sciences, Wuhan 430074, China
University of Chinese Academy of Sciences, Beijing 100049, China
Sino-African Joint Research Center, Chinese Academy of Sciences, Wuhan 430074, China

Andrew W. Gichira
Key Laboratory of Aquatic Botany and Watershed Ecology, Wuhan Botanical Garden, Chinese Academy of Sciences, Wuhan 430074, China
University of Chinese Academy of Sciences, Beijing 100049, China
Sino-African Joint Research Center, Chinese Academy of Sciences, Wuhan 430074, China

Jian Wang, Wenli Li, Fusheng Xu and Xumei Wang
School of Pharmacy, Xi'an Jiaotong University, Xi'an 710061, China

Yun Jia
Key Laboratory of Resource Biology and Biotechnology in Western China (Ministry of Education), School of Life Sciences, Northwest University, Xi'an 710069, China

Index

Printed in the USA
CPSIA information can be obtained
at www.ICGtesting.com
JSHW050846251023
50683JS00018B/86